MW00395297

PRAISE FOR *OF EPIDEMIC PROPORTIONS*

"Obesity is one of the sentinel epidemics of our time. In this humane book, Dr. Karasu offers a breadth of perspectives about the causes and consequences of the obesity epidemic, bringing to our understanding of the issue wisdom and imagination. We will not solve obesity through simple solutions alone. The ideas offered in *Of Epidemic Proportions* are an important step towards building a better intellectual architecture about the real scope of the challenges presented by the obesity epidemic."

— Sandro Galea, MD, DrPH, Dean, Robert A Knox Professor,
School of Public Health Boston University, author of
Healthier: Fifty Thoughts on the Foundations of Population Health

"What a brilliant and beautiful book! I think every physician and weight/nutritional counselor who deals with health and weight should read this."

— Sander L. Gilman, PhD, JD, Distinguished Professor of the Liberal Arts and
Sciences and Professor of Psychiatry, Emory University, and author of
Diets and Dieting *and* Obesity: The Biography

"*Of Epidemic Proportions: The Art and Science of Obesity* by Dr. Sylvia Karasu presents her many blogs in a single volume. These cover an eclectic series of topics dealing not only obesity, but also wellness, lifestyle, emotional health, and many other topic issues. Each blog is well researched, authoritative, and beautifully illustrated, including many great paintings. I am confident the reader will find this collection to be informative, entertaining, and both intellectually and visually pleasing."

— Antonio M. Gotto, Jr., MD, DPhil
Dean Emeritus, Weill Cornell Medicine
Provost Emeritus for Medical Affairs, Cornell University

"The book is remarkably thoughtful and rich in content, offering much to reflect on and peruse. The author, through a series of lively, well-researched and beautifully written essays, takes on pressing concerns for health and behavior. Moreover, she has been able to find works of art to illustrate and reinforce her messages. This is a visually and intellectually compelling read."

— Margaret A. Hamburg, MD, Past Commissioner of the
Food and Drug Administration (FDA) and President of the
American Academy of theAdvancement of Science (AAAS)

"Unique among books on obesity, *Of Epidemic Proportions* is a virtual tour-de-force. Visually stunning art images and well-researched scientific essays."

— Steven B. Heymsfield, MD, Professor of Metabolism & Body Composition,
Pennington Biomedical Research Center, Louisiana State University and President, The Obesity Society

"*Of Epidemic Proportions: The Art and Science of Obesity* is a remarkable work—visually and intellectually very compelling, the type of work that begs to be looked at. I am delighted to place it on display in the Oskar Diethelm Library, where scholars, practitioners, and historians will be able to peruse it."

— George J. Makari, MD, Professor of Psychiatry and Director of the De Witt
Wallace Institute for the History of Psychiatry at Weill Cornell Medicine and
author of Soul Machine: The Invention of the Modern Mind

"*Of Epidemic Proportions* is gorgeous, and I'm awestruck by how you did it. How did you find those illustrations and how extraordinary that you used them with your blogs? This is one of those situations where the whole is greater than the sum of the parts. Really, this book is a treasure, and I am honored to have a copy."

—*Marion Nestle, PhD, MPH, Paulette Goddard Professor of Nutrition, Food Studies, and Public Health, emerita, at New York University and author of* Unsavory Truth: How Food Companies Skew the Science of What We Eat

"This lavishly illustrated and beautifully written book brings new insights into the complex areas of eating, weight management and obesity. Dr. Karasu blends extensive knowledge of psychiatry, the humanities, and mind/body interactions to bring a fresh and most welcomed perspective to the multiple influences that interact in the complex area of obesity..."

—*James M. Rippe, MD, the Founder and Director of the Rippe Lifestyle Institute and Editor-in-Chief,* American Journal of Lifestyle Medicine *and author,* Lifestyle Medicine, *3rd edition*

"I love the intersection of art and science. This book reminds us of how much the inspiration and creativity of both can combine to motivate new insights. Dr. Karasu's images and blogs help us to put perceptions of our bodies in perspective and to understand the epidemic of obesity."

—*Barbara Jean Rolls, PhD, Professor and the Helen A. Guthrie Chair of Nutritional Sciences, The Pennsylvania State University and author,* The VolumetricsWeight-Control Plan

"With her wit, lucidity and deep insight, Sylvia Karasu's *Of Epidemic Proportions* is a gift to readers. This is a scholarly work like no other, providing her cumulative wisdom on the science of weight control in a collection that is as much a work of art as a popular science book."

—*Kenneth J. Rothman, DrPH, MPH, DMD, Distinguished Fellow, Research Triangle Institute and Professor of Epidemiology, Boston University and author,* Modern Epidemiology, *3rd edition*

"Dr. Karasu's writings are erudite, thought-provoking and engrossing, challenging us to engage our thinking at the intersection of science, history and culture. I am delighted to keep a copy of her extraordinary and beautifully illustrated book at my desk. It is a tour de force."

—*Judith A. Salerno, MD, MS, President, The New York Academy of Medicine and co-author,* The Weight of the Nation, *companion book to the 2013 Emmy-nominated HBO documentary series*

"A terrific read. Provocative and evidence-based—the best combination."

—*Cass R. Sunstein, Robert Walmsley University Professor, Harvard Law School, and co-author of* Nudge: Improving Decisions about Health, Wealth, and Happiness *and the author of the forthcoming* How Change Happens.

OF EPIDEMIC PROPORTIONS

THE ART AND SCIENCE OF OBESITY

EXPANDED EDITION

SYLVIA R. KARASU, MD

Front cover:
An obese man consulting his physician. Coloured etching. Wellcome Library, London.
Creative Commons Attribution-ShareAlike License 4.0. http://wellcomeimages.org.

Back cover:
Weighing Bars of Camphor, from *Tractatus de Herbis* by Dioscorides, Italian School, 15th century, Biblioteca Estense, Moderna, Emilia-Romagna, Italy. Primitive measuring device scale from the 15th century. Pedanius Dioscorides was an ancient Greek (1st century A.D.) physician and botanist and the author of *De Materia Medica,* a five-volume encyclopedia on the subject of herbal medicine. Bridgeman Images, used with permission.

Acknowledgment photo:
Lupines, Natallia Khlapushyna/Alamy Stock Photo, used with permission.

I have made every effort to obtain the appropriate rights and permissions for my use of any image not designated in the Public Domain.

sylviakarasu.com

Design by Alan Barnett

Author photo by Alan Barnett

Printed in the United States of America

CONTENTS

The more we learn about the world... the more conscious, specific, and articulate will be our knowledge of what we do not know, our knowledge of our ignorance... our knowledge can be only finite, while our ignorance must necessarily be infinite.

—Karl Popper, "Sources of Knowledge and Ignorance" in *Conjectures and Refutations*, p. 38, 1963.

...science can be seen as having two key complementary roles—dispelling false beliefs and creating new knowledge.

—David B. Allison et al, *Frontiers in Nutrition*, 2015.

In science, there is sometimes a fine line between healthy skepticism and misrepresenting and exaggerating scientific uncertainty.

—David B. Allison et al, *American Scientist*, 2018.

INTRODUCTION

Of Epidemic Proportions: The Art and Science of Obesity, initially released in March 2018, had been a collection of 93 scholarly blogs on the daunting science of weight control that I had written over the past seven and one-half years for the website of psychology-today.com. This first edition, of limited release and not for public sale, received considerable acclaim from those special friends and colleagues to whom I had sent my book. Many people had expressed a wish to purchase additional copies.

In response, I have decided to release this expanded version, now with 100 blog entries, an epilogue, and a comprehensive index, and make it available for wider distribution by offering it for sale on Amazon.com. Still by no means all-inclusive and undoubtedly idiosyncratic, this personal compilation focuses on many aspects of obesity that have intrigued me since I wrote, as senior author, *The Gravity of Weight: A Clinical Guide to Weight Loss and Maintenance* (2010, American Psychiatric Publishing, Inc.) I remain humbled, though, by the complexities and limitations of obesity research and gratefully acknowledge those exceptional investigators among you who emphasize the importance of scientific rigor, transparency, and reproducibility in this field and strive to rectify this challenging situation. It is to you that I dedicate my expanded version.

Though I have changed, and in numerous instances, supplemented many of the original art images that accompany each blog, I have still chosen to keep the essays, with a few exceptions, as initially published and maintain the order of their original date of publication. Of course, this means that some of the information, particularly from the earlier blogs, may need updating by the reader.

Sylvia R. Karasu, MD
Clinical Professor of Psychiatry
Weill Cornell Medicine
New York City
January 2019

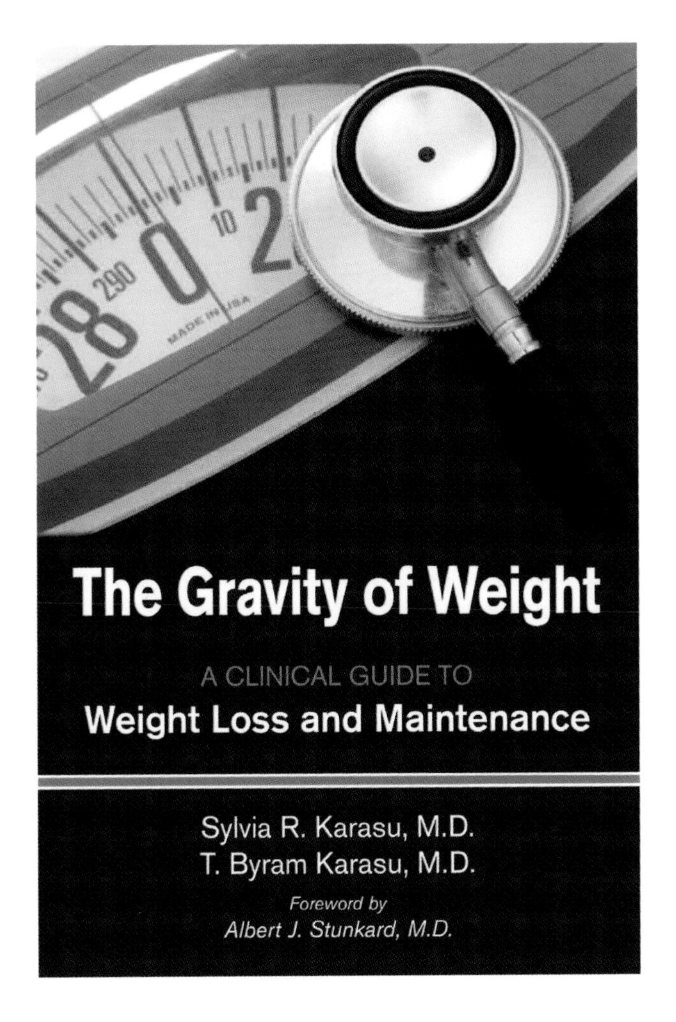

kroach, used with permission from istockphoto.com and from Tammy Cordova, designer of my book cover by American Psychiatric Publishing, 2010.

SORTING THROUGH THE INFORMATION EXPLOSION ABOUT WEIGHT

Casting a wider scholarly net amidst the expanding literature.

1 — Posted November 19, 2010

NOT ONLY HAVE AMERICAN (and even worldwide) waistlines expanded exponentially over the past three decades; so has the professional literature on diet and weight-related topics. According to a recent systematic review by Baier and colleagues, in the *International Journal of Obesity*, over 250 different professional journals include articles on the topic, though there are only three journals devoted exclusively to the subject of obesity. And these authors had not even included journals in such fields as economics and consumer affairs. Other researchers have described the situation of the information explosion as "lost in publication." In other words, if people want to keep up with the field, they have to cast a much wider scholarly net.

That was certainly my impression from writing my recently published book, *The Gravity of Weight*, (2010, American Psychiatric Publishing), a comprehensive text on all aspects of weight control written primarily for health professionals, such as psychologists, social workers, nutritionists, nurses, and physicians in all the medical and psychiatric specialties, as well as their intellectually curious patients. The book is a mind, brain, body integration that explores why weight control is so daunting for so many people. I culled research from over 900 sources, including articles from popular newspapers like *The New York Times*, and journals such as *The New England Journal of Medicine, Journal of Clinical Psychiatry, Plastic and Reconstructive Surgery, Journal of Clinical Endocrinology and Metabolism, Neuroscience and Biobehavioral Reviews*, and even the *Journal of Consumer Research*. In my book, I discuss some of the classic papers on obesity, such as Pavlov's early 20th century studies on satiety and sham feeding (in which Pavlov prevented the ingested food of his dogs to reach their stomachs) to cutting edge research on the newly discovered existence of substantial areas of brown fat in adults, which may be a "metabolic brake" that might lead eventually to genetic or pharmacological treatments for obesity.

What I have in mind for readers of my new blog, *The Gravity of Weight*, the daunting science of weight control, is to distill relevant information from scientific journals in diverse areas of expertise on weight-related subjects, from the "usual suspects" of diet and exercise to the more unusual, such as the connection between weight and circadian rhythms. My next blog posting will discuss the connection between "light pollution" and weight. ■

HAVOC WITH CIRCADIAN RHYTHMS: "LIGHT POLLUTION" AND WEIGHT

God said, "Let there be light," but not so much!

2—Posted November 19, 2010

How do environmental factors contribute to the increase in overweight and obesity in recent years? Most people focus on the fact that we have too much readily available food laden with addictive concoctions of fat, sugar, and salt. Furthermore, over time, portion size has increased so substantially that plates that were once around eight or nine inches have morphed into elliptical serving plate sizes of 12 or more inches, so much so that we have come to suffer from "portion distortion," as Brian Wansink and his colleagues have described. In other words, we no longer have appropriate "consumer norms" of what constitutes a single serving size. In his book, *The 9-Inch Diet*, for example, advertising executive Alex Bogusky tells how Americans, so used to buying super-sizes of everything, thought they were buying individual drinking glasses when they were actually purchasing flower vases from the company Ikea!

We are also more likely to live in temperature-controlled environments such that we expend fewer calories to maintain our body temperature than we had before we had heating and air conditioning. Other researchers focus on exposure to environmentally toxic chemicals, the so-called "endocrine disruptors," that are found in our plastic food and beverage containers or even our shampoos and nail polish that may be contributory to weight gain.

And of course, we are expending far fewer calories in our activities of daily living in a world where we have automated stairs, automobiles, televisions, and even energy-saving devices like washer-dryers and dishwashers than did our nomadic ancestors. It is estimated that these Stone Agers used about 1300 calories a day in this kind of physical activity where most of us, in comparison, are much more sedentary and on average, use only 550 calories per day. Even the use of email and instant messaging rather than walking to a colleague's office, may contribute to a slow incremental creep of weight gain over time.

Along those lines, though, researchers believe that it may not be the sedentary lifestyle exclusively that devices like computers or television create that are responsible for our increasing weight. It may be differences in light exposure that they and other modern devices, particularly in our 24/7 world, produce. Laura Fonken and her colleagues, for example, in a recent article in the *Proceedings of the National Academy of Sciences*, have suggested that over the years, our industrial societies have exposed us to nighttime lighting in all forms that may be playing havoc with our circadian rhythms and consequently our metabolic systems.

Circadian rhythms are the natural built-in biological clocks that are present in all light-sensitive organisms, from the lowly fruitfly to human beings. We now know that many cells and organ systems in the body can actually "sense" time. And many, if not most, of our hormones, such as the stress hormone cortisol, are secreted in a diurnal variation. Our biological rhythms are primarily synchronized by the natural light of the sun, but other environmental signals, including the timing of food intake or even the kinds of foods we eat, can affect these rhythms. Experiments have even demonstrated that drug-seeking behavior in rats can be affected by circadian rhythms such that cocaine cravings vary over 24 hours and cravings may last longer when attempts at extinction are carried out at certain times than others. Anecdotally, at least, the strength of cravings in humans can also vary over the course of a day.

Disorders of circadian rhythms can be caused experimentally, but the most common disorder is the one we know

The Gutenberg Bible, Lenox Copy, The New York Public Library.

Joseph M.W. Turner, *Venice, from the Porch of Madonna Della Salute,* circa 1835, Metropolitan Museum of Art.

Bequest of Cornelius Vanderbilt, 1899. Public Domain.

Nuremberg Chronicle, 1493, Hartmann Schedel, author.

Wikimedia Commons/Public Domain.

as "jet lag," also known as "circadian misalignment," a syndrome manifested by sleep disruption (including insomnia and/or excessive sleepiness) and other symptoms of malaise, due to a mismatch between a person's current environment and his or her usual sleep-wake pattern. The term "jet lag" specifically refers to a transient and time-limited syndrome brought on by jet travel across different time zones, but it can be used metaphorically for any shift in circadian rhythms. When there is a mismatch between a person's biological clock and his or her environment, we can speak of "social jet lag." Social jet lag is commonly seen in adolescents who stay awake all night and sleep during the day. In its more chronic and insidious form, it is seen in those shift workers who work night after night. These workers are more prone to metabolic disturbances, including obesity and overt diabetes, and there have even been reports of increased substance abuse among them. Interestingly, people who suffer from the night-eating syndrome, in which they tend to consume most of their daily caloric intake after dinner, tend toward overweight and even obesity.

The researchers Fonken and her colleagues, working with mice, found that even small changes in the magnitude and timing of exposure to light led to weight gain and metabolic abnormalities in their experimental animals. They suggest that their findings may have relevance to humans: seemingly "innocuous" environmental changes in exposure to light at night-"light pollution," as they call it, such as by prolonged use of computers or even television-may also lead to changes in daily patterns of food intake and may be one additional environmental factor to consider in our proneness toward weight gain. So when God uttered his first words in *Genesis*, "Let there be light," perhaps He did not mean so much! ■

"DID YOU EVER SEE A FAT SQUIRREL?"

Not why are many overweight, but why is anyone thin?

3 — Posted November 22, 2010

OUR BODIES PROCESS about a million calories each year. That's ten million calories in a decade. Given the typical environment in which most of us find ourselves, with an overwhelming array of caloric possibilities, "Why is anyone thin?" So asked Dr. Jeffrey Friedman, one of this year's recipients of the Lasker Award (often called the "American Nobel.") In other words, perhaps we shouldn't always be asking why there are so many overweight or obese people, but rather instead, given our environmental temptations, we should be asking how anyone remains at a normal weight. In an article in the *American Journal of Clinical Nutrition* from last year, Friedman acknowledged that the obese eat too much and exercise too little. For the most part, it is the First Law of Thermodynamics-that is, essentially, calories in, calories expended. The "deeper question" for Friedman, though, is why do the obese eat more and exercise less. And his answer is that it is sometimes "less about conscious choices" and "more about their biological makeup." To a large extent, this is determined by our genetics.

Used with permission/istockphoto.com. Credit: kumphel

Some people, no matter how little they eat or how much they exercise, will always have a weight problem; conversely, others, no matter what they do, will never gain much weight. And our bodies have evolved to preserve the status quo, or in biological parlance, homeostasis. So when we lose weight, our bodies are predisposed to regain the weight and conserve energy. And anecdotally, it seems the more rapidly we tend to lose the weight, as, for example, with severe calorie restriction, the more rapidly our bodies tend to regain.

"Did You Ever See a Fat Squirrel?" was the title of a popular diet book by Ruth Adams in the early 1970s. Perhaps not a fat squirrel, but have you ever seen an obese mouse, which has a genetic mutation that leads to a leptin deficiency, massive obesity, excessive overeating (hyperphagia), insulin resistance, and other metabolic abnormalities? An obese mouse with this genetic mutation is in the next photo, with a normal mouse for comparison. In the early 1990s, it was Friedman's laboratory at New York City's Rockefeller University that first isolated the hormone leptin, found predominantly in fat tissue and one of the hormones responsible for regulating food intake and energy balance, among its many other functions. An actual genetic leptin deficiency is quite rare in humans, which is unfortunate, since giving exogenous leptin can reverse the obesity and the accompanying metabolic abnormalities in these people. Would that the genetics were that simple! In fact, most obese people have a state of excessive, but ineffective leptin, analogous to the state of insulin resistance. Giving injections of leptin, in general, never became the panacea weight loss solution originally anticipated, though leptin may eventually have more of a role in preventing weight gain, particularly in combination with other medications.

We know from family, twin, and adoption studies that about 70% (and estimates up to 90%!) of our weight is genetically determined and perhaps as hereditable as our height. Actually, some researchers believe weight is more

Genetically engineered obese mouse, with mouse of normal weight for comparison
Wikipedia Commons/Public Domain.)

John Singleton Copley, *A Boy with a Flying Squirrel,* 1765.
Museum of Fine Arts, Boston, Wikipedia Commons/Public Domain.

hereditable than most other conditions, including heart disease, breast cancer, hypertension, or even mental illness. Many years ago, Claude Bouchard and his colleagues did some classic experiments on a metabolic unit in which different sets of identical twins were exposed to the same number of calories and same amount of exercise. The twin pairs themselves each gained about the same amount of weight, but, surprisingly, there were considerable differences in the weight gain among different pairs despite exposure to the same environment. Recently, Bouchard has emphasized that genetic variation "has much to do with the risk of becoming obese," even though it is clearly not the only cause.

In fact, though, genetics seems to determine even where our fat accumulates (e.g. whether around our abdomens-the so-called "apples" or around our thighs-the so-called "pears"), how we will respond to medications for weight loss, or even our reluctance to exercise (e.g. from differences in motivation and reward to differences in ability and coordination.) But we are not just talking about a few genes. Researchers studying the Human Obesity Gene Map found that there are over 300 separate trait areas (loci) that may be involved in weight control. Warden and Fisler, in a recent article in *Progress in Molecular Biology and Translational Science,* acknowledge the extraordinarily complex issues involved in the genetics of obesity. They believe, however, that an individual's genetic profile could eventually lead to greater "flexibility" in national recommendations for changes in lifestyle involving diet and exercise that are geared to preventing obesity. Even the USDA food pyramid and children's lunchboxes might be genetically individualized some day!

The point is that we can make our environment work with our genetics when it comes to weight control. So when you sit down to your Thanksgiving table this week, remember your genetic predispositions may weigh quite heavily, as it were, on you. But don't despair: The more you can appreciate and acknowledge the contribution of your biological makeup, the more you should be able to control your penchant to overindulge. ■

THE "HOLIDAY CREEP:" SEASONAL WEIGHT GAIN

During holidays, do we gain as much as we think?

4 — Posted November 28, 2010

Do we really gain as much weight during the holiday season from Thanksgiving to New Year's as we think we do? When surveyed, most people have the perception that they gain a great deal of weight during this time.

Some years ago Drs. Jack and Susan Yanovski and their colleagues at the National Institutes of Health conducted a well-designed study to investigate just that question. Their subjects were mostly NIH employees, with a range of occupations (from those who worked in maintenance to professionals engaged in research) and with both ethnic and racial diversity as well. Their sample of almost two hundred men and women ranged in ages from 19 to 82. Among other measurements taken, they measured their subjects' weights on four different occasions throughout the year. They masked the real purpose of the study by presenting it as one interested in seasonal changes in vital signs lest their subjects be tempted to change their eating patterns in response.

Santa Claus as illustrated by 1905 *Puck Magazine*.

image from the U.S. Library of Congress's Prints and Photograph Division, Wikipedia Commons/Public Domain

What they found is that people thought they were gaining four times as much as they actually had gained! Fewer than 10% of their sample actually gained over five pounds. Most had gained less than a pound. But those who were overweight or obese initially were the ones most likely to gain those five or more pounds, so their conclusion was that the holiday season "may present special risks" for those who do already have a weight problem. Furthermore, the researchers found that even though the average weight gain was fairly minimal, people tended to gain the most weight for the year around this time and not lose this holiday weight once the holidays were over. The result is a "cumulative" gain, or what we might call a kind of "holiday creep." In other words, even though for many, the actual weight gain this time of year is trivial, it may not be trivial over time. In general, as we age, if not watching carefully, we tend to gain one to two pounds a year as our metabolism slows, and particularly, if we lose muscle. So the holiday season may be a crucial time responsible in part for this cumulative gain year after year.

There are many reasons people tend toward weight gain during the holidays, including overeating as a reaction to the stress of being with family or even the stress involved in all the holiday preparations. Others develop a "What the hell?" attitude such that they decide to forgo their usual diets during this time. Furthermore, some people are sensitive to shorter days and have a seasonal affective disorder such that they may become depressed and tend to overeat in response. Sometimes, people are less physically active in the winter holiday months so they are less likely to burn off the excess calories. These are the "hibernators." Still others may find the tempting array of foods prepared specially for the holidays irresistible. The more variety and the more palatable the food choices given to people (or animals, for that matter), as for example, with a buffet style of eating, the more likely they will over-consume.

"Diet consistency," that is, not changing your eating style for holidays or weekends, has been shown to be an important factor to keep weight off in those who have lost weight and maintained that loss. This is one of the results of the National Weight Control Registry, a study begun by Rena Wing and James Hill in the early 1990s, of a non-random group of dieters who have lost at least 30 pounds and kept the weight off for more than a year. (Most in the study have lost much more weight and kept it off for a much longer time.) Another finding from this study is that successful dieters are accountable: they weigh themselves frequently (and often daily) and watch what they eat, often by counting calories or even keeping a food diary. People, even those who maintain their weight, are often notoriously inaccurate in remembering or tallying how many calories they eat in a day. This is called the "eye-mouth gap"—the difference between how much people think they are eating and how much they are actually eating. Obese people sometimes report only one-half to two-thirds of what they are eating, and even thin people may report only 80%. Of course, it is quite difficult to remember everything we eat. As food author Michael Pollan, in his book, *In Defense of Food,* has said, "I'm not sure Marcel Proust himself could recall his dietary intake... with the sort of precision demanded."

So what else should one do to prevent "holiday creep?" Well, Cora Craig, writing a few years ago in the *Canadian Medical Association Journal,* suggested that Santa Claus, though himself fairly obese, remains jolly and tries to keep himself as healthy as possible by his physically active lifestyle, that is, by jumping from rooftop to rooftop. For the rest of us, she suggests lots of "dancing and prancing!" ■

Portrait of Saint Nicholas of Myra, first half of 13th century), Saint Catherine's Monastery, Egypt

Wikipedia Commons/Public Domain

THE BODY'S DAMAGE CONTROL: THAT'S THE WAY THE COOKIES CRUMBLE

The fate of one chocolate chip.

5 — Posted December 5, 2010

MOST PEOPLE WHO KNOW SOMETHING about calorie counting can appreciate that adding a few extra calories every day over the course of time can have a substantial effect on our weight. After all, as I have said, "calories in, calories out." Brian Wansink, in his wonderful book, *Mindless Eating*, makes the point that eating three extra Jelly Belly jelly beans, equivalent to a mere 12 calories a day will add an additional 4380 calories for the body to process after one year. Since there are roughly 3500 calories* to one pound of weight gain, that will lead to over a pound extra by the end of the year, even if we change nothing else.

One of the most effective commercials, sponsored by the New York City Department of Health, called "Pouring on the Pounds," appeared just about a year ago and emphasized the same thing. This commercial, well worth watching (particularly if you never want to drink soda from a can again) on YouTube at www.youtube.com/watch?v=-F4t8zL6Foc, shows a young attractive man drinking down globules of fat. The message, "Don't drink yourself fat!" since, after all, one can of soda a day can lead to an extra ten pounds of weight a year later. But does that mean we can extrapolate that in two years you are twenty pounds heavier or in three years, thirty pounds heavier, and by ten years, 100 pounds heavier from that one daily can of soda? Well, no. Why?

The body has compensatory mechanisms, or as I would say, a kind of damage control, to prevent such dangerous changes to its balance (i.e. homeostasis). And that's where our chocolate chip cookie comes in. Earlier this year, in the *Journal of the American Medical Association*, Katan and Ludwig explored just this conundrum using the example of eating one extra 60-calorie chocolate chip cookie for life: by one year, the body may gain a few pounds from those extra calories. But as we all know by experience, this weight gain does not continue at the same rate indefinitely: even a decade later, no one will gain an extra 100 pounds from eating one extra 60-calorie cookie or three extra Jelly Belly jelly beans a day. Katan and Ludwig explain that over time, "an increasing proportion of the cookie's calories will go into repairing, replacing, and carrying the extra body tissue." And after a few years on this regimen, our body weight will stabilize, albeit at a higher level than initially. Unfortunately, though, when we lose weight, the amount of calories needed to maintain (i.e., repairing, replacing tissue, moving) our lower weight body also decreases. We can remember from physics that it takes fewer calories, for example, to move a lighter object than a heavier one. What this means is that once someone has lost a substantial amount of weight, he or she needs either to increase physical activity or decrease caloric intake. And Katan and Ludwig say that most people do the opposite: once they have lost weight, they return to their previous eating and exercising patterns before their weight loss, with the result that they will tend to gain their weight back.

A word of caution: many people who have a weight problem cannot stop at one chocolate chip cookie. Weight expert, Dr. Stephen Gullo, whose book, *Thin Tastes Better*, said it best, "Where there can be no moderation, there must be elimination." In other words, for some, it is easier not to take that first bite. ■

Chocolate chip cookies. For some people, it is easier not to take that first bite.

Kimberly Vardeman, Lubbock, Texas. Wikipedia Commons: Creative Commons Attribution.

* **Note:** Researchers no longer accept the simplistic rule that eliminating 3500 kilocalories will lead to a pound of weight loss. For more updated information, see my Blog 99, *Mathetical Models: Obesity by the Numbers.*

'TIS BETTER TO HAVE LOST AND REGAINED THAN NEVER TO LOSE AT ALL

The ups, downs, and spin on yo-yo dieting.

6—Posted December 11, 2010

THE YO-YO IS A TOY that dates back to Ancient Greece and may have originated even earlier in China. There is, however, nothing playful about our experiencing unintentional weight fluctuations that go up and down like a child's yo-yo. According to the *Oxford English Dictionary*, the first reference to "yo-yo dieting" appeared in an advertisement in the New York Times in the early 1960s. In the scientific literature, the term began appearing in the 1970s and 80s, where it is also referred to more technically as "weight cycling." Since about 75 to 95% of those who lose weight will tend to regain some, all, or even more than their initial weight before their weight loss, weight cycling is a common enough occurrence. In fact, some researchers like Foster and his colleagues believe "Weight regain remains the most common long-term outcome of obesity treatment."

Weight cycling can be an intentional pattern, such as seen with professional ballet dancers, wrestlers, or boxers (e.g. remember Robert DeNiro's considerable weight gain when he played the boxer, Jake La Motta, in Scorsese's *Raging Bull.*) It can also be a normal pattern, such as seen in the weight gain during pregnancy and subsequent weight loss after a baby's birth. Mostly, though, it is a pattern all too familiar to dieters: cycles of intentional weight loss and subsequent unintentional weight regain over time. For years, researchers have grappled with both the medical and psychological consequences of this pattern. Results, though, have been inconsistent, partly because there is no definition of what constitutes weight cycling: For example, there is no standard regarding how much weight loss is involved, over what period of time, or even how often or how many weight cycles do occur. Some studies don't even specify whether the weight loss is intentional (e.g. through dieting or exercise) or unintentional (e.g. through disease, medication, mood disorders). Furthermore, investigators may rely on self-reports of their patients, often years after the fact such that data are subject to the inevitable and natural distortions of memory. Typically, as well, we know that both men and women tend to overestimate their height and underestimate their weight when they self-report.

Those who have lost weight appreciate that it is often much easier to lose than to maintain that loss over time. After all, weight maintenance is much less reinforcing than weight loss. And often dieters are not realistic in terms of the amount of weight they can actually lose. Weight maintenance may involve accepting a shape and weight previously regarded as unacceptable or just "not good enough." When dieters fail at weight maintenance, many inevitably try again, often subjecting themselves and their bodies to countless weight fluctuations over time. Incidentally, a recent study by Erez and colleagues from Israel suggests that it may be possible to predict from both genetic studies and from levels of the hormone leptin which people are most susceptible to weight regain.

What about the clinical significance of this yo-yo pattern? After all, if there are negative medical or psychological consequences to weight cycling, should health professionals always be recommending weight loss again to those who

Yo-yo player, Altes Museum, Berlin.

Formerly in the Schloss Charlottenburg; Photographer: Bib Saint-Pol, 2008) Wikipedia Commons/Public Domain

have regained their previously lost weight? For example, there have been suggestions in the scientific literature, particularly in some (but not all) animal studies, that metabolism may be negatively affected such that less weight is lost after a subsequent regain. There has also been the suggestion that bone density may be affected by repeated cycles. Probably, though, it is more likely that those on diets may fail to get adequate amounts of supplemental calcium and vitamin D.

What about the psychological consequences of weight cycling? Clearly, for some predisposed to anxiety and depression, weight regain may lead to decreased self-esteem and self-confidence and particular feelings of loss of control. There is even the suggestion that weight cycling is associated with binge-eating.

Though weight cycling may not be a completely benign phenomenon, some researchers, such as Prentise and colleagues, have suggested that weight cycling may have been common in primitive times when our ancestors had to deal with intermittent cycles of famine and plenty.

The jury is still out on all the possible effects of weight cycling. A report some years ago from the National Task Force on the Prevention and Treatment of Obesity concluded that the "available evidence" of any adverse effects of weight cycling did not override all the potential benefits of weight loss. And even a weight loss of ten pounds can have substantial positive effects on cardiac health, diabetes, blood pressure, sleep, and sex. Drs. Mehmet Oz and Michael Roizen have had an effective campaign this past year of "Just ten"-indicating that even a weight loss of just ten pounds in those overweight or obese is worthwhile, a conclusion supported by weight expert, Dr. Louis Aronne, of Weill Cornell Medical College.

Bottom line: still 'tis better to have lost and regained than never to have lost at all! ■

Lady with a yoyo c. 1770 kanoriya. Indian miniature painting, Northern India.

Contributor: ephotocorp / Alamy Stock Photo / Brooklyn Museum

◀ Previous page: *Girl Playing with an Antelope,* Indian, Pahari, about 1840–50. For many, yo-yo dieting, i.e., the gaining, losing, and then re-gaining of weight, is a way of life. Whether this pattern is detrimental to health is open to question for some researchers though most do believe it is better to lose even if it means regaining the weight eventually.

STICKER SHOCK: CAN WE MAKE TEMPTATION LESS TEMPTING?

Liking fries, not in spite of, but because they're unhealthy

7 — Posted December 22, 2010

"We like fries not in spite of the fact that they're unhealthy but because of it." So said Malcolm Gladwell, author of *Blink* and *The Tipping Point*, in an article written some years ago in *The New Yorker.* He added, "... nothing is more deadly for our taste buds than the knowledge that what we are eating is good for us." Gladwell was discussing how McDonald's French fries came to be so perfectly and reliably cooked every time. The problem, though, is that they became not much more "than a delivery vehicle for fat." Over the years, the fries have become somewhat healthier as they are no longer cooked in trans fats or, to the delight of vegetarians, in beef tallow. But as we all know, Americans generally don't go to McDonald's to eat healthy. The hamburger, the McLean Deluxe, for example, explained Gladwell, was then an unmitigated failure, even though in blind taste tests, people actually thought the healthier burger tasted better. But once people knew it was healthy, they rejected it.

Ten years later, are we eating any better now that calories are clearly posted in our fast food chain restaurants? In fact, some of us are suffering from "sticker shock:" we had been notoriously poor at estimating the calorie count in many of our favorite foods despite an interest in watching our calories. For example, who had any idea that one piece of Starbuck's reduced fat Very Berry (even sounds healthy) Coffee Cake could have 350 calories and 10 grams of fat, a piece of pumpkin bread could have 390 calories and 15 grams of fat, or one raspberry scone could have 500 calories and 26 grams of fat? Has knowledge of these calorie counts had any effect on our consumption? Apparently, so far, not so much!

As Malcolm Gladwell says, we like fries because they are unhealthy.

Popo le Chien. Creative Commons Attribution Share Alike License, Wikipedia Commons.

A recent study, published in the *American Journal of Public Health*, by Tamara Dumanovsky and her colleagues, is the second part of an earlier study sponsored by the New York City Health Department. This previous study, before the mandated posted calorie counts, found that one in three customers in fast food restaurants was purchasing as many as 1000 calories for lunch alone—half of a full day's recommendations for adults. In 2008, New York City became the first state to mandate "clear and conspicuous" calorie postings. This most recent study, sampling around 1200 adults,(representing 15 different chains) compared consumption three months before the mandated calorie posting and then three months after. Though customers varied by fast food chain in how much they were actually cognizant of the posted calorie counts, (e.g. 87% of McDonald's customers but only 70% of customers at Starbucks, for example), in general, only about 27% of customers (and a substantially lower percentage in older customers) said that the calorie information actually affected their purchases. Though the percentage of customers who were aware of calorie counts had increased considerably after the mandate, the proportion of those using the information was still quite low (on average, from one in four to one in five, depending on age, location of the restaurant, etc.) The researchers concluded that calorie postings, while clearly increasing awareness of nutritional information, seem not necessarily to have had a major impact yet in getting the majority of us to make healthier choices.

Part of the problem is that we have so many options among our food choices. And we can make choices based on many factors, as for example, when we choose foods on the basis of cost or convenience. Cognitively, we are, of course, capable of choosing a less desirable alternative for reasons such as health. In that sense, we are different from other species. Imagine, for example, a carnivorous animal having an internal dialogue about whether to choose the grass (e.g. salad) rather than eat its kill or even whether to eat a second animal once it is full from the first.

One of the most puzzling questions, though, is why we humans choose to eat foods when we know they are unhealthy and even seem to prefer these unhealthy foods. In other words, why is providing nutritional information about the fat, sugar, and salt content of fast food not necessarily enough for the majority of people? The answer is not an obvious one. After all, humans are probably unique in the animal kingdom for being capable of extrinsic motivation in its full form. This is motivation as a means to something else, according to psychologist Roy Baumeister, who has his own psychologytoday.com blog, *Cultural Animal.* Our cognitive abilities enable us to visualize potential outcomes. This is different from intrinsic motivation-or motivation to satisfy our own immediate needs. And we are capable of self-regulation-we can reconsider and stop an action. Sometimes, though, this extrinsic motivation fails us: we either lose sight of or minimize future benefits (e.g. maintaining a healthy weight) in favor of immediate wishes to indulge. The challenge is to make temptation less tempting: align short-term and long-term goals to reduce any discrepancy between short-term and long-term consequences. ■

For those who tend to eat huge quantities of food, "sticker shock" of calorie counts may not be as much a deterrent to overeating as to those who are quite surprised to learn calorie counts of foods they enjoy. Flemish painter Jacques de L'Ange, *Gluttony,* circa 1642.

Milwaukee Art Museum. Wikipedia Commons/ Public Domain.

FROM SNAKE PIT TO SNAKE OIL: DIET REGIMENS

Are we still "beating the insane to keep them quiet?"

8—Posted January 13, 2011

Back in the late 1970s, obesity researcher, Jules Hirsch of Rockefeller University, suggested that the dietary regimens to which the obese were then subjected were "the modern day equivalent of beating the insane to keep them quiet." In other words, those regimens, analogous to "snake pit" asylum treatments, did not work very well and were perhaps even overtly cruel, and if they did work, the results were short-lived and hardly curative. Today, all these years later, many people still, looking for a quick fix, subject themselves to the cruel and unusual punishments found in nutritionally unbalanced crash diets, often with severe caloric restriction (e.g. 800 to 1000 calories/day), that promise substantial weight loss in a relatively short time. What these misinformed people end up losing is lean muscle and water rather than fat. There is no quick fix to dieting. For those who want to maintain weight loss over time, a diet must become a way of life, a "regimen," not a short-term endeavor.

More recently, Dr. David Katz, Director of the Prevention Research Center at Yale, and author of one of the most comprehensive books on nutrition, called *Nutrition in Clinical Practice,* conducted an extensive survey of the "seemingly limitless" possibilities for weight loss. Katz found a "prevailing gullibility" among the public who seemed "beguiled" and generally willing to suspend disbelief when it comes to promises of weight loss. In other words, many people, believing in the equivalent of snake oil, will accept any weight loss claim at face value, no matter how seemingly preposterous. In his review, Katz, incidentally, found no evidence that any diet approach, other than sensible calorie restriction, is superior to any other when we are talking about "sustainable weight loss." But, as most dieters appreciate, dieting by extreme calorie restriction is not only "intrinsically unsustainable" over time, but also downright dangerous. Incidentally, a recent long-term study by Sacks and his colleagues reported in *The New England Journal of Medicine* confirmed that calorie counting, rather than any specific percentage of protein, fat, or carbohydrates, is essential for weight loss. With that said, I do believe it is also likely that we will eventually isolate genetic factors that will determine what percentages of protein, fat, and carbohydrates work best for a particular individual.

Some general principles to maintain a healthy diet regimen and avoid either snake pit tactics or snake oil salesmen:

- Maintain a consistent caloric range (depends on weight goals, activity level, height, weight, frame, gender) on all days (weekends and holidays included)
- In general, avoid fasting and very low calorie (800 cal/day) diets unless under medical supervision; (for some, the faster the weight loss, the more likely weight regain)
- Eat foods high in fiber (about 25 grams/day)
- Maintain low fat intake (25 to 35% of daily calories) with particular restriction of saturated fats (e.g. red meat) and avoidance of trans fats that are particularly heart unhealthy (often found in processed foods to prolong shelf life). Some diets suggest very low fat intake (about 10% of daily calories), but if a diet is too low in fat, it is hard to maintain over time
- Eat low glycemic foods: that is, carbohydrates that release insulin slowly (e.g. brown rice, steel-cut, slow-cooked oatmeal rather than white rice or instant oatmeal) (note also the actual percentage of carbohydrates in a food, the so-called glycemic load) and avoid processed carbohydrates that come in packages
- Avoid foods sweetened with high fructose corn syrup (unnatural glucose-fructose combination)

Snake Charmers, a chromolithograph by Alfred Brehm, c. 1883.
Wikipedia Commons/Public Domain.

Alphonse-Étienne Dinet, *The Snake Charmer,* 1889. Art Gallery of New South Wales.

Wikipedia Commons/Public Domain.

Snake Oil Cure All in Masonville, Colorado.

Shea Oliver. Used with permission, Alamy Stock photo.

- Avoid foods with chemicals, even if labeled "natural—read all labels
- Eat lean protein (certain proteins more satiating than others and digestion of protein uses more calories than either carbohydrates or fats)
- No more than 2 glasses of wine/day (red wine has anti-oxidant resveratrol), but alcohol can increase appetite, as well as make someone lose track of how much eating (and alcohol is preferentially oxidized so fat is stored rather than burned)—Avoid foods that trigger you to eat more
- Eat breakfast daily* and aim for three meals/day (preferably at same time each day); dinner should be smallest meal and avoid eating later in the evening
- Eat slowly (to increase sense of fullness)
- Always remember portion control (avoid "portion distortion" (Wansink))
- Monitor food intake daily (keep food diary)
- Weigh self daily or at least weekly—importance of accountability
- Exercise regularly (to maintain substantial weight loss, more than one hour/day of moderate aerobic exercise for cardiovascular fitness; in addition, flexibility for balance; and resistance, strengthening for preservation of lean muscle
- Remember all physical activity burns calories, including maintaining posture, fidgeting, even chewing gum
- Know your own zones of vulnerability: multi-task eating (eating while working or driving); overeating in particular situations (e.g. restaurants or parties); bingeing at certain times of day
- Sleep more than 6.5 hours/night but less than 9 hours

For more specific details, please see my book, *The Gravity of Weight*. (American Psychiatric Publishing, Inc. 2010) ■

* **Note:** Not all researchers believe in the importance of eating breakfast as a means of weight control.

STRICTLY FROM HUNGER: THE ABCs OF INSUFFICIENT FOOD

"Is there a right to food?"

9—Posted February 11, 2011

"THE HISTORY OF MAN *is in large part the chronicle of his quest for food,*" wrote Ancel Keys, an early researcher in obesity. Even *the Bible* is replete with references to fears of inadequate food supplies. Remember Pharaoh's dream of the seven fat cows and seven lean cows that Joseph interpreted to mean seven lean years and seven years of plenty?

But "Is there a right to food?" So ask Drs. John Butterly and Jack Shepherd provocatively in their new book, *Hunger: The Biology and Politics of Starvation.* Though it has been reported that obesity may now be more common in much of the world today than is hunger, these authors report that there are still over 1.02 billion people worldwide who suffer from chronic hunger. According to these authors, there is a *"nutritional crisis,"* not only among those chronically hungry but paradoxically among those who suffer some form of malnutrition now in combination with obesity. Even in the U.S., where childhood obesity and overweight continue to increase at an alarming rate, Butterly and Shepherd report that *almost 15% of all households "do not have enough to eat daily and suffer from recurring hunger."*

Hunger is a complex physiological signal, often "compelling and unpleasant" that creates an urge to eat. The word 'hunger," has come to have psychological connotations as well. Hilde Bruch did early work on anorexia nervosa, a disorder where some sufferers claim they don't experience normal pangs of hunger. She believed that awareness of hunger sensations are not present at birth but develop either "accurately or distortedly" in the context of an infant's interactions with caretakers. As a result, some people seem not to be able to distinguish "*hunger—the urge to eat—from signals of bodily discomfort that has nothing to do with food deprivation*" and reflective of states of emotional tension. In effect, this is psychological hunger. It is probably that hunger that led to our slang expression *"strictly from hunger,"* for something "very bad," "undesirable" or "lousy." When people, though, are subjected to chronic physical hunger, as well as malnutrition, they become more susceptible to disease. According to Butterly and Shepherd, as a result of its effects, chronic hunger becomes the leading cause of death throughout the world. When there are severe food shortages that lead to an actual breakdown of society, we call that *"famine."* Ultimately, though there are many causes of starvation or famine (e.g. weather, political conflict or corruption, reliance on a single crop, etc.), Butterly and Shepherd believe that poverty is the underlying cause. The rich rarely starve—except, of course, by choice!

What are the physiological and psychological effects of hunger? Ancel Keys was one of the first to conduct a long-term scientific study of hunger and semi-starvation. During World War II, while some countries like Holland were in the throes of famine, (and American parents were telling their children to eat everything on their plates because children were starving in Europe), Keys devised an experiment with more than 30 male conscientious objector volunteers. These men were willing to restrict their food intake substantially for several months for the purpose of providing detailed information on the psychological and physiological effects of caloric restriction. Called the *"Minnesota semi-starvation experiment,"* Keys documented his results in an enormous two-volume tome. The men in the study, of course, were undergoing semi-starvation "under the best possible conditions"-comfortable living conditions, warm clothing, and no threat of attack. Over the course of the experiment, the men were to lose about 25% of their weight by caloric restriction and exercise and then spend three months in rehabilitation. Humans, incidentally, can tolerate a weight loss of up to 10% of their weight without much functional disorganization that starvation can cause;

Famine Memorial, Dublin, Ireland, commemorating the Great Irish Potato Famine of the 1840s.

Wikipedia Commons/Public Domain. Photo taken by AlanMc, 2006.

Jean-Baptiste Carpeaux, French sculptor, *Ugolino and his Sons,* 1865–67. Ugolino was given the choice between cannibalism and starvation. Rarely, during extreme conditions of famine, those are the only two desperate options.

Metropolitan Museum of Art, NYC. Credit: Purchase, Josephine Bay Paul and C. Michael Paul Foundation, Inc. Gift, Charles Ulrick and Josephine Bay Foundation Inc. Gift, and Fletcher Fund, 1967. Public Domain.

Grzegorz Stec, *Hunger*, 2010, oil on canvas.
Photo by the artist, who gives permission under Creative Commons Attribution-Share License.

a severe famine might result in substantial weight losses of up to 35% of weight. Any weight loss beyond that percentage is usually incompatible with life.

Keys' study, though published over 60 years ago, can give us some clues to why dieting by food restriction can be so difficult for so many people. What he found was that with prolonged under-nutrition, hunger sensations become "progressively accentuated," unlike in total fasting where feelings of hunger may dissipate after a few days. Keys, incidentally, reported that Jews in the Warsaw Ghetto in 1942 were barely subsisting on 600 to 800 calories a day. People in his own experiment were given almost twice as many.

During the experiment, Keys' subjects became totally preoccupied with food. They would often dawdle for hours over their meal, and they developed "striking changes" both physically and mentally. The men became depressed, listless, unable to concentrate, socially withdrawn, and apathetic. They also began to neglect their personal appearance: they no longer brushed their teeth, combed their hair, or shaved. And they lost all interest in sex. Keys called the syndrome *"semi-starvation neurosis."* Incidentally, some researchers more recently believe that hunger causes cognitive changes in people such that those who go on "hunger strikes" have serious mental changes that result in impaired decision-making skills.

The 3-month rehabilitation period in Keys' study, was even harder for some than the period of caloric restriction. The men ate thousands of calories (some up to 10,000 calories) a day and still felt unsatisfied-what was called a "post-starvation hyperphagia." The speculation was that their bodies were attempting to recover not just the fat they had lost but the loss of their lean body tissue-often gaining more weight than their initial weight—a process called "overshooting." This study has some clinical implications for today's dieter: those who lose a substantial amount of weight on caloric restriction may eventually gain back even more and it may also explain why exercise that preserves lean muscle becomes so important in the maintenance phase after substantial weight loss. ■

A GLASS HALF-FULL, A GLASS HALF-EMPTY: HOW MUCH WATER TO DRINK?

Water pollution and debunking the rule of eight.

10—Posted March 14, 2011

Devastation from the recent earthquake and tsumani in Japan has brought subsequent threat of contamination from a potential nuclear meltdown, but more immediately, there have been reports of severe food and water shortages. Depending on our nutritional and body fat status, for example, we can survive without food for weeks or even months as long as we have water. According to researchers, though, we can live only 6 to 14 days, depending on the rate of water loss, without any water. Our bodies are, after all, about 50 to 70% water; even the brain, according to physiologist Heinz Valtin, is 75% water. We have water within and outside our cells, including in our blood, which is 85% water. When body fluids fall below an optimal level, we enter a toxic dehydrated state: there develops an imbalance of the vital electrolytes sodium and potassium and a disturbance in brain function. Dizziness, mood changes, lethargy, brain swelling, delirium, coma, and even death can result. As a result, our body has an exquisitely precise homeostatic system for water regulation that involves the kidneys and specific hormones of the endocrine system primarily, though the small intestine is the primary site of water absorption in our body.

So how much water do we need to drink daily for optimal health? Conventional wisdom is we require eight 8-ounce glasses of water daily, particularly since we lose water when we breathe, sweat, or excrete. Valtin takes issue with this common admonition. Writing some years ago in the *American Journal of Physiology*, he attempted to trace its origin by conducting a comprehensive search of the literature. In fact, he could not find any scientific validation or convincing evidence for this well-known recommendation given to healthy adults! And he even went so far as to suggest that too much daily liquid may be dangerous for some people and may lead to hyponatremia (low sodium levels) or even unnecessary exposure to pollutants. And he noted that all liquid consumption should count in our daily tally, whether water, juice, coffee, soda, and even beer in moderation, though we know that many liquids such as sweetened sodas can add many additional calories daily. The point is water is essential for life but our water requirements vary with climate (including temperature and humidity), gender, diet, exercise, age, health, etc. In general an average healthy adult requires about one and one-half quarts daily to replace normal losses. According to researchers Jéquier and Constant, in a recent article in the *European Journal of Clinical Nutrition*, our level of thirst can usually determine our intake of water and it may vary substantially from person to person. This mechanism, though, may not be accurate in the elderly or in infants, who are particularly sensitive to the effects of dehyration. Furthermore, infants have immature kidneys that cannot concentrate urine well, a high metabolic rate, and a limited ability to indicate thirst. Certain disease states, such as those causing fever or diarrhea, may increase our usual daily needs.

Other researchers have suggested that many things may influence the effect of water on us: even the speed with which we drink may determine how much we retain. For example, drinking a large amount in a few minutes is excreted rapidly whereas the same amount over several hours is largely retained. And some have suggested that water in food makes us feel more satiated, though it is not clear how long that effect lasts or how much liquid is required.

What about the use of bottled water? There have developed fairly common sightings, at least in New York City, of people who cling to huge bottles of water, not unlike the security blankets of infants, as if they were crossing the Sahara alone and far from mother rather than the city streets. Bottled water, incidentally, is not necessarily

Tarutao National Park, Thailand.

Used with permission/istockphoto.com. Credit: Utopia_88.

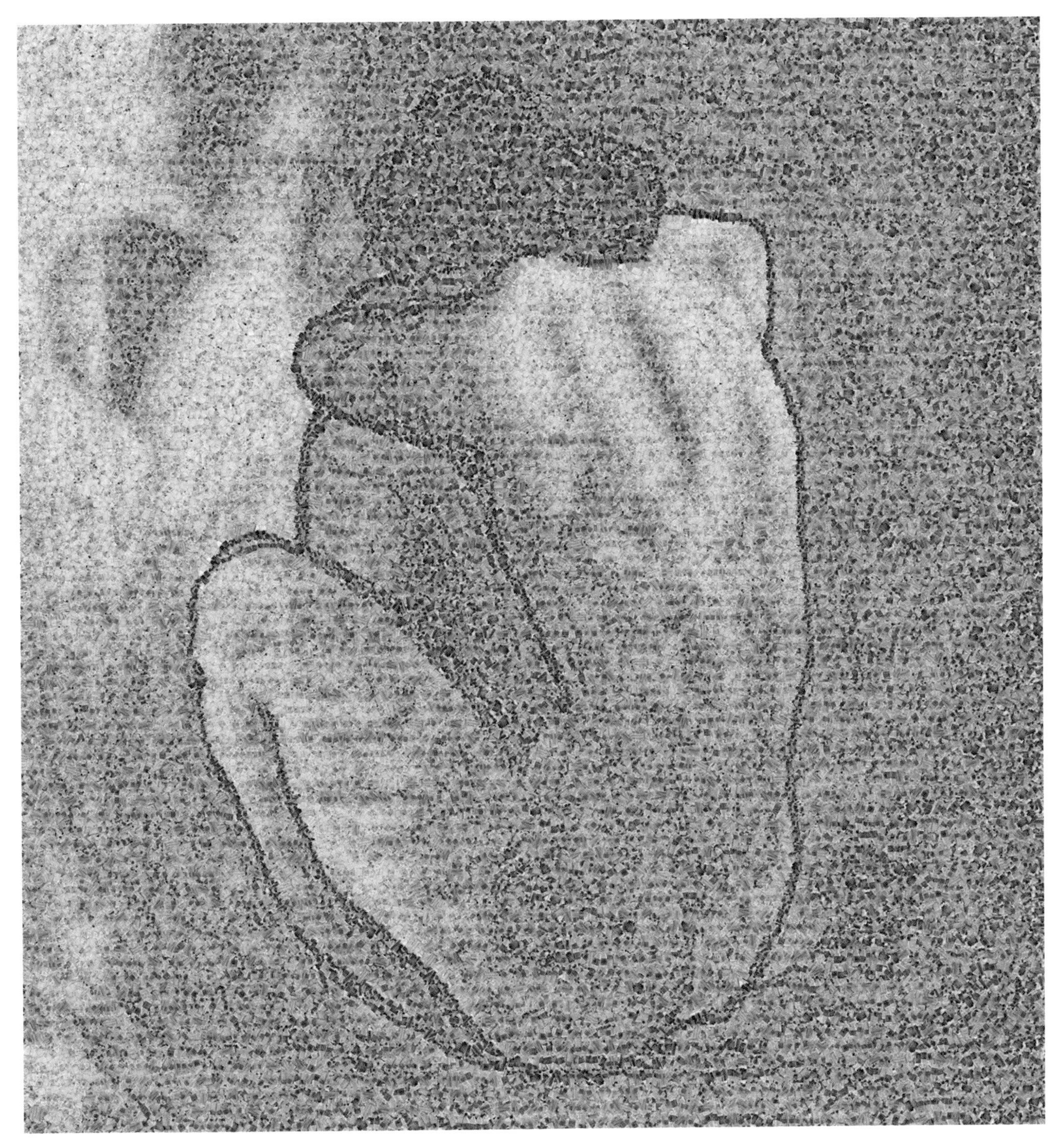

Chris Jordan, *Blue,* 2015. Depicts 78,000 plastic water bottles, equal to 1/10,000th of the estimated number of people in the world who lack access to safe drinking water.

Johannes Vermeer, *Young Woman with a Water Pitcher,* circa 1662. Metropolitan Museum of Art, NYC.

Marquand Collection, Gift of Henry G. Marquand, 1889. Public Domain.

safer than most city tap water. It may also carry its share of industrial waste, sewage, bacteria, and chemical contaminants. In fact it was reported that at least one company carried the label, PWS, for "public water source!" And writer Ian Williams, in an article, aptly titled, "Message in a Bottle," questioned whether water from melted glaciers is really so healthy, particularly when this water may have been "lying around from the last Ice Age" with such pollutants as lead, dioxin, and even "polar bear poop and...the occasional dead Inuit or Viking." The other problem with bottled water is the havoc all these plastic bottles are causing on our environment. For those ecologically-minded, I would strongly recommend the recent extraordinary book, *Moby Duck*, by Donovan Hohn, which vividly describes the "toxic goulash" created by plastic debris, among other pollutants, floating in our oceans. Hohn's own search for the almost 30,000 bath toys that fell overboard in a 1992 transit from Asia led to his "archeology of the ordinary" and a voyage of profound awakening to the problems of environmental pollution for both himself and for the reader. Hohn found that one beachcombing couple scavenging for debris had retrieved seventy-five different brands (many from other countries) of polyethylene water bottles!

Instead of those plastic bottles that can pollute the environment for hundreds of years, people may want to consider getting a water filter for their own tap water. And remember, though water is essential to our existence, there is no science to the rule of eight! ■

Used with permission/istockphoto.com. Credit: winterling.

FRAYED, FRAZZLED, AT THE END OF YOUR ROPE: STRESS & WEIGHT

Trying to cut one Gordian Knot of weight control

11—Posted April 23, 2011

BACK IN THE 1930S, Hans Selye, an Austrian born physician, described a characteristic adaptation syndrome that occurred when an organism was exposed to "diverse noxious agents," or "stressors." These stressors could be excessive exercise, surgical injury, drugs, or even temperature changes like exposure to cold. Selye emphasized that the body produced a classic, patterned response that was actually independent of the actual stressor. For Selye, there were three stages to this stress response: general alarm, a stage of resistance, and a stage of exhaustion or even death, if the stress is severe enough. And though the responses were highly specific, the actual stressors could be non-specific. Selye also made the point that stress did not have to be negative: it could, in fact, be positive or even healthy, such as when someone feels good when working on challenging or creative work. Selye considered stress the "salt of life." As such, he felt it was neither advantageous nor even possible to eliminate it from life. Instead, the challenge is to contain any stress and channel it into feelings of mastery.

Stress, incidentally can be either chronic or acute. When chronic, the body can develop two forms of adaptation: *habituation*, in which repeated exposure diminishes the effect of the stressor or *facilitation* (also referred to as sensitization) in which repeated exposure intensifies the effect of the stressor.

Over the years, as the word "stress" has entered common parlance, we now know so much more about the many complex physiological and behavioral responses that occur when an organism undergoes stress, whether acute or chronic. For Rockefeller University researcher Bruce McEwen, stress occurs when an individual's homeostasis or internal equilibrium is either actually threatened or perceived as threatened. McEwen emphasized that some of the most powerful perceived stresses for humans are psychological and experiential, such as those involving novelty, withholding of a reward, or anticipation of punishment, sometimes even more than experiencing the punishment itself.

Among the many reactions that occur, the body reacts to stress with a specific feedback system involving the brain's hypothalamus and pituitary gland, as well as the body's adrenal glands, called the HPA (hypothalamic-pituitary-adrenal) axis that leads to a release of a "cascade" of hormones, including cortisol. But neurotransmitters, such as dopamine, epinephrine, norepineprine, serotonin, and acetylcholine, among others, are also released. And our brain's mesocorticolimbic system, involved in anticipating, recognizing, and even remembering danger, as well as those systems involved in motivation and reward, also participate in the stress reaction.

Ironically, the process of eating, itself, can be a stress on our bodies for it is simultaneously both necessary for maintaining the body's homeostasis and a threat to its homeostasis. Researchers Power and Schulkin, in their book, *The Evolution of Obesity*, speak of this as the "paradox of eating." And conversely, when stressed, some people find themselves eating more, particularly the so-called "comfort foods" often high in fat, sugar, and salt, whereas others find themselves unable to eat at all.

In recent years, results from animal studies have appeared that have suggested that stress itself can lead to obesity. One mechanism described is that insulin, in the presence of increased secretion of corticosterone (analogous to our human cortisol) leads to an accumulation of increased visceral (i.e., the particularly dangerous) fat.

A very recent study in the journal *Obesity* by Wardle and her colleagues reports on a systematic literature review conducted to determine whether psychosocial stress actually does lead to weight gain in humans. From over 600 "potentially relevant" studies, these authors were able to include results in their final meta-analysis from only 14 research papers that met their stringent criteria of longitudinal (from one year to 38 years of follow-up), well-controlled studies from Europe and the U.S. published in peer-reviewed journals. Furthermore, the authors restricted their sample to studies that relied on actual physical measurements rather than self-reports of weight. Acknowledging that there are many types of stressors, as for example, physical, physiological, and emotional, these authors focused on "external stressors" such as "life events, caregiver stress, or work stress." They also noted that previous epidemiological research on the connection of stress to weight gain had led to "inconsistent results."

What Wardle and her colleagues found is that psychosocial stress can lead to weight gain, particularly demonstrated in those more methodologically sound studies and those with longer follow-up, though the effects were "modest and smaller than assumed in the lay literature." In fact, 69% of the studies included found "no significant relationship between stress and adiposity." The authors did find, however, that men seem to have "stronger physiological responses" (e.g. increased cortisol levels, etc.) to stress than women. Interestingly, the authors did not find that visceral fat accumulation was more "sensitive" to stress than general fat accumulation, though they acknowledge that more studies might be warranted.

A Rope Maker, circa 1425, anonymous. Many people feel "at the end of their rope" by stress.

Hausbuch der Mendelschen Zwölfbruderstiftung, Band 1. Nürnberg 1426–1549. Stadtbibliothek Nürnberg, Amb. 317.2, via http://www.nuernberger-hausbuecher.de. Wikipedia Commons/Public Domain.

The connection between stress and weight accumulation is obviously a complex one and there are many other variables, including age, genetics, other psychological states (e.g. anxiety or depression), and even eating behavior, involved. In fact, the connection may lie somewhere between being at the end of one's rope and trying to cut a Gordian Knot! ■

Man Resting on a Spade, Engraved by Pierre Millet (French), 1874, Woodcut on this laid paper. Gift of Dr. Van Horne Norrie, 1917.

Metropolitan Museum of Art, NYC. Public Domain.

CALLING A SPADE A SPADE: SHOULD A PHYSICIAN'S WEIGHT MATTER?

Steering between the Scylla and Charybdis of weight discussions

12—Posted May 24, 2011

In a recently published article in the journal *Obesity Reviews,* Zhu and colleagues examined the relationship between a health professional's own weight and its effects on his or her own "health-promoting" recommendations. The authors note that research has already established that health professionals' own smoking and exercise habits do, in fact, influence how they practice and what they recommend to patients. But what about a doctor's or nurse's own weight? Does it have any relevance?

New York pediatrician and writer Perri Klass, MD sensitively tackled that issue in an article, *When Weight is the Issue, Doctors Struggle Too,* some years ago in *The New York Times.* Dr. Klass admitted she has herself had chronic difficulties controlling her own weight, and asked, "How on earth…am I supposed to give sound nutritional advice when all they have to do is look at me to see that I don't follow it very well myself?" She added, "…how am I supposed to help stem the so-called epidemic of childhood obesity when not a week goes by that I don't break my own resolutions?" After talking about nutrition and exercise with an adolescent patient who had been gaining weight, she describes looking the mother in the eye and adding quite spontaneously, "If this were easy, I would be thin and fit." But not all physicians are apparently as psychologically-minded and compassionate as Dr. Klass.

According to Zhu et al, overweight and obesity are not uncommon in doctors and nurses. For example, the authors refer to a study by Miller and colleagues that noted that in a sample of over 750 nurses in the U.S., 54% were overweight or obese by body mass index measurements. And in a sample of 1.2 million staff in the National Health Service in the United Kingdom, 700,000 were classified as either overweight or obese.

In their own systematic, in depth review of nine studies, Zhu and colleagues found that those nurses and physicians who were themselves of normal weight were more likely than their overweight colleagues to provide overweight or obese patients with guidelines and advice to achieve weight control and "use strategies to prevent obesity in their patients. " But weight status of the physicians and nurses apparently did not seem to influence either their general assessment of their patients' weights or their referral practices. Perhaps not surprisingly, "specialty was one of the strongest independent predictors for weight management practices," such that those involved in primary care like pediatrics or family practice were more likely to discuss weight than surgeons. The authors recommend further research since there are so few studies that address this issue directly and the relationship between health practitioners' own weight and their behavior toward patients is "complex." Their research, though, suggests that physicians' and nurses' weights seem to be "less strongly and consistently" associated with how they practice than smoking and exercise behaviors.

The Doctor by Luke Fildes (1843–1927), 1891.
Wikipedia Commons/Public Domain.

Whether overweight or obese or not, though, how should health professionals address these issues in weight-challenged patients? An earlier study by Wadden and Didie, aptly titled, *What's In a Name?* found that obese men and women rate the word "fatness" significantly the most undesirable description for their weight, but also did not like the words "obesity," "excess fat," or "large size." The researchers found that obese patients preferred more neutral words like "weight problem," "BMI," (body mass index), "excess weight," and "unhealthy body weight."

Wadden and Didie cautioned physicians that use of those more undesirable terms could be "hurtful or offensive" and even "derogatory" to patients. Like Odysseus, then, those professionals who work with these patients have to steer clearly between the Scylla of using terms that may seem derogatory and the Charybdis of avoiding any discussion of weight control whatsoever. But sometimes a spade is just a spade. ■

King of Spades, The German, from Harlequin Cards, 2nd Series (N220), Issued by Kinney Brothers (American), 1889, Commercial color lithograph.
The Jefferson R. Burdick Collection, Gift of Jefferson R. Burdick. Metropolitan Museum of Art, NYC. Public domain.

Mount Rushmore with American Flag Background.

iStock.com, used with permission. duckycards.

WE HOLD THESE TRUTHS TO BE SELF-EVIDENT: THE OBESE CANDIDATE

Should morbid obesity disqualify a politician from seeking higher office?

13—Posted October 2, 2011

Should morbid obesity be a disqualifier for higher office such as the presidency? This question has surfaced in the media recently as there is a political movement to encourage New Jersey Governor Chris Christie, an obese person, to join the 2012 Republican race against President Obama. According to an article in the *Huffington Post* by reporter Sam Stein, Peggy Howell, public relations director for the National Association for the Advancement of Fat Acceptance (NAAFA), has apparently come to Mr. Christie's defense.

This organization, founded in the late 1960s, under the name National Association to Aid Fat Americans, began as a nonprofit human rights organization to defend against discrimination. Over the years, it has functioned as a clearinghouse for lawyers who advocate for the rights of the obese, as well as provide support groups, a social network, and workshops. NAAFA has rejected the word *obese* because it "pathologizes heavier weights" but, analogous to gay activitists who have claimed the word "queer," have reclaimed the word *fat*. For those in NAAFA, fatness is just "body diversity," analogous to racial, ethnic, or sexual preference diversity.

Prejudice and overt discrimination against the obese have a long history. Early Christian teaching labeled gluttony and sloth among the seven deadly sins. Albert Stunkard found, for example, that even in a scroll from 12th century Japan, men appear to be laughing at an obese woman who had to be carried by others. And even Shakespeare called Falstaff "fat-kidneyed" and a "huge hill of fat." There is also evidence that health care professionals, including physicians and nurses who work in the field of obesity, maintain certain prejudices against their patients, with some viewing their patients with typical stereotypes such as they are lacking self-control, worthless, and lazy. And one study found that those caring physically for the obese were repulsed by them and preferred not to have any physical contact with them. Even Ancel Keys, one of the foremost researchers in obesity and the first to popularize the use of body mass index (BMI) as a standard for classifying levels of obesity, called obesity "disgusting as well as a hazard to health." And Wadden and Stunkard have described how some have called obesity an *aesthetic crime*—namely, "it is ugly."

Further, there are many studies that indicate that the obese are discriminated against in regard to education (e.g. college acceptances), employment opportunities and compensation, marriage, and even jury selection. In one study, children were asked whether they would like to befriend an obese child or one who is handicapped and in a wheelchair, they rated befriending an obese child as the least desirable among several other options, including the handicapped child.

Obesity is multi-faceted and involves far more than self-discipline. About 70 percent of our weight is determined by our genes, (with hundreds of gene loci regulating weight) but environmental (from portion control to circadian rhythms), neurochemical, gender, perinatal, developmental, and psychological factors also are involved.

"Chicago Nominee: 'I knew him, Horatio; a fellow of infinite jest..." *Hamlet*, Act 1V, (sic), Scene 1. This satire refers to the 1864 American presidential campaign.

Artist: Justin H. Howard (American, active 1856–1880), Probably published by Thomas W. Strong (New York), 1864. Wood engraving. Gift of Georgiana W. Sargent, in memory of John Osborne Sargent, 1924. Metropolitan Museum of Art, NYC, Public Domain.

So are the obese physically and emotionally capable to withstand the considerable stress of higher office such as the presidency? That depends, and the situation is clearly a complex one. While there is a category of metabolically benign obesity in which an obese person exercizes regularly and has not demonstrated any evidence of disease, most with obesity are much more highly likely to have the metabolic syndrome, a cluster of abnormalities including abdominal fat (the most dangerous place for fat accumulation), hypertension, abnormal glucose levels, insulin resistance, and overt diabetes, as well as abnormal blood lipid levels and even sleep apnea that make them more susceptible to chronic (and often debilitating) illness. Furthermore, obesity has not only been associated with cardiovascular disease, but also orthopedic problems, and many kinds of cancer. And there is evidence that obesity is associated with increased mortality, with the higher the BMI, the more likely an earlier death. Studies that have not demonstrated this connection are often those that have not controlled for smoking so that a higher body weight seems erroneously protective against death.

Further there is no specific personality found in all who are obese. And though it is also not clear which is cause and which is consequence, obesity has been associated with increased levels of anxiety and depression.

So while obesity, and particularly morbid obesity, with a BMI of 40 kg/m^2 (Class 3 obesity) itself is not a contraindication to seeking higher office, it certainly should be taken into consideration. ■

A TIME FOR EVERY PURPOSE: THE SCIENCE OF CHRONO-PHARMACOLOGY

They say timing is everything, but for taking medication?

14—Posted November 14, 2011

Whether you cannot read "To everything there is a season, a time for every purpose under the sun..." without hearing Pete Seeger's music and the voices of The Byrds from their song "Turn, Turn, Turn" or whether you know these words from the *Bible's Ecclesiastes* (3:1–8), they may be able to teach us something about the importance of our body's natural rhythms as they relate to the timing of ingesting medication prescribed to us.

In recent years, researchers have come to appreciate that medications can have vastly different effects, from beneficial to toxic, depending on the time of day when they are administered. Sometimes this timing may be more important in determining a drug's effect than its route of administration (e.g. oral or intravenous) or even how quickly the medication is eliminated from the body. This is the science of chronopharmacology, and it is based on the fact that we, as well as all light-sensitive organisms (including some bacteria and fungi), have internal biological clocks, i.e. circadian rhythms, that regulate many, if not most, of our physiology. Originally it was thought that there was one "master clock" (i.e. "the pacemaker" or suprachiasmatic nucleus) that is located in the brain's anterior hypothalamus and is synchronized by the earth's 24-hour light-dark cycle as the sun's light hits our retina. We now know that many other environmental signals called zeitgebers ("time givers"), such as eating (even specific foods such as chocolate or other carbohydrates), drinking alcohol, or exercise can affect these natural rhythms and phase-shift them one way or the other. One of the most common causes of the phase-shifting of our circadian rhythms, of course, occurs with jet lag. Those who have experienced this syndrome that is brought on by jet travel across time zones know how uncomfortable it can make us feel, and how our patterns of eating and sleeping, as well as our sense of well-being, can be severely compromised, at least transiently.

Marinus van Reymerswaele (1490–1546), German (1541) at the Museo del Prado (Madrid). Jerome (pictured) produced a 4th-century Latin edition of the Bible, known as the *Vulgate* that became the Catholic Church's official translation.

Wikipedia Commons/Public Domain.

We also now know that most cells have "clock" genes and can sense time; these genes control our sleep-wake cycle, body temperature, blood pressure, and many metabolic processes such as digestion and the release of many of our hormones (e.g. cortisol, melatonin, and insulin), as well as our intake of medications.

According to a 2010 article in the journal, *Annual Review of Pharmacology and Toxicology*, researchers Georgios Paschos and his colleagues at the University of Pennsylvania School of Medicine note that circadian rhythms can affect absorption, distribution, metabolism, and elimination of many of our commonly used medications. For example, benzodiazepines, calcium channel blockers, acetaminophen, and antidepressants are all absorbed more rapidly when given by mouth in the morning rather than at night. And the anticoagulant heparin is twice as effective when given during the day, whereas the antibiotic gentamcyin is best tolerated when given in the afternoon. Furthermore, chemotherapeutic agents against cancer can have more devastating toxic effects (e.g. peripheral neuropathy) when cells are undergoing division, a process also under the control of circadian rhythms. Many medical conditions, such as arrhythmias, acute myocardial infarctions (e.g. more common in early morning hours), stroke, and even sudden death "exhibit prominent circadian patterns" that affect when they are more likely to occur and how severe the symptoms are. Even symptoms of allergic rhinitis and bronchial asthma are more severe in the morning.

Researchers believe that it makes sense from an evolutionary perspective that the administration of medication would be under circadian control as a defense system—what is called "xenophobic (from the Greek, "fear of strangers") detoxification." In other words, it may have evolved to protect us from any potentially noxious substances, including foods.

Eating, incidentally, is also affected by our circadian rhythms. For example, those who eat only one evening meal for the 24-hour period are more prone to weight gain and even obesity, as are those who have the night eating syndrome (e.g. eat most of their calories after 6 pm, frequent awakening during the night to eat, avoidance of breakfast, and even abnormal patterns of hormone secretion.)

Unfortunately, the science of chronopharmacology is still in its infancy, but it is certainly worthwhile to ask your physicians whether information is known about the importance of timing when taking medication prescribed to you. ■

Portrait of Christiaan Huygens, oil on canvas, 17th century. The Dutch polymath and horologist Christiaan Huygens, the inventor of first precision timekeeping devices (pendulum clock and spiral-hairspring watch).

Wikipedia Commons/Public Domain.)

Richard Wallingford pointing to a clock, his gift to the Abbey of St. Albans. Wallingford's face is disfigured by leprosy. Title of work: *Golden Book of St. Albans*. 1380.

Wikipedia Commons/Public Domain.

The astronomical clock face, Prague. Photo by Andrew Shiva, 2013.

Wikipedia Commons. Creative Commons Attribution-Share Alike 4.0 International.

MEASURE FOR MEASURE: A MADNESS IN THE METHOD FOR WEIGHT CONTROL

Why don't we know more about obesity than we do?

15 — Posted December 24, 2011

Emperor Jahangir (reigned 1605–1627) weighing his son Shah Jahan on a weighing scale by artist Manohar (A.D. 1615, Mughal dynasty, India). 1615.

Artist: Meister der Jahângir-Memoiren, British Museum, London. Wikipedia Commons/ Public Domain.

In this season of excessive consumption and overindulgence, we might want to consider why control of our weight continues to seem so daunting for most people. Over the years, we have been inundated with news reports almost daily of new studies about obesity and overweight that either contribute little to our existing knowledge or provide information that overtly contradicts previous studies. While there have been literally thousands of published reports, ranging from large epidemiological research to small case studies in peer-reviewed journals, why don't we have more answers and a greater understanding about obesity than we do? Researchers clearly have been trying.

The answer lies in the fact that obesity research, unfortunately, lends itself particularly well to methodological difficulties. Admittedly, there are no impediments that are specific to obesity research. In fact, in his classic 1970s paper, Canadian Professor Dr. David Sackett, delineated 55 categories of potential bias that can occur at any stage of a clinical research study, from conducting an initial literature review and selection of a sample population to measuring and interpreting data collected, and even publishing a study's results.

What exactly is bias in research? It is any systematic error (as opposed to an error by chance) in the design or implementation of a study that can interfere with its validity. In other words, validity, whether internal or external, is essentially the degree to which a study is free from bias. Obviously, when the validity of a study is compromised in any way, researchers must become much more cautious in making inferences or issuing recommendations to clinicians and patients.

Major impediments to obesity research are of particular importance in the areas of measurement of body composition (particularly of adipose tissue); food and caloric consumption; and physical activity. For example, most obesity studies rely on self-report data that are notoriously inaccurate: people tend to overestimate their height and underestimate their weight, particularly when they are obese. Furthermore, the use of body mass index (BMI) as the standard measure of obesity (weight in kilograms divided by height in meters squared, with obesity defined as a BMI of 30 kg/m^2 or greater) is controversial. Since BMI measures not only adipose tissue but also muscle and bone, it can be highly inaccurate in many populations, including athletes (with increased muscle mass) or the elderly, as well as in very tall or short people or even in children. Measuring adipose tissue by use of skin calipers is also potentially inaccurate: it depends on the skill of the examiner and may vary from one examination to another or from one observer to another.

The Ancient Egyptian *Book of the Dead* depicts a scene in which a scribe's heart is weighed against the feather of truth.

National Geographic, Ancient Egyptians, circa 1285 B.C., British Museum. Wikipedia Commons/ Public Domain.

Measurement of food and caloric consumption, other than on a metabolic hospital unit, whose artificial setting has its own set of complications, is also potentially highly inaccurate. Researchers collect information through food diaries, 24-hour recall, or food-frequency questionnaires. These are subject to distortion either inadvertently due to subjects' faulty memory or intentionally due to subjects' embarrassment over their behavior so that they may tell researchers what they think they want to hear, rather than how they actually behave. Further, subjects may also change their style of eating while they are in the study and fail to adhere to protocol, particularly over the course of a prolonged time-period and so bias their answers. Sometimes, even the control population changes its behavior to follow the protocol of the experimental group.

Likewise, measurement of physical activity also lends itself to inaccuracy: for example, devices such as a pedometer, capture only certain kinds of movement, and people often overestimate how much physical activity they are engaging in or even fail to judge whether an activity is vigorous or moderate. Standard activity tables are actually only general estimates of caloric expenditure, as no two people engage in the same activity similarly.

Of course, inaccuracy in measurement is not the only source of bias. Non-randomization of the sample population is another. For example, those who volunteer for a study may be different from those who don't. Typically, they may be more health-conscious. One of the most publicized long-term studies, for example, is the National Weight Control Registry. Even though this study, begun in the 1990s, has yielded considerable information regarding those successful at weight maintenance, its original sample began as a non-random population solicited through advertisements and not representative of the typical American population.

High attrition (i.e. drop-out) rates, as well, are common in obesity studies, particularly when data collection extends over many years' duration. Typical attrition rates hovering around 50% are not uncommon and can severely compromise the integrity of clinical research. These are only some examples of the limitations involved in obesity research.

Remarkably, given these impediments, we know as much as we do. ■

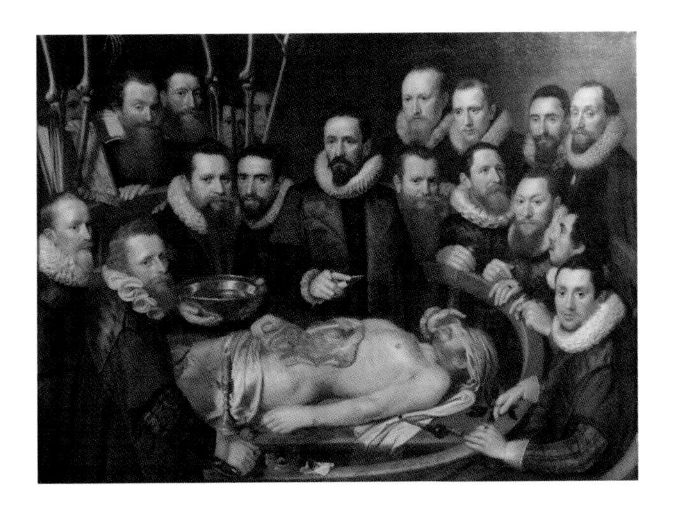

THE UNBEARABLE HEAVINESS OF BEING: CONSIDERING BARIATRIC SURGERY

When diet, exercise, and even medication are not enough

16—Posted January 23, 2012

As the population of the U.S. grows heavier and heavier, obese patients have turned to major abdominal surgery as an alternative treatment when they have been unsuccessful at losing significant weight (and maintaining that weight loss over time) by a combination of diet, exercise, and even medication alone. Currently, surgical treatment is recommended only when a person has at least a body mass index (BMI) of 40 kg/m^2 (considered "extreme," "morbid," or Class 3 obesity) or a BMI of 35 kg/m^2 (class 2 obesity) when there are substantial, significant accompanying medical conditions such as diabetes, hypertension, cardiac disease, or obstructive respiratory disease.(e.g. sleep apnea).

Bariatric surgery has considerable benefits besides substantial weight loss. Even before weight loss (and sometimes within days of surgery), some patients have improved metabolic profiles, including resolution of their diabetes and hypertension.

There have been many techniques in use over the last 50 years, but the two most common ones in use today are the gastric bypass (also known as the Roux-en-Y) and some form of gastric banding. Gastric banding is considered a restrictive technique in which the stomach capacity is made smaller by a band, placed either vertically or horizontally. Over time, though, the smaller stomach pouch may enlarge or the staples used to hold some bands in place may deteriorate so that a second surgical procedure is required. Some banding techniques can be done laparoscopically, with minimal hospital stays.

Anatomy lesson of Dr. Willem van der Meer, 1617, painting by Michiel van Mierevelt and Pieter van Mierevelt, Museum Prinsenhof, Delft. Bariatric surgery, such as the Roux en Y technique, involves rearrangement of the abdominal anatomy.

Wikipedia Commons/Public Domain.

Gastric bypass involves more extensive surgery and has a restrictive component, a malabsorption component, and a hormonal component. The surgery involves creating a small pouch in the stomach, but also "rearranging gastrointestinal anatomy," as researchers Cummings and Flum have described, by bypassing parts of the small intestine. This procedure can be done either abdominally or laparoscopically but does not allow, as do the banding techniques, for adjustments, should they become necessary over time so it has been considered a "one size fits all technique." Significantly, though, gastric bypass surgery inhibits the secretion of several hormones including ghrelin, the hormone that makes us feel hungry, and this may be responsible for the decreased appetite that may occur after surgery. Typically, bypass surgery leads to greater weight loss than the banding techniques and is less likely to require a second surgery.

Bariatric surgery is not necessarily a benign procedure. In an article written a few years ago in *The New York Times*, Stephen J. Dubner and Steven D. Levitt, who wrote the book *Freakonomics*, suggested that surgery for weight loss is what economists call a "commitment device." In other words, once started, "retreat is not an option." Dubner and Levitt jokingly suggested that in lieu of such a drastic strategy, the obese carry around their necks a Ziplock bag with a towelette "infused with…a deeply disgusting" smell to deter them from eating too much.

Over the years, mortality and complication (morbidity) rates from the various bariatric procedures have decreased, but still obviously depend on the skill and experience of the surgeons and their surgical teams. Serious complications, though, can still occur after surgery, including wound infections, pulmonary embolism, bowel obstruction, hemorrhage, deep vein thrombosis, and even the catastrophe of peritonitis. But even uncomplicated gastric bypass surgery requires postoperative follow-up as vitamin D, B12, calcium, and iron deficiencies should be anticipated. Furthermore, up to 70% of those who have had gastric bypass develop the "dumping syndrome"—symptoms that include light-headedness, nausea, flushing, sweating, abdominal pain, palpitations, and diarrhea after eating sugary foods. Vomiting is not uncommon with any bariatric procedure and can last for years after surgery. Though patients can lose from 40 to over 100 pounds over time, about 20 to 30% will regain considerable weight in the two years after surgery. The mechanisms for weight regain include increased intake of calories as the stomach pouch may enlarge over time, decreases in metabolic rate, and even changes in hormonal levels over time. Sometimes, as well, patients develop disordered eating patterns, including a binge eating disorder, which may contribute to weight regain.

Determining the best candidates for bariatric surgery is not always easy. Most centers today require a psychological evaluation (or at least a personal interview) prior to surgery though there is no agreement among mental health professionals about how this evaluation should be conducted. In some studies, psychiatric disorders, particularly depression, but also personality disorders, were common among bariatric patients. There are relatively few contraindications, though, for surgery, as there is a consensus that after surgery, most patients have greater self-esteem, less self-consciousness, and greater assertiveness. ■

▲ During the Middle Ages and the Renaissance obesity had been seen as a sign of wealth. Now this man might be a candidate for bariatric surgery. *The Tuscan General Alessandro del Borro*, attributed to Charles Mellin, circa 1630. Gemäldegalerie, Berlin.

Wikipedia Commons/Public Domain.

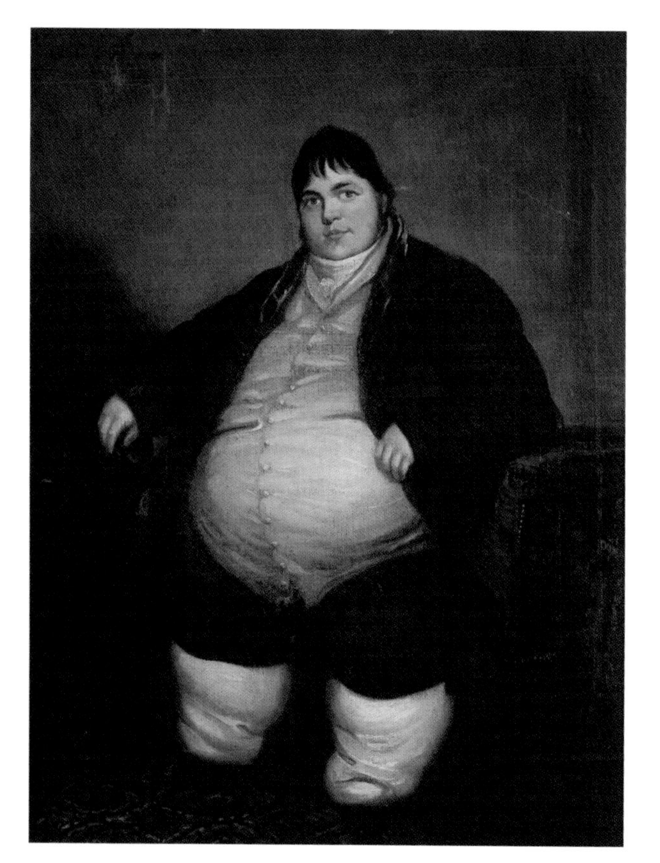

◀ Example of extreme obesity. Most bariatric surgeons would hesitate to operate on someone with such extreme obesity. At his death, at age 39, Lambert was reported to be over 700 pounds. Portrait of Daniel Lambert, circa 1806–07, artist unknown. This is not the same portrait of Daniel Lambert by Benjamin Marshall that is exhibited at the Royal Academy in 1807 and later in Leicester Museum and Art Gallery, United Kingdom. It is apparently similar to the composition of an engraved portrait of Lambert "published for the Proprietors by Bell & De Camo, 1 August 1809."

Wellcome Collection/Creative Commons, licensed by 4.0

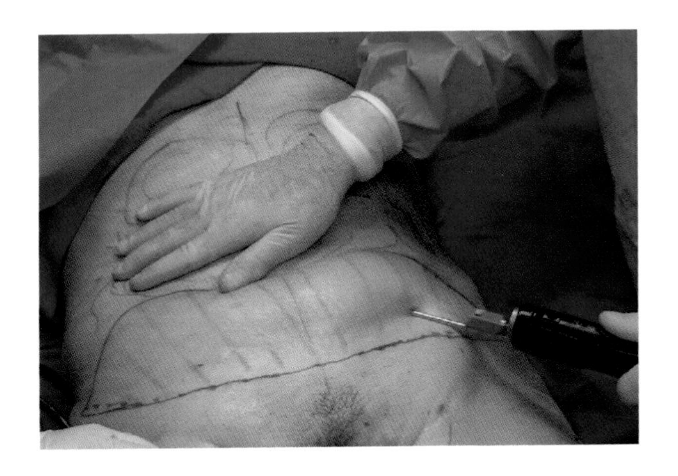

LIPOSUCTION: THE MOST UNKINDEST CUT OF ALL?

When to consider liposuction.

17 — Posted February 11, 2012

Woman receiving a more complicated procedure than liposuction. Surgeons are combining the removal of her *subcutaneous fat* around her abdomen by power-assisted liposuction with the technique of abdominoplasty, a procedure that is performed after liposuction, to tighten the excess skin that has lost its elasticity.

Marc Antony, in Shakespeare's *Julius Caesar,* speaks of the extraordinary betrayal that Brutus inflicted on Caesar—"the most unkindest cut of all"—when Brutus joins the conspirators and assassinates Caesar. Sometimes, our bodies seem to betray us, as well, no matter how well we treat them, when we accumulate fat in places that seem immune to diet and exercise. It is then that some people may want consider liposuction.

Liposuction is one of the most common surgical procedures performed in the United States, with hundreds of thousands done yearly on both men and women (though much more commonly in women.) It is essentially "body contouring." The technique consists of making small holes in specific areas of the body, injecting a fluid, such as saline or epinephrine to constrict the blood vessels and minimize blood loss, and then removing subcutaneous fat (i.e., the fat beneath the skin as opposed to visceral, i.e., fat around internal organs.) The fat can be removed by various methods, including power suction or laser. Depending on the quantity of fat removed, it can be an office procedure, or if "large scale" liposuction (involving removal several pounds of fat) is performed, it may require a hospital setting and general anesthesia.

Joseph A. Rabson, MD, Chief of Plastic Surgery at Montgomery Hospital in Plymouth Meeting, Pennsylvania, emphasizes that liposuction should definitely not be considered a procedure for weight loss *per se*. What he has seen, though, is that those who undergo liposuction are often more motivated to lead a healthy lifestyle and continue to diet and exercise after surgery, with subsequent weight loss and maintenance over time. Sometimes, surgeons will recommend weight loss prior to liposuction. But Dr. Rabson explains there is no such thing as "spot reducing," no matter how many sit-ups a person does. In other words, exercise can tighten and firm, but when someone loses weight, he or she loses weight all over the body, not just in certain areas. In some people, however, there are those problem areas that are just genetically predisposed to hold onto the fat and that's where liposuction can be most effective. Incidentally, liposuction alone will not be effective when a person has lost considerable weight because the skin has lost its elasticity; other surgical techniques may be required in addition.

Adipose tissue (fat), though, is a highly active endocrine organ that secretes hundreds of substances called adipokines (including leptin and adiponectin) that are involved in regulating fat accumulation. It is generally now believed that the number of fat cells (adipocytes) in the body remains constant and is not affected by the calories we eat. When fat cells are lost such as by liposuction removal, the body produces more. And the body can compensate, as well, by the enlargement of the remaining adipocytes. As a result, weight gain can, of course, occur after liposuction in other areas of the body so that diet and exercise must be considered essential after surgery. Obese people, especially those who have been heavier since childhood, incidentally, have many more and considerably larger fat cells than those who are lean.

Further, there is controversy regarding whether liposuction can benefit a person's metabolic profile, such as the abnormalities seen in the metabolic syndrome. Unlike liposuction, bariatric surgery (see blog 16, *An Unbearable Heaviness of Being: Considering Bariatric Surgery*) clearly has been shown to have considerable benefits in reducing

The Assassination of Julius Caesar, by Vincenzo Camuccini, circa 1804–05, Galleria Nazionale d'Arte Moderna e Contemporanea, Rome. Caesar was assassinated on the Ides of March (15 March) 44 B.C.
Wikipedia Commons/Public Domain.

glucose levels (and even the need for medication for diabetes), hypertension, and abnormal lipid levels. But bariatric surgery deals with the more metabolically active (and dangerous) visceral fat accumulating around the waist and encasing internal organs whereas liposuction aims at decreasing subcutaneous fat accumulation. (For those who are interested in some references regarding this controversy, please see my book, *The Gravity of Weight*, pp. 438–455.)

The overwhelming majority of those who undergo liposuction, especially the majority who do not gain weight after the procedure, feel there are many positive psychological benefits, including increased self-esteem and confidence, and increased comfort in wearing their clothing.

Plastic surgery, though, has been called *body image surgery* by Bolton and colleagues. As a result, physicians must be sensitive to those patients who have serious misperceptions about their body and even overt psychopathology, such as body dysmorphic disorder, and distinguish them from those whose dissatisfaction is specific and realistic. Years ago, researchers Napoleon and Lewis suggested that one screening technique plastic surgeons can employ is to ask patients to list, in order of importance, the five major areas of body dissatisfaction that they would like to improve. If the surgeon's own list is substantially different from and does not even include areas on the patient's list, that is a red flag and cause for concern (and even possible need for a psychiatric consultation.) ■

WHAT EXACTLY IS CELLULITE? "COTTAGE CHEESE" THIGHS

When this cottage cheese is definitely not "fat-free"

18—Posted March 1, 2012

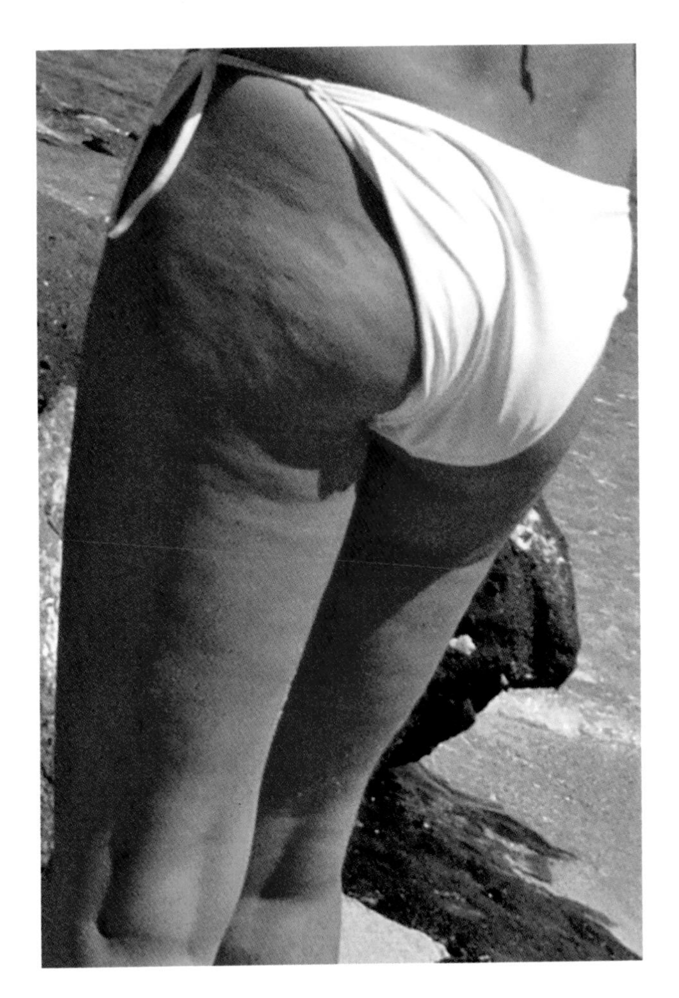

The dimpled skin appearance of cellulite. Some have compared it to the texture of cottage cheese.

Gazing (Flickr.com), Author: tata_aka_T, Tokyo, Japan, 2007. Creative Commons Attribution 2.0 Generic. Wikipedia Commons/Public Domain.

CELLULITE, known medically as gynoid lipodystrophy, is considered a localized metabolic disorder of the subcutaneous layer of skin, the dermis, and its small blood vessels. Over years, these vessels become compressed, resulting in inflammation, poor lymph drainage, proliferation of fibrous tissue, tissue damage, unevenness of the thickness of the skin, and the protrusion (herniation) of fat through layers of the skin. Some researchers have called cellulite a "condition of altered connective tissue" or an "abnormality of the architecture of the skin." Khan and his colleagues, in a 2010 comprehensive two-part article in the *Journal of the American Academy of Dermatology*, describe cellulite as "a result of several ultrastructural, inflammatory, histochemical, morphologic, and biochemical changes." The primary areas involved—the thighs, abdomen, and buttocks—acquire the uneven, dimpled, and textured skin of an orange (peau d'orange) or the lumpiness of cottage cheese. Though considered by some an abnormality, cellulite is extraordinarily common and found, to some degree after puberty, in from 85 to 98% of women (though more often in Caucasian than Asian or Black women.) Researchers emphasize that it is an aesthetic condition, though highly embarrassing to most women, and not associated with mortality or morbidity.

Cellulite was first described in the 1920s and was thought to be related to a water imbalance. Over 90 years later, we still don't know too much more about its pathophysiology or even exactly why it is seen primarily in women. There are many factors that seem to predispose women to develop cellulite, including genetic factors, hormonal factors (e.g. particularly estrogen), and lifestyle choices (e.g. lack of exercise, pregnancy, high carbohydrate diet with increased insulin levels, increased weight, excessive salt intake, protracted intervals of sitting or standing that interfere with blood flow, etc.) Though more prevalent and more extensive in obese women, cellulite is also seen in women of normal weight (i.e., normal body mass index, BMI) as well. Weight gain, itself, can make the appearance of cellulite worse. Weight loss and particularly significant weight loss, can make its appearance better.

Khan and his colleagues emphasize that all the treatments that are currently available offer only "mild" and at best, only temporary improvements in the appearance of cellulite. Treatment modalities include various topical creams (including ones that increase collagen formation such as the peroxisome proliferator-activated receptors (PPARs), massage, radiofrequency waves (using a combination of electrical and heat stimulation that can be painful), ultrasound, and even lasers. Intense pulsed light (i.e., light in the visible spectrum) that uses thermal energy, in combination with retinyl-based cream, has also been tried, with some success, to increase collagen production

Raphaelle Peale, *Still Life with Oranges*, 1818, Toledo Museum of Art, Ohio. The dimpled skin of cellulite has been referred to as *peau d'orange*—the skin of an orange.
Wikipedia Commons/Public Domain.

and decrease the irregular *peau d'orange* skin of cellulite. Liposuction, though, in general, is not recommended as a treatment because it has to be performed too close to the skin's surface and may result in a poor aesthetic result and even more likely, lead to complications. Other invasive techniques that require surgically cutting into the skin or injecting various compounds (e.g. caffeine, theophylline, hormones, vitamins, herbal substances) into the subcutaneous layers of skin are also not recommended as they may lead to localized edema, bleeding, tender nodules, infection, allergic reactions, and even irregularities in the external look of the skin.

It is probable that treatment will eventually entail the combination of different modalities. For example, ultrasound, which can lead to damage of fat cells, by itself, has not proved efficacious in treating cellulite but may have a role eventually when combined with other therapies. Likewise, there may be a role for the use of carboxy therapy in which carbon dioxide is injected into the skin's subcutaneous layers, or even the selective use of cryolysis, i.e., the destruction of cellulite by exposure to cold, although all the ramifications of this technique on adipose tissue are not known. Another new theoretical approach involves use of the phosphodiesterase inhibitors that increase circulation by vasodilation. The most familiar one, of course, is sildenafil, more commonly known as Viagra. Some researchers suggest that a topical form of sildenafil may eventually be developed and have a therapeutic role in treating cellulite.

Despite its ubiquitous presence in most women, cellulite remains an enigma for physicians, not only in terms of its pathophysiology, but also in terms of effective treatment. ■

LEAD US NOT INTO TEMPTATION: SCIENCE & THE MARSHMALLOW TEST

What 40-year-old marshmallows can teach us about dieting

19—Posted March 26, 2012

In the 1960s and 1970s, Walter Mischel and his colleagues began a study with 500 nursery school children of faculty and graduate students at Stanford University. These researchers were interested in "demystifying" the concept of "willpower" in their 4-year-old subjects. "What began as a set of experiments with preschoolers turned into a life-span developmental study," said Mischel et al. The investigators devised a "delay-of-gratification paradigm," popularly known over the years as the famous "marshmallow test," and even described in Daniel Goleman's book, *Emotional Intelligence*.

The researchers were interested in how children were able to resist temptations (i.e., delay gratification). There were several variations on the actual experiment, but essentially it involved presenting these preschoolers with particularly desirable treats, such as cookies, pretzels, and marshmallows (hence "the marshmallow test") and apprising them of the advantage (e.g. "a larger, later reward") of resisting their temptation to eat them immediately and delaying until the experimenter returned, usually only about 15 minutes later.

A tasty, gooey, freshly roasted marshmallow may be hard to resist. For Mischel's experiments, though children were given the opportunity to eat just plain marshmallows, some could not resist the immediate temptation.

Creative Commons Attribution 2.0 Generic. Source: Flickr by Nina Hale 2006. Wikipedia Commons.

What they found is that those children who could delay their gratification were, in general, ten years later as adolescents, more "socially and academically competent" than their peers, as well as "more able to cope with frustration and resist temptation." Unbelievably, Mischel et al reported, "seconds of delay time in preschool also were significantly related to the SAT scores when they applied to college." The researchers also suggested that a family environment where "self-imposed delay" is "encouraged and modeled" may give children "a distinct advantage" to deal with frustrations throughout life.

How were some preschoolers able to delay gratification? They used several strategies, such as avoiding looking at the treats in front of them, covering their eyes with their hands, talking to themselves, singing, and even trying to go to sleep during the brief waiting period (i.e., all "directing their attention and thoughts away from the rewards.") Paying attention to the rewards "consistently and substantially" interfered with the children's ability to delay.

Over time, the researchers also found, though, that stimuli can be what they call either "hot" ("consummatory") or "cool" (i.e., abstract or "non-consummatory"). In other words, even though "attention is the crux of self-control," it can have either a "facilitating" or "interfering" effect on resisting temptation, depending on cognitive features such as whether our attention is "arousing or abstract." For example, older children and adults can develop more sophisticated strategies: instead of thinking of the appealing qualities of a marshmallow (e.g. toasted, warm, gooey, and delicious), they can redirect and refocus more on its shape and color ("non-consummatory," cool properties) and imagine the marshmallow as a tasteless ball of cotton.

Most recently B.J. Casey, PhD, Sackler Professor and Director of the Sackler Institute for Developmental Psychobiology at Weill Cornell Medical College in New

Cotton plant, Texas. Photo taken by employee of the U.S. Department of Agriculture, 2006. Instead of imagining a delicious gooey marshmallow, imagine a more neutral stimulus like tasteless balls of cotton.

Wikipedia Commons/Public Domain.

York, and her colleagues have followed up on nearly 60 of the original sample of preschoolers, now in their 40s, to study gratification delay and two brain systems—"hot" (emotions and desires) and "cold" (cognitive control) involved. Their article can be found in the September 2011 issue of the *Proceeding of the National Academy of Science* (USA) What they found is that the ability to resist temptation is fairly stable over the lifecycle and predictive of behaviors 40 years later! Of course, since marshmallows and cookies don't quite have the same tempting quality as they once did when these subjects were nursery age, Dr. Casey and her colleagues devised experiments using social cues of faces with different emotional (e.g. happy, sad, neutral) expressions to study impulse control in their middle-age subjects.

For the first time, though, Casey et al were also able to use functional magnetic resonance imaging (fMRI) to study a subset of 26 of the original preschoolers. They were able to identify and confirm that different areas of their brains become more active in the so-called "high delayers" (prefrontal cortex) and "low delayers" (ventral striatum). They found, as well, that resisting temptations varies not only by cognitive control mechanisms but also by the context and the salience (i.e., the "compelling nature") of the stimulus. For example, these cognitive mechanisms can be "hijacked" by the more emotional and primitive areas of the brain (e.g. limbic system), especially at certain vulnerable times (e.g. adolescence) or with exposure to certain environmental cues (e.g. drugs, tempting foods). Even the behavior and attitudes of other people, i.e., social influences, can influence the ability to delay gratification.

This of course, has relevance for dieters, and restaurants seem to know this intuitively when they, knowing that many people will be less likely to resist, bring diners a dessert tray with an assortment of delicious offerings. So try to distract yourself by averting your eyes from the tray (a kind of "out-of-sight, out-of-mind" cognitive behavioral technique) or else imagine some neutral or perhaps even some truly disgusting substitute there in front of you. ■

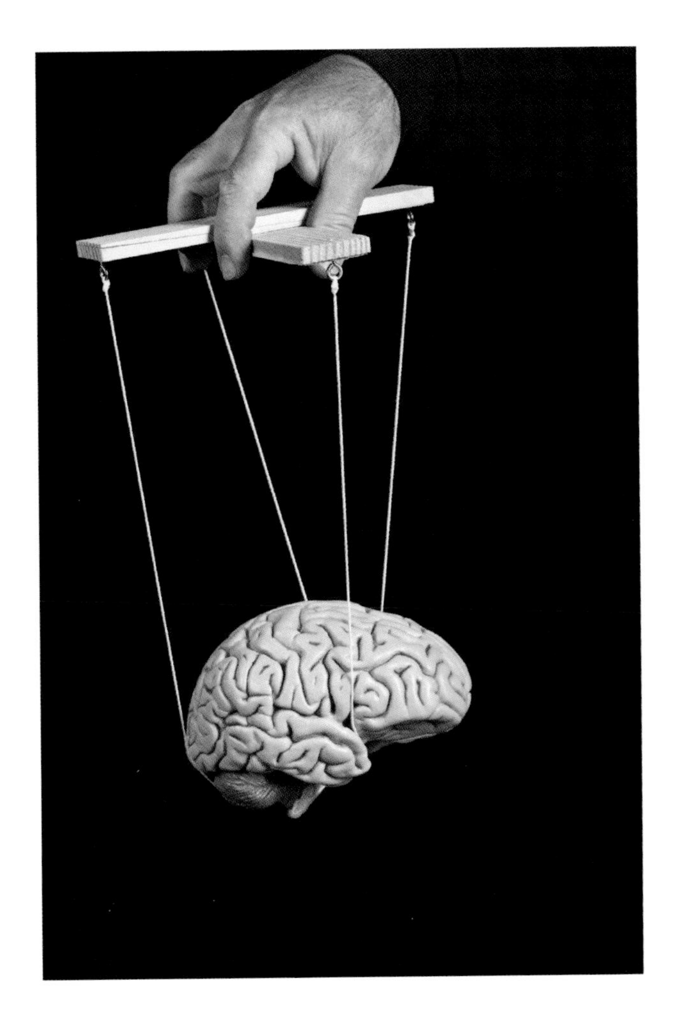

Used with permission, istockphoto.com. Credit: DebbiSmirnoff.

STRING THEORY: ATTACH INCENTIVES TO ASSIST IN WEIGHT CONTROL

Generating motivation to lose weight and maintain the loss.

20—Posted April 16, 2012

SINCE WEIGHT LOSS and maintenance of that loss require constant surveillance, most people need all the help they can to stay on track. Not only must they have a sense of self-efficacy, i.e., that they can be successful in accomplishing their goals, but they must be motivated to initiate and follow through with healthy lifestyle changes that may need to be continued indefinitely.

Motivation can be either *internal*, (called "self-motivation" or "autonomous" motivation), such as developing a sense of mastery or competence for its own intrinsic sake, or *external*, such as arising from a desire to gain (or avoid) some reward or punishment. This reward (or punishment) can be either quite concrete, such as money, or more abstract, such as gaining someone's praise and respect (or disapproval.) External motivation is the basis of the popular NBC television show, *The Biggest Loser*, now in Season 13 and paying over $250,000 to the grand prize winner this year. Here incentives may include fame and fortune, but also the pride and praise involved in winning competition with others and or even avoiding considerable humiliation and public embarrassment. In general, internal motivation leads to a sense that one's behavior is more under his or her own control, rather than contingent on external factors.

In her new book, *Strings Attached: Untangling the Ethics of Incentives*, Ruth W. Grant speaks of incentives in the field of behavioral psychology as a "particular kind of motivation—an extrinsic prompt, deliberately designed to elicit a desired behavior..." In other words, incentives used in this context mean "how to get people to act in accordance with their true interests, when their settled habits tend in the opposite direction." Grant further differentiates legitimate from illegitimate incentives; for example, incentives can be "irresistible" or "coercive" such as when a person does not really have a free choice or the incentives are not voluntary. Incentives should "maximize choice" and provide "tangible benefits." When assessing an incentive, Grant says, we have to consider what the "reason for the incentive" is, i.e., whether there is "undue inducement."

Incentives, though, can backfire: once there is no longer an incentive, one's motivation can stop as well. In other words, incentives work in the short-term. Grant makes the point that incentives have to be transformed into "habitual behavior and become part of accepted expectations" for them to have long-term impact. Further, incentives can send a message that we don't expect much from a person—that we have generally low expectations for that person to have self-directed behavior. And says Grant, "Why we do things impacts how well we do them." Grant also makes the point that incentives don't "address the root causes of the problem" and hence are "often limited as a tool for long-term improvements." Further, incentives, though, must have meaning and value to a person. If they don't work effectively, it is possible that they were not significant enough for that person.

John and colleagues, in a recent article in the journal *Preventive Medicine,* describe two types of monetary incentive systems for use in those who want to lose weight. The first, the deposit contract incentive system, has participants put their own money down to motivate them to lose weight; they will forfeit their own money if they fail to meet specific weight loss goals over time. Here, two principles are involved: over-optimism, or a belief people have that they can achieve their goals, and loss aversion, or people's unwillingness to lose their own money.

The second monetary incentive system is the lottery-based system, whereby people receive someone else's money (e.g. NBC sponsors of TV's *The Biggest Loser)* if they achieve their weight loss goals. John and colleagues found that both systems can produce weight loss, but there was "substantial" weight regain once the incentives were removed. What these researchers found, though, was that a pure loss is worse for most people than the idea of losing a deposit of their own money. The "peanuts effect" is the failure to appreciate that small deposits add up over time. In other words, aggregating all the small deposits that people stand to lose may motivate them to lose weight. The researchers also conclude that one size does not fit all: incentive programs may work best when they are "customized" to individuals. They also suggest that making a large first deposit may have greater consequences and contribute to continued larger deposits over time and they even suggest that some people may do well to switch from one form of incentive plan to another.

Bottom line: incentives may have a place in weight control in the short-term but they work best when they are individualized and when they lead people to develop intrinsic motivation and long-term changes in behavior. Otherwise, their effects may be short-lived. ■

Money is often used as an external motivation, arising from a desire to gain (or avoid) some reward or punishment.

American money, Federal Reserve, U.S. Government. Wikipedia Commons/ Public Domain.

Auguste Rodin, *Iris, Messenger of the Gods,* Circa 1895. The new hormone in skeletal muscle is named after this Greek god, Iris.

Metropolitan Museum of Art, NYC (Public domain.)

SPECIAL DELIVERY: WHAT CAN BROWN (FAT) DO FOR YOU?

Newly discovered muscle hormone, irisin, has exciting potential for obesity.

21 — Posted May 12, 2012

At least since the days of Hippocrates in Fifth century B.C. Athens, we have known the value of exercise for health, but not until very recently have researchers begun to comprehend precisely how exercise actually exerts its beneficial effects. Only in the past ten years, for example, have we begun to think of skeletal muscle as an endocrine (i.e., secretory) organ that is capable of communicating with other organs by hormones—myokines—that muscles can release into our bloodstream.

A new discovery in the cell biology lab of Professor Bruce Spiegelman, his postdoctoral fellow, Dr. Pontos Boström, and their colleagues, who are researchers at the Dana-Farber Cancer Institute and Harvard University, undoubtedly brings us closer to understanding the importance of exercise on a molecular level. Reported in the January 2012 issue of the prestigious journal *Nature*, these researchers have isolated a new hormone found in skeletal muscle that they have named, "irisin," after the Greek messenger goddess Iris. This goddess Iris, "swift-footed" and "golden winged" is mentioned by Homer in the *Iliad,* and there are depictions of her on Ancient Greek vase painting. The famous French sculptor, Rodin, depicted Iris erotically or pornographically, depending on your perspective.

For our purposes, here, though, irisin seems to be part of a communication (i.e., messenger) system between muscle and fat and may explain why physical activity and specifically exercise are so advantageous. Levels of irisin increase significantly in the blood of mice and humans with exercise, and irisin seems to offer a protective effect against many diseases (e.g. diabetes and diet-induced obesity) so far, at least, in mice. Interestingly, irisin found in the mouse is structurally (i.e., genetically) equivalent to that in human muscle. In contrast, human leptin, the hormone in fat involved in energy balance and regulation, shares only about 83% of its genome with mice (and insulin, only 85%). Most importantly, irisin also seems to have the effect (i.e., acts as a signal) of "browning" white adipose tissue. This is significant because white adipose tissue is much less metabolically active than brown adipose tissue, which can generate heat, (i.e., has a thermogenic effect and important for animals that hibernate, for example), and hence can cause an increase in energy expenditure (i.e., burning calories.) In other words, exercise seems to activate this "browning process," at least so far in mice.

For years, researchers believed that brown adipose tissue was found predominantly only in human infants and was responsible for their ability to regulate temperature without shivering. The assumption was that brown fat remained only in very small vestigial amounts as we age, unlike animals like rodents which keep their brown fat throughout life. It is now believed, with the advent of higher-resolution imaging, that adults have more brown fat than originally thought (and cold temperatures make it more visible on scans.) Brown fat, which has a large supply of mitochondria, (i.e., the "power plants" of the cell so-called because, among other functions, they are responsible for creating

Luca Giordano, *Iris,* 1684–1686, Palazzo Medici-Riccardi, Florence.

Wikipedia Commons/Public Domain.

Human anatomy discus thrower.

Used with permission, istockphoto.com. Cosmin4000.

chemical energy) gets its color from its rich vascular supply. There is speculation that brown fat actually protects against obesity by regulating thermogenesis, one of the ways we burn calories. (For those interested in more details about thermogenesis please refer to my textbook, *The Gravity of Weight*, pp. 57-63.) Brown fat is seen most prominently in human adults in neck and shoulder (cervical and supraclavicular) areas and is more prominent in younger adults who have lower body mass indexes, as well as those who have never smoked. Researchers in Spiegelman's lab now believe there are actually two different types of brown fat: so-called classical brown fat and now brown fat (called "brite cells or beige cells") originating from the browning of white fat.

Those who maintain their weight by strenuously watching their caloric intake and exercising regularly (as much as an hour a day, especially after substantial weight loss) know how very difficult it is for most people. We also know that physical inactivity seems to increase our risk for many chronic diseases, including cardiovascular disease and cancer. The idea of "exercise in a pill" (or even in an injectable form) seems only an implausible fantasy and clearly too good to be true. We are unquestionably a long way off, but perhaps irisin can hold some promise eventually for obesity treatment. Researchers do believe that brown fat may play a role in energy homeostasis and provide a "metabolic brake" that might one day be manipulated genetically or pharmacologically to treat obesity. Further, writing in a recent April issue of *The New England Journal of Medicine*, though, Pedersen ("The Muscular Twist on the Fate of Fat") says that it is possible that patients with diseases that compromise their ability to exercise may also benefit from the discovery of irisin. ■

A BITTER PILL TO SWALLOW: GRAPEFRUIT JUICE AND MEDICATION

Some fruit juices may not be for breakfast anymore.

22—Posted May 31, 2012

We have all heard of the benefits of fresh fruit as part of a healthy diet. Fruits are rich in antioxidants that help protect against cell damage caused by exposure to unstable compounds called free radicals, and they have fiber that is naturally filling and may aid in digestion. Sometimes, however, fruit and fruit juices may interfere with a drug's pharmacokinetics and create a food-drug interaction. In other words, sometimes they can affect the bioavailability of a medication and may adversely interfere with blood levels, causing either lower levels of medication that are ineffective or higher levels that may be toxic. The juice most commonly implicated is grapefruit juice, but pineapple, pomegranate, and even some kinds of orange juice have also been found to interact with medications.

Canadian researcher David G. Bailey and his colleagues had discovered the interaction of grapefruit juice and medication serendipitously in the early 1990s. They had been experimenting with the effects of alcohol on felodipine, one of the calcium channel blockers, medications used to treat hypertension. Wanting to mask the taste of the alcohol, Bailey said they tried "every juice in a home refrigerator one Saturday evening." White grapefruit juice from frozen concentrate proved the most effective to disguise the taste, but unexpectedly, they found that their patients developed an increase in their heart rate, lower standing blood pressure, and orthostatic hypotension, as well as medication blood levels that were fivefold higher with grapefruit juice than when the medication was given with water. It turns out that the primary pathway by which grapefruit juice interferes with medications is by way of CYP3A4, an isoenzyme of the cytochrome P450 enzyme family, that is found in the small intestine. This cytochrome P450 family, incidentally, is the same one involved in the metabolism of the selective serotonin reuptake inhibitors such as fluoxetine (Prozac), and is, in fact, responsible for the metabolism of the majority of medications currently in use. As a result, when less of the medication is metabolized, more remains in the blood. In grapefruit juice, furanocoumarins are the main chemicals that are responsible for the toxic interaction.

Giuseppe Arcimboldo's *The Summer,* (from *The Seasons*) 1563, Kunsthistorisches Museum, Vienna. Grapefruit is most commonly associated with potential interactions with medications, but many other fruits may decrease or increase medication blood levels.

Wikipedia Commons/Public Domain.

The actual effect of the medication-grapefruit juice interaction is, though, quite variable among people, suggesting a genetic component. Furthermore, some patients are more susceptible to the effects because of preexisting (and particularly chronic) medical conditions such as hepatic insufficiency that predispose them to drug sensitivities and abnormal drug effects. In those susceptible, one glass of juice can be enough to affect blood levels and these effects can last for 24 hours and even become cumulative over time. Researchers believe that juice that is more concentrated and of a greater quantity causes "more marked interactions," but factors like storage and preparation of the juice can factor in as well.

Since that initial chance association, researchers have found the grapefruit juice-medication interaction is far more common than originally thought and as noted, occurred not just with grapefruit juice but with other juices

Jan Matejko, *Alchemist Sendivogius,* 1867. Museum of Art in Lódz, Poland. The old alchemists or apothecaries would mix together compounds to make primitive medications.
Wikipedia Commons/Public Domain.

and foods as well. For example, in a more recent study by Methlie and colleagues reported in the *European Journal of Endocrinology* (2011), both grapefruit juice and licorice increased cortisol levels in patients given exogenous cortisol for treatment of Addison's Disease, a disease (from which President John F. Kennedy suffered) characterized by a deficiency of the body's own cortisol. Increased medication levels have been seen with many medications, including beta blockers, cardiovascular drugs, statins for lowering cholesterol, benzodiazepines, antihistamines, anti-epileptics, anti-depressants, and immune-suppressants. Most interactions result in increased blood levels. Commonly used medications, such as methadone (a synthetic opioid), cyclosporine (used for psoriasis and rheumatoid arthritis), midazolam (Versed, used as a mild anesthetic for relaxation and sedation prior to surgery), triazolam (Halcion, used for sedation), verapamil (a calcium channel blocker used to treat hypertension), and sertraline (Zoloft, an anti-depressant), have been shown to have their blood levels affected by grapefruit juice. For the statins, (e.g. lovastatin (Mevacor), atorvastatin (Lipitor), and simvastatin (Zocor), all used to lower cholesterol, grapefruit juice seems to be associated with adverse effects such as muscle pains (myalgias) and rhabdomyolosis (rapid destruction of skeletal muscle.) With the medication fexofenidine (Allergra, used to treat allergies), grapefruit juice, orange juice, and even apple juice have been reported to reduce blood levels and hence its effectiveness.

Bottom line: Consult your physician and pharmacist about any known interactions your medication may have with common foods, juices, or even other medications. Some will be completely contraindicated, whereas others may suggest cautious use, particularly because there can be such individual variation in response. Sometimes medications will come with warnings, including a warning to wait four hours between drinking juice and taking a drug. For those interested in a more complete (but far from exhaustive) list of potential interactions, please see the review article by Seden and colleagues in the journal *Drugs*, 2010, "Grapefruit-drug Interactions." ■

SUPERSIZING AND THE TYRANNY OF THE SODA POLICE?

"Consumption norms," "portion distortion," and the "completion compulsion."

23—Posted June 13, 2012

"How far will society go to regulate 'healthy behavior'?" So asked physician Faith T. Fitzgerald in an editorial, "The Tyranny of Health," in *The New England Journal of Medicine* almost twenty years ago. Fitzgerald wrote that people tend to see "failures of self-care," including obesity, as "crimes against society," and evidence that others have "misbehaved." As a result, they blame these others for their own illnesses, particularly because society bears so much of the burden in health care costs when people get ill. In the past twenty years, obesity rates have continued to skyrocket in the U.S. and most of the rest of the world. Those who study obesity use words like "pandemic" and "epidemic" to describe the situation and continue, rightly so, to warn of the dire consequences of increased incidences of type II diabetes, hypertension, abnormal blood lipid levels, and fat in all the wrong places, i.e., dangerously around our internal organs. Ironically, there are three phases to a classic epidemic, writes Dr. Katherine Flegal, who has conducted epidemiological studies on obesity and its prevalence: initially there is a reluctance to appreciate what is happening; the second phase consists of finding some framework to explain the events and may include blaming the victim; the third phase consists of creating pressure and a sense of urgency for the community to respond. We may be hovering between the second and third phases.

Then NYC Mayor Michael Bloomberg, designated the "nanny mayor," created a stir when he suggested that soda sizes be regulated to combat the obesity epidemic.

Used with permission, istockphoto.com, esolla.

Fitzgerald's question came to my mind during the recent controversy raised by New York Mayor Michael Bloomberg, who recommended, in an attempt to curb the burgeoning rates of obesity, that society regulates the sale of sugared soft drinks. Instead of thinking of obesity as a "crime against society," though, many were quick to deride our "nanny" Mayor for his police tactics and think of themselves as the victims. The press has had a field day. Frank Bruni, while supporting Bloomberg's recommendation in his recent editorial in *The New York Times*, pointed out the "random and absurb" nature of the ban on sodas greater than 16 ounces. After all, someone could still buy a 20-ounce milkshake with far more calories and considerably more fat. Bruni so aptly added, "Man cannot balloon on Mountain Dew alone." Jon Stewart, meanwhile, debated the potential existential conundrum when he mused on melted frozen hot chocolate and the physicality of the Slurpee. Is it a liquid? a solid? Stewart pondered. When a comparison to cigarette smoking was made (and how our attitudes toward smoking have changed over the years), Stewart retorted, "Yes, but there is no such thing as second-hand carbonation."

Obesity is a complex disorder that results from a combination of genetic, psychological, neuro-endocrinological, and environmental factors. Our genetics, though, have not changed in the past twenty years. Bruni rightly noted that what is most likely driving the obesity epidemic is the overproduction of inexpensive and increasingly available highly caloric (often processed) food rich in salt, sugar, and fat. This fact has been emphasized by Kelly Brownell, PhD, Director of the Rudd Center for Food Policy and Obesity at Yale University, and a major advocate for changes in food regulations and policy.

As a result, we actually do need some policing, since we cannot seem to police ourselves. Several psychological factors are involved: First, humans have what's been

Police Superintendent's Party: A Gift of Food and Drink by Tsukioka Yoshitoshi, Japanese, 1877, polychrome woodblock print, Gift of Lincoln Kirstein, 1960.

Metropolitan Museum of Art, Public Domain.

called a "completion compulsion," that is, we tend to eat in units. This was first noted in the 1950s by psychologist Paul Siegel when he found that people don't leave a fraction of a cookie. Many people understand that mentality when they cannot seem to stop eating until they reach the divider of a package (or even, if so inclined, to finish the entire package.) The recent development of 100-calorie snack packs is a step in the right direction.

Further, we lose sight of what are called "consumption norms," according to Dr. Brian Wansink, Director of Cornell University's the Food and Brand Lab. In other words, we begin to think whatever we are being served is reasonable and appropriate, no matter what the size. We develop what Wansink calls, "portion distortion."

How did supersizing begin? Greg Critser, in his book, *Fat Land*, explains that those in the food industry realized that people did not want to buy two boxes of popcorn when they were in a movie theater—because they didn't want to be seen eating two boxes. Eating two boxes just seemed too gluttonous—somehow one jumbo-sized box of popcorn seemed less "piggish." For just a little more money, retailers could sell much more—and with it, supersized Coca-Cola. Alex Bogusky, who came from an advertising background, says this is "value marketing"—increase the size of a product and increase its desirability for the customer. Bogusky's book, *'The 9-Inch' Diet*, tells how Americans have become so used to supersizing that some people thought they were buying "normal" drinking glasses when they were buying flower vases from Ikea!

Not only are we fat, but if you want the proverbial second opinion, we are generally lazy too. Wansink's research has demonstrated that we are more apt to eat candy that has its wrapper removed or even placed just next to us. Want to inhibit eating? Just keep the wrapper on and move the candy several feet away, preferably to a locked cabinet. Cognitive behavioral therapists have known of these techniques for years.

So, of course, we can go back and order that second soda—but most of us won't—nor will most want to be seen as gluttonous and take two sodas simultaneously. If we cannot do it for ourselves, let's welcome the soda police, at least as a start. That's no longer the tyranny of health. ■

FATAL FLAWS: DETERMINING WHO IS OVERWEIGHT AND WHO IS OBESE

How flawed measurement may underestimate the true prevalence of obesity

24—Posted July 11, 2012

THE OBESITY CRISIS has reached epidemic proportions, not only in the U.S. but worldwide. An article last year by Boyd Swinburn and his colleagues in the prestigious British journal *The Lancet* has called it a "global pandemic." In the U.S. alone, for example, almost 17% (12.5 million!) of children and adolescents are now considered obese; about 1/3 of adults are obese and another 1/3 are overweight.

How do we determine who is obese and who is overweight? Typically, in adults, we use body mass index (BMI), which is our weight in kilograms divided by our height in meters squared. It is an equation, popularized by Ancel Keys in the 1970s, that dates back to the Belgian 19th century mathematician (and father of modern statistics), Adolphe Quetelet. By today's standards, established in the late 1990s, a BMI of 25 kg/m^2 to 29.9 kg/m^2 is overweight, and a BMI greater or equal to 30 kg/m^2 is obese, with further divisions into Class I, II, III (morbid or extreme obesity), and Class IV (super morbid obesity), depending on increasing levels of BMI. As I have noted previously, researcher Jeffrey Friedman, from Rockefeller University, explains that obesity is a "threshold" measurement. In other words, it is defined when BMI "exceeds a defined threshold" such that "a relatively small increase in average weight has a disproportionate effect on the incidence of obesity." BMI is only an approximate measure of fatness, and most researchers today believe it is an imprecise measurement, even in adults, because it does not factor in muscle and lean body mass. An athlete with considerable muscle may be considered obese whereas an elderly person with muscle wasting (sarcopenia) may be considered of normal weight when he or she is actually overweight or even obese.

Is the obesity crisis even worse than we have imagined? Recently Shah and Braverman, writing in the journal *PLoS One* (2012) have suggested that BMI is so inaccurate a measure that it significantly underestimates the actual prevalence of obesity. Because increasing levels of obesity are associated with major health morbidity (e.g. many cancers, osteoarthritis, type 2 diabetes, dyslipidemias, cardiac disease, stroke, sleep apnea, etc), these researchers contend that we need to identify clinically useful biomarkers and clarify what exactly we are measuring when we use BMI. BMI has become so popular and widespread as a diagnostic tool because of its "convenience, safety, and minimal cost," but its cutoffs are "arbitrary." Instead, Shah and Braverman suggest that a more accurate and direct measure of body fat can be obtained by using DXA (dual-energy X-ray absorptiometry), the same machine that measures bone density. Because this machine uses X-rays, it obviously cannot, though, be used with pregnant women. They also suggest that fasting leptin blood levels should be used to indicate body fat percentages. Leptin, as I have mentioned previously, is the hormone (isolated in Jeffrey Friedman's lab) that is secreted primarily by fat cells (adipocytes) and among its many functions, regulates energy balance. Just as obese people who have the metabolic syndrome, (e.g. abdominal obesity, abnormal blood lipid levels, hypertension, and abnormal glucose levels) have high, though insensitive levels of insulin (i.e. a state of insulin

Standards of beauty, including body sizes, have changed over time since the Renaissance. Anonymous Italian painter, Portrait of a Woman as Cleopatra, Second half of the 16th century.

Walters Art Museum, Baltimore, Maryland. Wikimedia Commons/Public Domain.

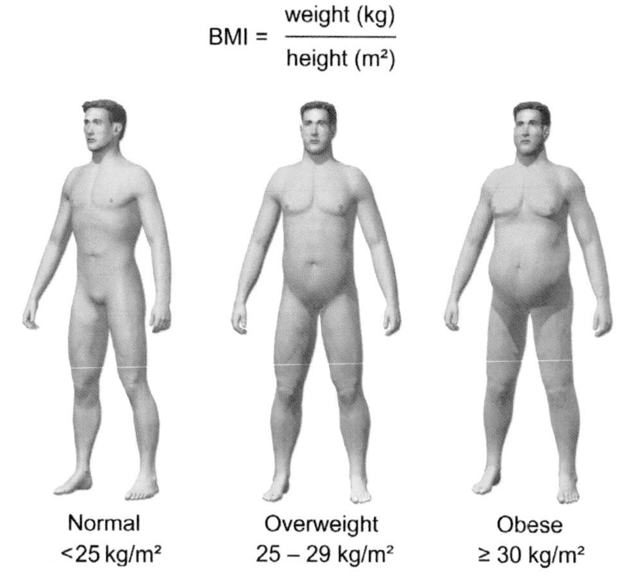

Body Mass Index (BMI) has become the standard, though highly flawed, measure of determining who is overweight and obese.

Creative Commons Attribution-Share Alike 4.0. Photo by BruceBlaus, 2015.

English carnival poster advertising Leo Singer's midget troupe called the *Singer Midgets,* circa 1915, author unknown. Sometimes, BMI measurements are not accurate in athletes who have excessive muscle or in very short or very tall people.

Wikimedia Commons. Public Domain.

resistance), so they can have high blood levels of leptin and a state of leptin insensitivity (i.e., resistance.)

What about diagnosing obesity in children? Measurement of weight and specifically fat in children is even more "complex and confusing," according to researchers Katherine Flegal and Cynthia Ogden, writing in the journal, *Advances in Nutrition* last year. In children BMI varies not just with height but with age as well.

Flegal and Ogden note that both terms "overweight" and "obesity," though used extensively, "can be ambiguous" and sometimes even used interchangeably. In other words, the "terminology is far from standard." Though there are some variations, "overweight" is considered a BMI from the 85th to 95th percentile above age and gender and "obesity" as above the 95th percentile for children, but "these cutoff values are not necessarily exact." These are statistical, though, rather than clinical distinctions. They also note that we don't even have "any well accepted standards for body fatness" or "even strong evidence for any precise definition" for children. In other words, say Flegal and Ogden, these numbers still do not necessarily tell us which children are necessarily at risk for future adverse health consequences, even though, for example, higher levels of BMI in children have been associated with morbidity such as increased blood pressure or abnormal lipid profiles, etc. that have been associated with cardiac disease in adults.

Bottom line: BMI is a screening, rather than diagnostic, imprecise measure of fat for children, adolescents, and adults, and its cutoff levels are arbitrary and statistical rather than clinical. Consider requesting blood leptin levels and DXA scans for more accurate assessment of body fat.

Note: A method using two-dimensional digital photographs to measure body composition as a "viable alternative" to BMI in both children and adults may eventually have widespread use. See the paper by Allison, Heymsfield and colleagues (Affuso et al, *PLOS ONE,* 2018). ■

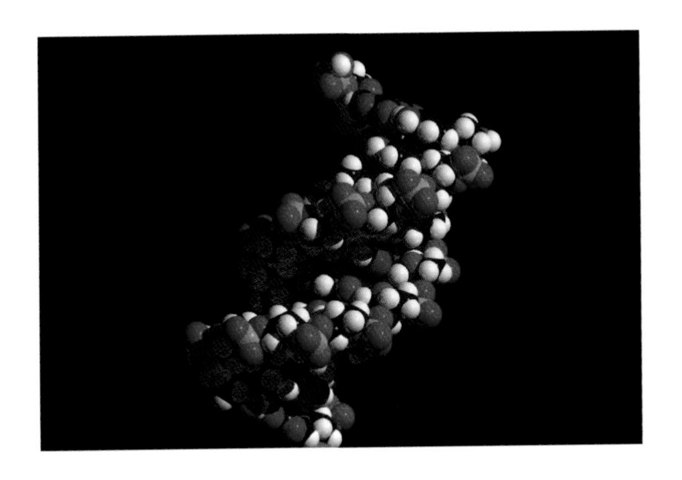

DOUBLE-CROSSING THE DOUBLE HELIX: WEIGHT & GENES

Metabolic imprinting and drifting through the "epigenetic landscape"

25 — Posted August 5, 2012

Model of the DNA's double helix structure. We are a product of our genes and the environment (i.e., epigenetics).

OpenStax Anatomy and Physiology. Wikipedia Commons/Public Domain.

MOST RESEARCHERS BELIEVE that there is considerable genetic input in determining who will become obese—perhaps up to 70% or so in some studies, if you are unlucky enough to have two obese parents. Scientists, though, from an evolutionary perspective, do not believe that our genes have changed significantly in such an infinitesimally short period of time. So why has the prevalence of overweight and obesity become epidemic or even globally pandemic in the past thirty years? That is where the science of epigenetics enters our picture.

Epigenetics and the concept of the *epigenetic landscape* are terms that were first used by C.H. Waddington in the early 1940s (well before the field of molecular biology) to describe all the biological processes that can occur during development that interact with genes and result in how an organism actually appears (i.e., its phenotype.) Choudhuri, in an 2011 article on the history of epigenetics in the journal *Toxicology Mechanisms and Methods*, explains that originally, epigenetic mechanisms were almost seen as "metaphysical and without any understanding of their molecular underpinnings" and became a "default explanation" when genetics failed to explain a particular phenotype.

More recently, *epigenetics* is defined as any long-term, persistent change in gene function that does not actually involve any alterations in a gene sequence or structure. Essentially, though, epigenetics, while *not changing a gene's sequence or structure,* does involve modifications to a gene that can be inherited from one cell to another or even from one generation to another. These modifications, for example, may lead to either silencing or activating genes and may be adaptive or nonadaptive. We now know they may *include* adding or subtracting methyl groups (a carbon atom bonded to three hydrogen atoms) called "methylation," changing the configurations of histones, (proteins around which the genetic material DNA winds in a cell's nucleus) or even producing small (called micro) strands of RNA. (RNA is the nucleic acid involved in protein synthesis and many biological reactions including controlling genes but unlike the double-stranded double-helix configuration of DNA, it is single-stranded and has different chemical components.) Epigenetic modifications can be reversible or stable, as well as occur randomly ("stochastically") or induced by changes in the environment.

Choudhuri describes epigenetic mechanisms as "an editorial hand that edits and modifies the language of DNA," but adds that we still don't understand what "regulates the regulator," i.e., "how signals trigger epigenetic changes."

So even though we have been successful in mapping the human genome, we have been less successful in assessing the contribution of any exposures in the body's internal as well as external environment, whenever those exposures appear. The *exposome* (a term first used by Christopher Wild) is the "totality of exposures" received from conception throughout life. Howard Slomko and his colleagues at Albert Einstein College of Medicine, in a recent "mini-review" of epigenetics for the journal *Endocrinology*, note that although a person's genome is fixed at conception, his or her internal chemical environment is constantly changing because of changes in both a person's internal and external environments. Exposure to chemicals, smoke, drugs, radiation, diet, and even inflammation, stress, infection, etc. may all have an impact on our DNA. Today, epigenetic mechanisms have wide-ranging implications for research in aging, cancer, and obesity.

This epigenetic landscape, from prenatal and early postnatal development and throughout childhood and adulthood,

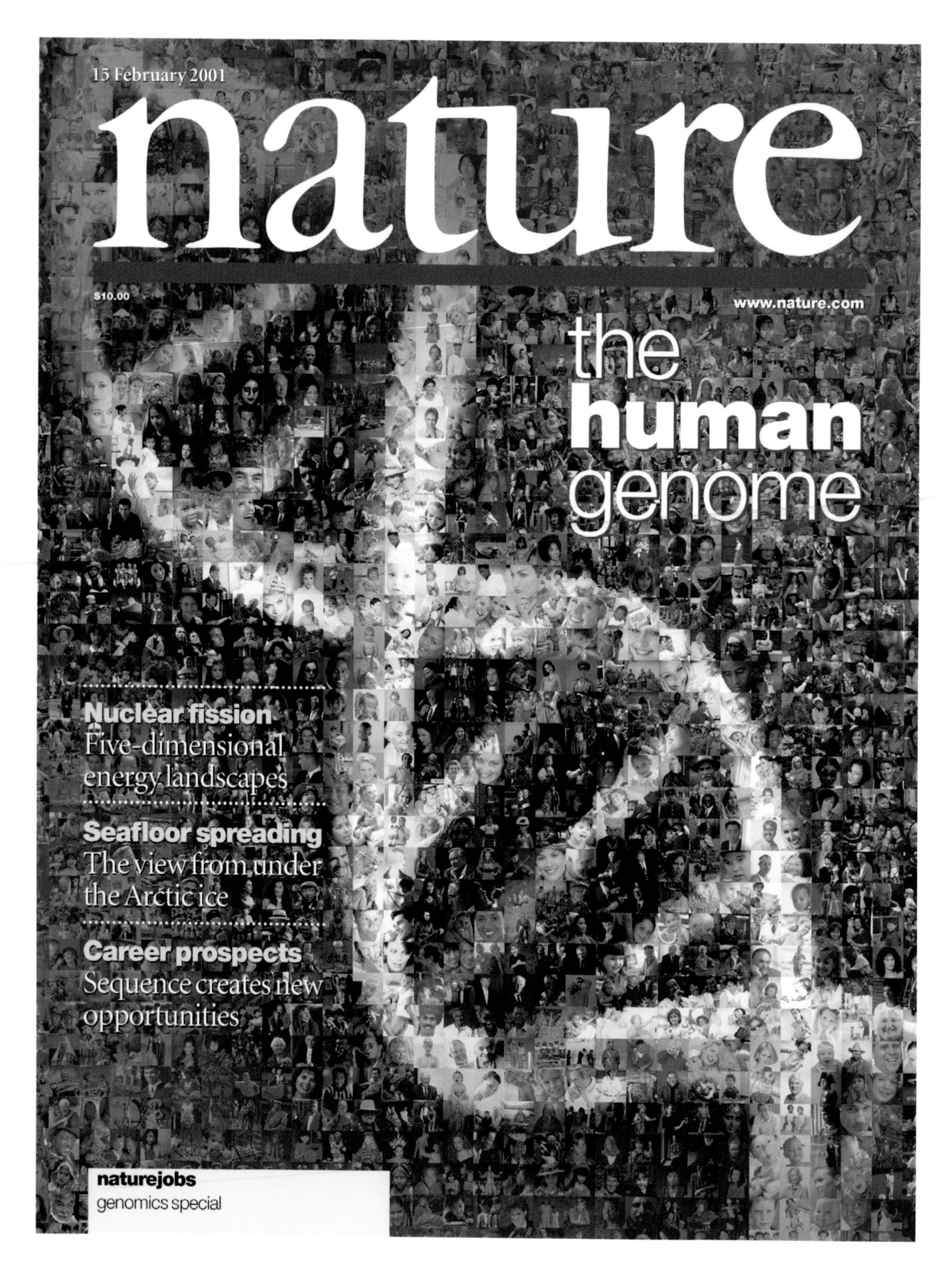
15 February 2001
nature
$10.00
www.nature.com
the
human
genome
Nuclear fission
Five-dimensional energy landscapes
Seafloor spreading
The view from under the Arctic ice
Career prospects
Sequence creates new opportunities
naturejobs
genomics special

◀ Previous page: When scientists, with groundbreaking efforts, were able to crack the code for the human genome, *Nature* featured this extraordinary cover for its February 15, 2001 issue. We now know that our personal genetics interface with the environment. Epigenetics, while not changing a gene's sequence or structure, involves modifications to a gene that can be inherited from one cell to another and even from one generation to another. Epigenetic mechanisms have been described as "an editorial hand that edits and modifies the language of DNA." (Choudhuri, 2011)

*Springer/*Nature*, Cover, Copyright, February 15, 2001, used with permission.*

Paul Klee, *The Pathos of Fertility,* 1921, drawing.

The Berggruen Klee Collection, 1984. Metropolitan Museum of Art, NYC, Public Domain.

can have a signficant and sustained impact throughout life. In fact, pathologist George Martin has noted that exposure to different environments over time has been postulated to result in *epigenetic discordance* or rather poetically, *epigenetic drift*, even between monozygotic (identical) twins, who share all their genes, as they age, and may explain, for example, why one twin develops Alzheimer's Disease in his 60s and the other not until his 80s.

We are all familiar with the importance of a nontoxic uterine environment for the development of a growing fetus. The teratogen, thalidomide, for example, taken years ago by unsuspecting pregnant women, resulted in major congenital malformations for the fetus. Less dramatic, but perhaps no less ultimately significant effects, though, may result from a pregnant woman's diet and may be responsible, at least in part, for an increased susceptibility to obesity and differences in regulation of fat and glucose in her offspring much later in life. A metabolic "obesogenic environment" *in utero*, for example, may expose the fetus to increased levels of glucose, as well as to increased levels of the hormones insulin and leptin. This has been called *metabolic priming* or *metabolic imprinting* such that nutritional exposure prenatally results in a kind of *endocrinological memory* with potentially far-reaching consequences. A gene-environmental interaction occurs: Not only you are what you eat, but you may be what your mother has once eaten! ■

Pieter Bruegel the Elder, *The Tower of Babel,* 1563, Kunshistorisches Museum, Vienna.

Wikipedia Commons/Public Domain.

A TOWERING BABEL: STRUCTURAL FRAMEWORKS FOR WEIGHT

Can we all speak the same language about obesity?

26—Posted September 1, 2012

Most of us are familiar with the Biblical story of the Tower of Babel from Genesis (11: 1–9): the whole world had one language and a common speech and all the people began to build a city and a tower to reach the heavens. The Lord came down to see the city and the tower that the people were building and said, "If as one people speaking the same language they have begun to do this, then nothing they plan to do will be impossible for them…..let's go down and confuse their language so they will not understand each other." So the Lord scattered them from there all over the earth, and they stopped building the city. That is why it was called "Babel" (sounding like the Hebrew word for "confused")—because there the Lord confused the language of the whole world.

No one really knows whether there was a real Tower of Babel or obviously what it looked like. For art lovers, the best depiction of the Tower of Babel comes from the 16th century Pieter Bruegel painting of the same name that now hangs in the Kunsthistorisches Museum in Vienna. Visit the museum's website, http://www.khm.at.) There is even a glorious rock formation called the "Tower of Babel" in the Arches National Park in Utah.

How is all this, though, relevant to the study of obesity? Researchers in the field cannot agree on how to think about obesity and we can indeed sometimes find ourselves in a confusing muddle of vastly different conceptual frameworks. As I have mentioned in my previous blogs, obesity is defined medically as a threshold based on body mass index, i.e., weight in kilograms divided by height in meters squared. The higher the body mass index, the worse the level of obesity. Simplistically, obesity is an energy imbalance due either to increased consumption of food and/or decreased expenditure of energy.

But is obesity a medical disorder or disease, particularly since it is so prevalent worldwide? Louis Aronne, a major obesity researcher, and his colleagues, acknowledge this is a "controversial" though hardly new question. They believe that obesity "meets all the criteria of a medical disease, including a known etiology, recognized signs and symptoms, and a range of structural and functional changes that culminate in pathological consequences." Along those lines, obesity has been considered a metabolic (i.e., endocrine) disease, a neurochemical disease, a disease of inflammation, a genetic disease, a viral disease, and even a brain disease. It has also been considered an impulse disorder as evidenced by a failure of self-control. Currently, in the upcoming ICD 10 (International Classification of Diseases by the World Health Organization), it will be classified under the "Endocrine Nutritional and Metabolic Disease" category. It will not, however, be included in the upcoming Diagnostic and Statistical Manual (DSM-V) for psychiatric disorders. Researchers Bachman and Histon (2007)summed it up well, "Before adding the stigma of "brain-damaged" to the high physical and social burden obese persons already bear," they would like to see more compelling data.

Not everyone, though, uses a disease model. Power and Schulkin, in their book, *The Evolution of Obesity* (2009) maintain that obesity, rather than a pathological condition, is an example, in terms of evolutionary theory, of "inappropriate adaptation" to our environments that are now laden with energy-dense foods (e.g. sugar, fat, salt) not seen until recent generations and that provide fewer opportunities for physical exertion. Swinburn and colleagues, writing in the British journal *The Lancet* (2011) have called obesity the "result of people responding normally to the obesogenic environments they find themselves in." Psychiatrist Michael J. Devlin (2007) has called obesity "a cultural disorder of sorts." The most different language of all, though, comes from those in the National Association for the Advancement of Fat Acceptance (NAAFA), who believe that obesity is an example of body diversity, analogous to sexual, ethnic, or racial diversity, that should be celebrated rather than treated.

Albert Stunkard, one of the field's most renown researchers, wrote back in 1959 that obesity is not a single disease with a single etiology. Rather, he saw it as "the end stage of a variety of different conditions with different etiologies." Because of its complexity, we may never all be able to speak the same language when it comes to the study of obesity. What likely regulates fat accumulation in humans is determined by many factors, including genetic, environmental, neural, endocrinological, psychosocial, and behavioral. Perhaps, though, we do not need to speak the same language as long as we are open to translation of the many languages we now have. ■

Tower of Babel rock formation in Arches National Park, Utah.

Used with permission, istockphoto.com. Credit: narawon.

Flemish painter Lucas van Valckenborch, *The Tower of Babel*, 1594, Louvre Museum, Paris.

Wikipedia Commons/Public Domain.

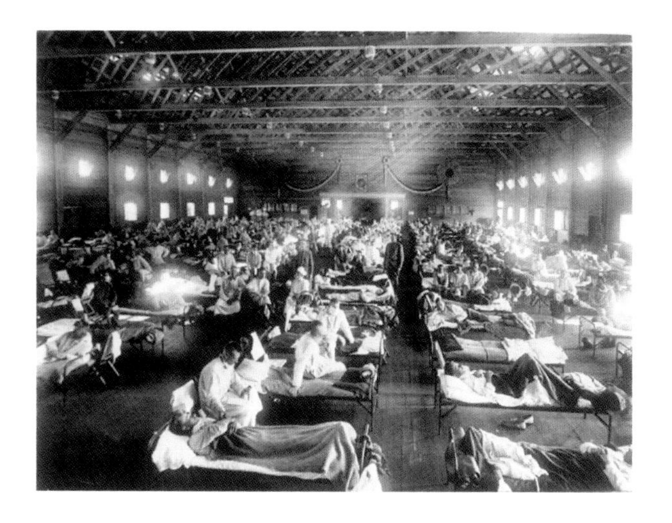

WEIGHT: A FLU SHOT IN THE ARM, A SHOT IN THE DARK

Infection, the flu, and the immune response in obesity

27 — Posted September 23, 2012

AS THE WEATHER BECOMES cooler and fall has begun, we are entering flu season that typically runs from October to May. Though there are a few contraindications, the flu shot is recommended for everyone older than 6 months of age, but particularly for the elderly, pregnant women, and the young who are most vulnerable, as well as for those with chronic diseases such as asthma, diabetes, or chronic lung disease. It takes two weeks after the shot for immunity to build up, and since the flu virus is constantly mutating (called "antigenic drift"), last year's immunization is not effective against this year's flu. Some hospitals are asking their entire staff to be immunized.

The flu can be deadly. Anyone who wants a true page-turner about the horrific and devastating effects of the flu, not only on individuals afflicted but on societies worldwide, should read John M. Barry's *The Great Influenza,* about the 1918 pandemic, published several years ago. The Centers for Disease Control (CDC) reports that an estimated 20 to 50 million people worldwide died in that particular pandemic. In fact, the CDC is currently studying the genetic make-up of that 1918 strain since the biological properties that make one flu strain more virulent than another are still not well understood. For those interested in this research conducted by the CDC, visit its website, http://www.cdc.gov/flu/about/qa/1918flupandemic.htm.

Could another flu pandemic occur? Yes, of course, but it is not possible to predict when or even how it will start.

Soldiers from Fort Riley, Kansas, ill with Spanish influenza at a hospital ward at Camp Funston. Influenza can have deadly consequences; the pandemic of 1918 killed millions of people worldwide. Photograph taken (before 1923) by a U.S. army photographer.

Wikipedia Commons/Public Domain.

The CDC believes, though, if it does, it will be caused by a subtype of influenza virus for which humans currently have little or no natural immunity. According to the CDC, "The single best way to protect against the flu is to get vaccinated each year." According to its statistics for the years 1976 through 2007 (i.e., 31 flu seasons), the CDC estimates that from a low of 3000 to a high of 49,000 people in the United States alone have had flu-related deaths. About 90% of deaths have occurred in those older than 65 years, but even healthy people can develop serious cases of the flu. According to Sheridan and colleagues, 250,000 to 500,000 people die from influenza worldwide each year!

Where does obesity fit into this picture? Evidence has been gathering that the obese are, in general, more susceptible and vulnerable to infections, including an increase risk of sepsis, as well as respiratory and urinary tract infections, especially after surgery, and excess fat seems to have a negative impact on an obese person's general immune functioning. This becomes significant when the World Health Organization (WHO) reports that about 500 million adults and almost 43 million children below the age of 5 are considered in the obese range worldwide. Milner and Beck, in a recently published (2012) review article in the *Proceedings of the Nutrition Society*, "Obesity and the Immune Response to Infection," note that obese individuals, in general, are more likely to have longer stays in intensive care units and are even more likely to die in the hospital. Furthermore, they may even have different responses (e.g. poor antibody response) to vaccinations, such as tetanus and hepatitis B immunizations, as well as to treatment with antibiotics. Not only does this impact the obese individual, but may also have more general public health consequences in terms of the spread of disease. In fact, recent studies have indicated that obesity was found to be an independent risk factor for increased morbidity and even for greater mortality in the 2009 H1N1 flu pandemic. Milner and Beck believe more research is clearly

Egon Schiele, *Seated Woman, Back View*, 1917. Schiele died of influenza in 1918 at age 28. Metropolitan Museum of Art, NYC.

Bequest of Scofield Thayer, 1982. Copyright Metropolitan Museum of Art. Image source, Art Resource, NY, Used with permission.

Norwegian artist Edvard Munch, *The Sick Child,* 1896. Gothenburg Museum of Art, Sweden.
Wikipedia Commons/Public Domain.

needed but also believe that it is a "cause for major concern" that certain medications and vaccines "may not function as intended in obese individuals." Researchers Louie and colleagues, writing in the journal, *Clinical Infectious Diseases,* in 2011, found that 50% of those 534 in their California study who had been hospitalized during the 2009 H1N1 epidemic were obese, and those with extreme obesity (body mass index greater than 40 kg/m^2) had increased odds of death. Their conclusion was that obese individuals "should be treated promptly" and given priority for both vaccine and antiviral medication when there are shortages.

Bottom line: along with all the other medical and psychological morbidities associated with obesity (e.g. diabetes, cardiac disease, certain cancers, sleep apnea, etc.), we can add that those who are obese may have less of an antibody response to the flu vaccine and may have increased morbidity and mortality if they get the flu. In other words, a shot in the arm may be a shot in the dark! ■

STRENGTHENING TIES THAT BIND: WEIGHT CONTROL

What determines who are the people successful at weight loss and maintenance?

28—Posted October 10, 2012

In Book XII of Homer's *Odyssey*, the sorceress Circe tells Odysseus of the dangerous but enchanting sound of the Sirens. "There is a great heap of dead men's bones lying all around, with flesh still rotting off them," for those who don't heed her warning about the lure of their singing. She cautions that Odysseus put wax in his men's ears, but if he wants to hear their extraordinary song for himself, his men must tie him to the mast of their ship. And if he begs for them to untie him, they must bind his cords more tightly. The story has been depicted in art throughout the ages, including on Greek black figure and red figure pottery and a 19th century painting by English artist John William Waterhouse, "Ulysses and the Sirens."

Metaphorically for some, the lure of food has the same effect as the Sirens' song and our "obesogenic" environment, the same potentially deadly consequences. Johnson et al, in a recently published article in the *International Journal of Obesity,* revisit the concept of dietary restraint and self-control to distinguish when restraint may lead to "effective weight control" and when it might lead to undermining that control. They explain that so-called "restraint theory" for years held that restraint in eating led to "counter-regulatory responses," reduced ability to monitor satiety, and even "dis-inhibited, binge-like eating patterns." On the other hand, those who were more "relaxed" about food were more likely to have better weight control, healthier eating patterns, and even a better body image. The problem, say Johnson et al, is that restrained eaters are not a homogeneous group and some are more vulnerable than others to succumbing to their food temptations. In other words, some dieters find restraint ineffective and even counterproductive and prefer what is called an "undieting," approach; others find using restraint a helpful cognitive means for maintaining control of their food intake. Furthermore, researchers can break down dietary restraint into "rigid" restraint (i.e., with an "all-or-none approach to weight control) and "flexible" restraint (i.e., a "more graduated" approach such that some tempting foods can be eaten in small quantities while compensating by eating less of other foods) as long as the person doesn't cross his or her own "eating boundary." What distinguishes successful restrained eaters from unsuccessful ones? Johnson et al report that successful restrained eaters are able to respond to food cues by being able to think long-term and focus on a 'dieting' goal" whereas those not successful activate a more immediate pleasurable 'eating' goal. These authors acknowledge that genetics likely plays a role in determining differences among those dieters who can exert more self-control from those who have more difficulty.

Homer's Odysseus tied to the mast to resist the beguiling voice of the Sirens. Detail from Attic red-figure stamnos, circa 480–470 B.C.

Photograph by Jastrow, 2006, British Museum, Wikipedia Commons/Public Domain.

Along those lines, Lorraine G. Ogden and her colleagues, writing recently in the journal *Obesity*, found four distinct subgroups among their successful dieters in their National Weight Control Registry. This ongoing "observational" study, begun in the early 1990s has followed over 5000 self-referred dieters who have been able to lose at least 30 pounds and keep the weight off for at least a year (and most, for much longer.) In previous publications, the researchers who began the study have described that successful dieters, in general, were more likely to eat breakfast regularly, exercise about one hour a day (mostly walking) and expend 1000 to 2000 calories/week, monitor their weight and continue to do so even during weight maintenance, eat low calorie, lower fat foods, and maintain diet consistency not only on weekdays but also weekends and holidays. In this most recent article, the authors studied

English painter John William Waterhouse, *Ulysses and the Sirens,* 1891, National Gallery of Victoria, Melbourne, Australia. *Wikipedia Commons/Public Domain.*

over 2200 participants from their larger sample. They found that about 50% of their sample fell into Cluster 1, "a weight-stable, healthy, exercise-conscious group who are very satisfied with their current weight." Almost 27% fell into Cluster 2, a group who have "continuously struggled" with their weight since they were children, required the most resources and strategies to lose and control their weight, and experienced greater levels of stress and depression. Almost 13% fall into Cluster 3, a group who are successful with weight control on their first try, ("participants with immediate and long-term success") more likely to have maintained their weight loss for the longest time, have the least difficulty with weight control, and least likely to have been overweight during childhood. Almost 10% fall into Cluster 4, a group who tend to be older, eater fewer meals, less likely to exercise to control their weight, and more likely to have health difficulties.

Bottom line: Even among those successful at weight loss and maintenance over time, "one size does not fit all" and some will struggle more with the process than others, whether with food restraint specifically or weight maintenance in general. ■

WEIGHT CONTROL: THE BIOLOGICAL BRAIN, THE PSYCHOLOGICAL MIND

How mind matters in weight control: piecing together puzzling questions

29—Posted November 4, 2012

WEIGHT CONTROL is the result of a complex integration of many factors, including environmental, genetic, neuroendocrinological, and psychosocial. There is no question that much of weight control is biological, but how does the psychological mind work with the biological brain and body?

Neuroscientist Antonio Damasio has written of our "minded brain." In other words, our human brain and the body "constitute an integrated organism." For Damasio, all the images and thoughts that constitute our minds—our "mental phenomena"—are "biological states that occur when many brain circuits operate together."

Drs. John Monterosso, associate professor of psychology and neuroscience at the University of Southern California and Barry Schwartz, a professor of psychology at Swarthmore College, explain "All psychological states are biological ones." In a *New York Times* article this past year, Monterosso and Schwartz wrote of the "misguided" belief people have in what these researchers call "naive dualism," namely that psychological causes are distinct from biological ones. In an earlier (2005) article in the journal *Ethics and Behavior*, Monterosso and his colleagues designed a series of experiments on the nature of responsibility. They found when their subjects have this belief, they see behavior as voluntary only when it seems to come from the mind (or soul), but when there is a physiological explanation (and "when participants tended to view the body as the cause of the behavior,") these subjects perceived this behavior as less voluntary and hence they were apt to attribute less responsibility to those actions.

The brain is a "minded brain."

Used with permission, istockphoto.com. Credit: wildpixel.

How does all this relate to weight control? Our minds can be very powerful and persuasive when it comes to weight. For example, our highly developed prefrontal cortex, i.e., our cognitive brains, enable us to plan our meals, think about food rationally, remember what we have eaten, and even remember what foods made us sick many years earlier. We are able to appreciate the consequences of our behavior such as the importance of watching what we eat and exercising regularly. Researchers Lowe and Butryn write of how we can be "restrained eaters"—that is, we are capable of eating "less than we want, rather than less than we need." We are also capable of choosing a less appealing food because it is healthier or making food choices based on extrinsic factors such as cost, brand, convenience, or even what other people are eating or the television commercial we just saw.

It is our "minded brain," though, that can also seem to sabotage our efforts to maintain our weight, despite our best intentions to do so. We can be overwhelmed by the sugary, fat concoctions in our environment or our own cravings for a particular food and give in to our temptations and dismiss any long term goal of maintaining our weight. The technical term for this is *delay discounting*: we devalue

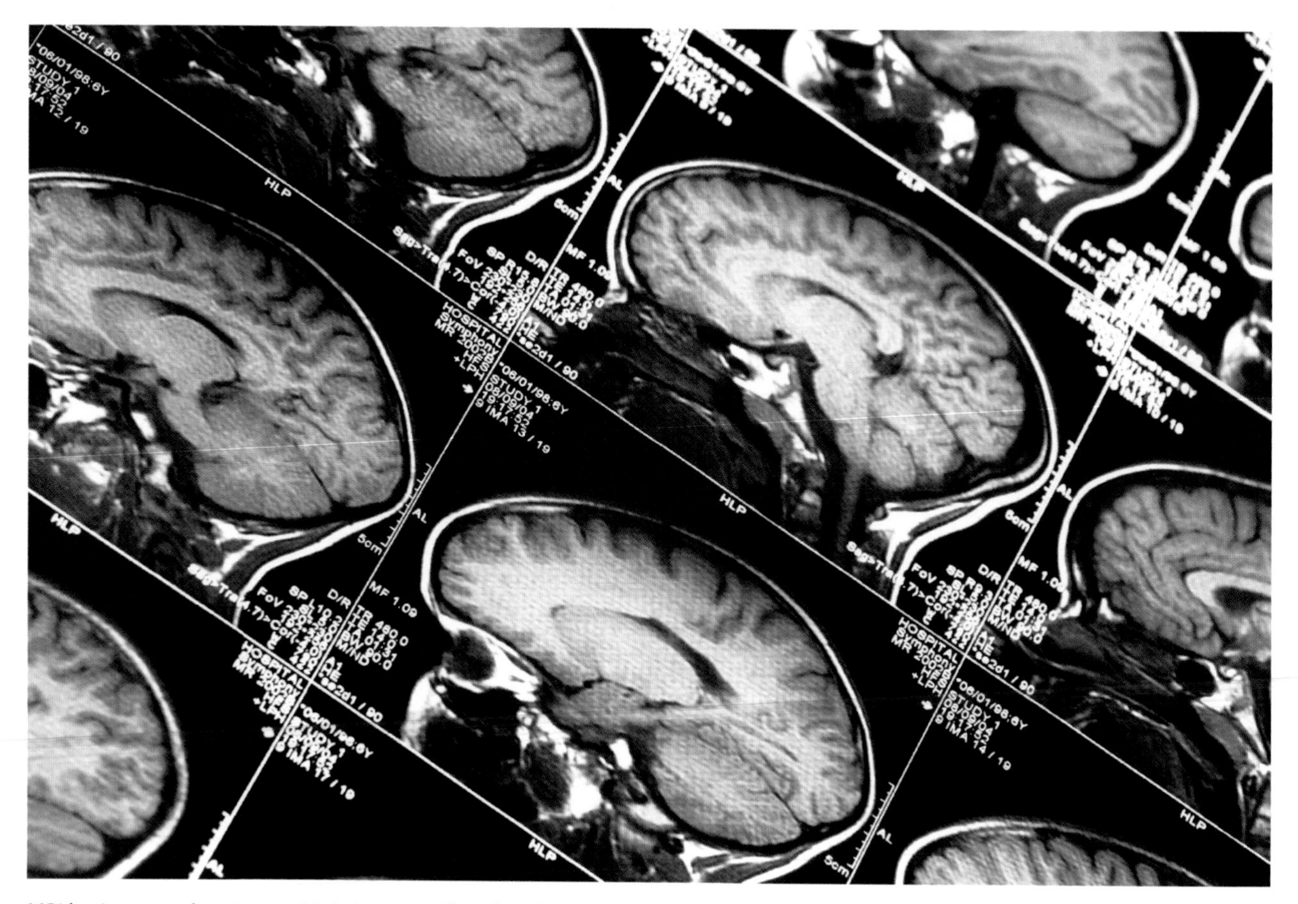

MRI brain scans showing multiple images of head and skull.
Used with permission, istockphoto.com. Credit: haydenbird.

or even discount something that may happen in the future (e.g. gaining weight or developing medical consequences from obesity) for the sake of some immediate reward or gratification (e.g. eating those chocolate chip cookies). We can also develop what researchers Herman and Polivy call "the perverse logic of the dieter" or the "what-the-hell" effect when we eat one food we shouldn't and then give up our diet entirely—in an all-or-none fashion. We make value judgments about foods and ourselves—they are good or bad—and we are good or bad for eating them.

Whether we have a biological set point for weight regulation is controversial but some believe in a *cognitive set point*, first described in the 1970s. It is a point in the perception of our own weight, shape, or size that involves a more deliberate control over our eating. Though this point can change over time, as for example, when our "acceptable" weight on the scale creeps up slowly or we choose to buy the next size in clothing, most who care about weight have a limit. In other words, there is a personal "diet boundary" we dare not cross.

For maintaining our weight, we also need a cognitive sense of self-efficacy, i.e., the sense or confidence that we can bring it about. Psychologist Roy F. Baumeister has written extensively on self-control (and the broader term, self-regulation). He notes that few impulses are truly irresistible (e.g. such as breathing, sleeping, urinating). Most of the so-called "irresistible" impulses are, in reality, rationalizations for our failures to maintain self-control. Baumeister notes that self-control enables humans to have flexibility in our responses and the ability to stop what we are doing in the middle.

We may have more voluntary control than we sometimes think. Some researchers believe that our complex biological systems that have evolved for weight control have been hijacked and overwhelmed by our current obesogenic environment. So even though there is a biological substrate, we need our cognitive controls—the "minded brain"—more than ever if we want to control the burgeoning development of worldwide obesity. ■

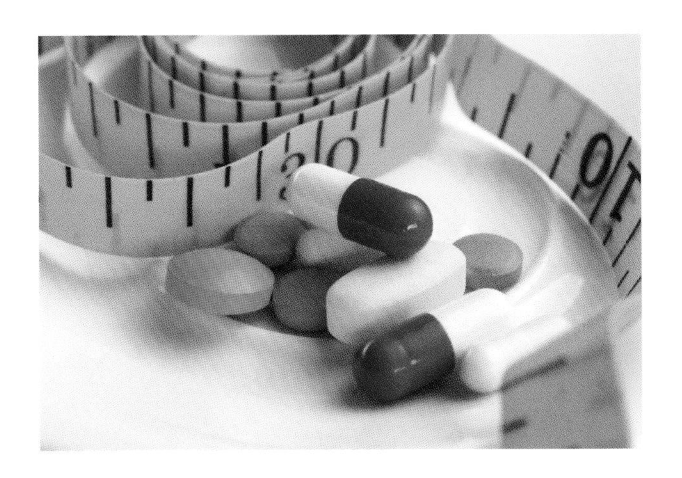

SEARCHING FOR MAGIC BULLETS: WEIGHT-CONTROL MEDICATIONS

Hope continues to spring eternal for magic elixirs that will lead to weight loss

30—Posted November 27, 2012

FOR THOSE PREDISPOSED to excessive weight gain, weight control is a lifelong preoccupation. Any medication prescribed, therefore, will likely need to be continued indefinitely, analogous to those drugs used to treat chronic conditions like diabetes or hypertension.

Medications for weight control have had a particularly inauspicious past. Not only have there been extremely few medications available, particularly compared to those approved for other chronic medical conditions, but even these few have had remarkably short commercial half-lives.

Most recently, sibutramine (Meridia), approved for weight control, was withdrawn from the U.S. market by the Food and Drug Administration when reports of serious side effects appeared. Years ago, the combination fenfluramine and phentermine (fen-phen) was withdrawn when unexpected reports of deleterious cardiac valvular disease and deadly pulmonary hypertension developed in some patients over time. Furthermore, rimonabant (Accomplia), a medication that blocks cannabinoid receptors and decreases rather than increases hunger (i.e., the opposite of the marihuana "munchies") was also withdrawn when scattered reports of depression and suicidal ideation occurred even in patients without a prior history of psychiatric illness.

Dietary supplements, often sold in health food stores, have also been used, sometimes with dire consequences. The herbal compound Ephedra, for example, had been reported to lead to cardiac side effects, including myocardial infarctions, seizures, and even death. The major problem with these dietary supplements is that they are not FDA-controlled and hence not subject to any regulation. Several years ago, a *New York Times* expose, reported that many of the so-called diet pills sold under various names contained medications not listed among their ingredients and even medications for which one would need a prescription (e.g. diuretics, anti-epileptics, etc.) and that could lead to serious complications (e.g. dehydration, hypotension, etc.)

Furthermore, medication has been prescribed only for obesity (i.e., a body mass index—BMI greater or equal to 30 kg/m^2 or 27 kg/m^2 in those with serious medical morbidity related to excessive weight.) There has never been a medication developed or recommended for those who want to lose and keep off those proverbial 5 to 15 pounds, and medication is always prescribed in the context of the recommendation for lifestyle changes of regular exercise and dietary management (usually including lowering fat intake and caloric restriction.)

While there are many medications currently in development, two new medications have recently won FDA approval (at least for the time being): Belviq (lorcaserin) and Qsymia (combination of phentermine/topiramate in extended release.) These are not innocuous medications and significant side effects have been reported with each. Lorcaserin, an appetite suppressant (i.e. promotes satiety), is a drug that affects serotonin levels and as a result cannot, for example, be used with the many other medications that also affect serotonin levels because a life-threatening serotonin syndrome (e.g. agitation, hallucinations, incoordination, vomiting, and even coma) may develop. Furthermore, cognitive impairment (e.g. difficulty with memory and attention) and psychiatric symptoms (e.g. euphoria and even hallucinations) have been reported. Even concern about cardiac valvular disease, blood cell counts, and the risk of pulmonary hypertension have been mentioned, as has priapism (prolonged erection) in men.

To date, there are no magic bullets for weight control.

Used with permission, istockphoto.com. Credit: zaretskaya.

Jean-Léon Gérôme, French artist, *Cafe House, Cairo (Casting Bullets)*, 1884 or earlier. Would that finding "magic bullets" for weight loss were as easy as casting bullets for guns. Metropolitan Museum of Art, NYC.

Bequest of Henry H. Cook, 1905. Public Domain.

Medication for weight loss must be in the context of lifestyle changes, including a sensible diet and exercise.

Used with permission, istockphoto.com. Credit: chang.

The combination drug Qsymia comes in different strengths and the dose is titrated up over time. The most common side effects are insomnia, constipation, dry mouth, dizziness, and parethesias, but cognitive impairment (including word-finding difficulties and problems with concentration) have also been noted. It should not be used with alcohol, and it is contraindicated in pregnancy as fetal abnormalities (e.g. cleft lip and palate) have been reported. Furthermore, weight loss is not dramatic with either medication and may not be sustained over years.

For those who are obese, it is likely that more than one medication may be needed, analogous to the need for polypharmacy in hypertension or type II diabetes. One medication may be useful for weight loss while another may be necessary for weight maintenance over time. Eventually, it is possible that there will be available genetic screening to enable "personalized" medicine.

Unfortunately, there is no magic bullet for weight control. Michael Fumento, a medical journalist and author of *The Fat of the Land*, very wisely wrote, "...Americans by and large don't want a drug that makes them eat less. They want a drug that allows them to eat more but not gain weight." (p. 249) ■

TOP TEN REASONS WHY WE MAY ALL BE GETTING FATTER

Will everyone in the U.S. eventually be either overweight or obese?

31 — Posted December 17, 2012

AT THIS TRANSITION TIME of the year, 'tis the season for top ten lists. Lists appear, for example, of the top ten films, theater, and books of the past year. Late night comedian David Letterman has made popular his "Top Ten" list each evening, but there is nothing funny about why we as a nation are increasingly overweight and obese. There are dire predictions, in fact, from some researchers, that if we don't get a handle on the problem, eventually all Americans will be obese or at least overweight in another 35 years. In the U.S. alone, according to the most recent statistics about ⅔ are overweight and of this number, ⅓ are obese. Excess fat has been linked to cardiac disease, hypertension, type II diabetes and other endocrine disorders (e.g. the metabolic syndrome), as well as sleep apnea, sexual dysfunction, complications of pregnancy, osteoarthritis, and many types of cancer such as those of the colon and breast. There is even the suggestion that obesity may be linked to Alzheimer's Disease. But why are we getting fatter? No one really knows for sure, but there are many promising leads, including a "top ten" list of reasons.

David B. Allison, PhD, from the Department of Biostatistics at the University of Alabama in Birmingham, oversaw a project that involved some of the most prestigious names in the field of obesity, including Louis J. Aronne, MD of Weill Cornell Medical College, to explore "ten putative contributors to the obesity epidemic." Their concern was that this epidemic had become a "global issue" and showed "no signs of abating." Their report was published several years ago in *Critical Reviews in Food Science and Nutrition* in a comprehensive article that included extensive evidence from both human and animal studies (and almost 500 references). More recently, their findings have been presented in a considerably abridged version by Dr. Aronne and his colleague Suzanne B. Wright, from the Comprehensive Weight Control Program at Weill Cornell, and published in 2012 in the journal *Abdominal Imaging*.

Here is a list of ten explanations, though not necessarily in any order of importance, to account for the burgeoning worldwide obesity epidemic:

1. the current food environment: increased availability of relatively inexpensive, highly caloric food (particularly laden with fat, sugar, and salt) served in enormous portions
2. decreases in physical activity: increased time spent in sedentary activities such as watching television or using computers; advanced technology enables less caloric expenditure (e.g. escalators, TV remote controls, electric garage door openers, etc.)
3. decreases in sleeping time; trend over the years of decreased amount of sleep in the general population(from around 9 hours before WWI to less than 7 hours more recently), with many people chronically "sleep deprived"; "sleep debt" exerts "profound effects on metabolic hormones" that may lead to changes in circadian rhythms, increased food intake, weight gain, and even diseases such as type II diabetes and cardiac disease

The famous "Top Ten List" popularized by late-night comedian David Letterman.

Used with permission, istockphoto.com. Credit: pagadesign.

The toxic food environment is one of the major reasons for the exponential rise in obesity prevalence.

Creative Commons Attribution-Share Alike 4.0. (no author information. Wikipedia Commons.

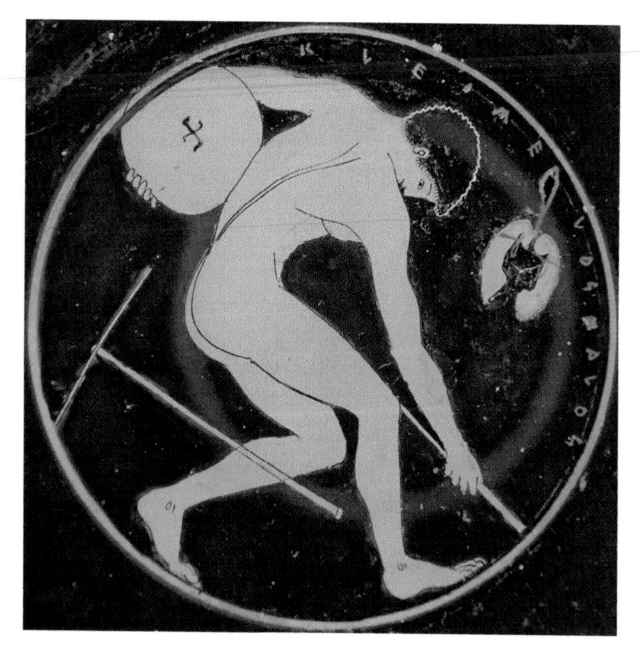

Young boy throwing a discus by Kleomelos, Greek painter. Red-figure ceramic, between 510 and 500 B.C. Another main reason for the increase in obesity is most people do not get enough regular exercise.

Collection of Giampietro Campana di Cavelli, 1861. Louvre Museum, Paris. Wikipedia Commons/Public Domain.

4. drug-induced weight gain: many more people taking medications that lead to weight gain (e.g. medications for depression, anxiety, psychosis, hypertension, diabetes, contraception). Even antihistamines can lead to weight gain; lithium for mania can lead to a 35 lb weight gain in some
5. decline in cigarette smoking: smoking can lead to weight control and giving up smoking can be associated with often considerable weight gain
6. exposure to "endocrine disruptors:" industrial chemicals ubiquitous in our environment have been associated with increased weight gain (e.g. phthalates, pesticides, flame-retardants, bisphenol A); these disrupt hormonal functioning and adipose tissue regulation
7. infections: several viruses have been associated with increased weight gain, both in animals and humans; adenovirus 36 is most commonly implicated; even changes in bacteria in the gut can lead to weight gain
8. intra-uterine effects: so called "epigenetic effects;" maternal weight (either overnutrition or undernutrition for the fetus) may lead to obesity later in life
9. increased maternal age: women postponing childbearing because of improved contraception and more women in the workforce; obesity rates in children reportedly higher with older mothers
10. temperature control and decreased variability in ambient temperatures with air conditioning and central heating: may lead to decreased metabolic rate and weight gain

This is not an exhaustive list. Even the huge genetic contribution of "assortative mating" (e.g. fat people tend to marry fat people) may be a factor. For those particularly interested in the details behind the "top ten" list, please consult Dr. Allison's original article. ■

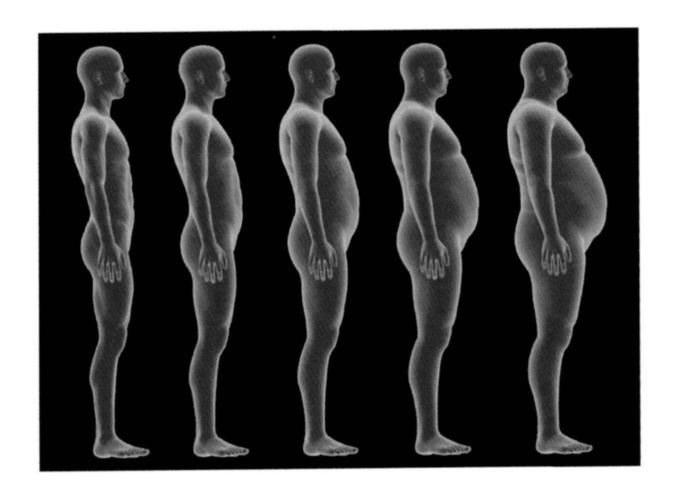

THE OBESITY PARADOX: IS THERE AN IDEAL WEIGHT FOR HEALTH?

Can that "lean and hungry look" actually be detrimental to your life expectancy?

32—Posted January 5, 2013

FOR YEARS, we have been told that overweight and obesity are associated with serious medical morbidity, including certain cancers like breast, prostate, and colon, metabolic disturbances such as diabetes and abnormal blood lipid levels, as well as heart disease, orthopedic disability, sleep apnea, and hypertension. Most, but not all, researchers have led us to believe that overweight and obesity, as defined by body mass index (BMI), can substantially shorten our life expectancy, and the worse the level of obesity, the more likely, an earlier demise. This week, though, a new study received considerable media coverage and raises questions about some of these assumptions. Can being overweight or even somewhat obese actually be protective against mortality? For most in the field of obesity, this notion certainly seems counter-intuitive, surprising, and even nonsensical, and hence it has been called the "obesity paradox."

Katherine M. Flegal, PhD, from the Centers for Disease Control and Prevention (CDC), and her colleagues, writing in the *Journal of the American Medical Association,* (and extending their original 2005 study) conducted a rigorous, systematic review of almost 100 other studies involving a combined sample of 2.88 million people and 270,000 deaths to assess whether overweight and all levels of obesity are in fact related to increased mortality risk as most believe. Dr. Flegal et al believe that knowing "relative mortality risks" associated with different levels of weight "may help to inform decision making in the clinical setting." They note that in the U.S., for example, almost 40% of adult men and almost 30% of adult women are considered overweight by today's BMI standards, while 36% of adults are considered obese. Significantly, though, more than half of those in the obese category are in the grade 1 category (BMI of 30 kg/m^2 to 34.9 kg/m^2). In general, mortality-weight statistics follow a U-shaped curve, with extremes of being severely below normal weight or severely above it are at highest risk. What these researchers found is that Grade 1 obesity was not associated with higher mortality and what's more, being overweight (BMI of 25 kg/m^2 to 29.9 kg/m^2) was actually protective and associated with significantly lower all-cause mortality! Flegal and her colleagues, though, acknowledge that many of the studies included in their review relied on notoriously inaccurate self-reporting of height and weight that could result in misclassifications of BMI and bias their results. They also note that their study is limited in not addressing morbidity, "cause-specific mortality," and issues of body composition (e.g. where fat is distributed on the body.)

Human body, from slim to overweight. The relationship between weight and mortality is a complex one.

Used with permission, istockphoto.com. Credit: angelhell.

In an Editorial accompanying the study, Drs. Steven B. Heymsfield and William T. Cefalu raise the provocative question, "...are the concerns about overweight as currently defined unfounded?" Professor of Law and author of *The Obesity Myth: Why America's Obsession with Weight is Hazardous to your Health,* Paul Campos, writing an editorial in the *New York Times,* in response to the Flegal study, certainly thinks they are and believes "our current definition of 'normal weight' makes absolutely no sense." He asserts that we are in an "absurd situation" in which we are serving "the economic interests of...the weight-loss industry and large pharmaceutical companies" and "baselessly categorizing at least 130 million Americans—and hundreds of millions in the rest of the world—as people in need of 'treatment.'"

Heymsfield and Cefalu, though, note that the Flegal study does confirm that higher levels of obesity are associated with increased mortality, but they also explain that the situation is far more complicated when we consider

Renoir, *Ambroise Vollard,* 1917, private collection. Vollard was a famous French art dealer.
Wikipedia Commons/Public Domain.

Artemisia Gentileschi, *Self-Portrait as a Lute Player,* between 1615 and 1617. Wadsworth Atheneum, Hartford, Ct.
Wikipedia Commons/Public Domain.

lower levels of obesity and even overweight. First, BMI (the ratio of weight in kilograms divided by height in meters squared) is not a perfect measure of fat: it does not account for differences in sex, race, or age or even a person's cardiopulmonary level of fitness, as well as other risk factors such as blood pressure, glucose and lipid levels, and waist circumference. For example, fat around our internal organs is more dangerous than subcutaneous fat so that the "apple" distribution of fat around the middle is worse than the so-called "pear" distribution. Furthermore, these authors note that physicians may be "increasingly aggressive" in treating the risk factors associated with obesity and hence may be affecting mortality statistics, particularly in these lower weight categories, such that "overweight and grade 1 obesity might lead to greater morbidity that is not captured when evaluating associations between all-cause mortality and BMI." They also point out that when there is a chronic wasting disease, being overweight and even slightly obese can be somewhat protective in providing "needed energy reserves" and even protection against trauma. There have been, for example, earlier reports that those overweight and mildly obese patients with chronic heart failure and coronary artery disease had decreased mortality compared to those of normal weight. It is always important to consider whether weight loss is intentional or unintentional.

Bottom line: One size does not fit all. The relationship between weight and mortality is a complex one. For most people, being of normal weight is still healthiest but for some, particularly with advancing age and chronic wasting diseases, those extra pounds may not be so detrimental. ■

ARE WE SUGAR-COATING SUGAR SUBSTITUTES?

There may be more artifice in artificial sweeteners than we realize

33 — Posted January 28, 2013

READ THE LABELS on our packaged foods, and in all likelihood, you will find they are laced with added sweeteners. Ng et al, according to a recently published article (2012) in the *Journal of the Academy of Nutrition and Dietetics*, report that of over 85,000 processed foods tested for the years (2005 to 2009) they studied, 75% contain sweeteners of some form. Most commonly, these products contain caloric varieties of sugar (e.g. corn syrup, fruit juice concentrate, honey, etc.), but increasingly, manufacturers are turning to artificial sweeteners, used either alone or even in combination with caloric sweeteners. For example, the researchers found that noncaloric sweeteners were added in more than a third of yogurts, 42% of flavored waters, and most diet drinks. And there is reason to suspect that far more people are using these sugar substitutes today than even a few years ago, though actual use is hard to quantify since "purchase" is not the same as "use."

Today the technical term is "nonnutritive" because essentially these substances provide no calories. (Artificial sugars used to sweeten gum, though, such as sorbitol, mannitol, maltitol—the so-called sugar alcohols—actually have two calories per gram.) Currently there are five nonnutritive sweeteners available in the U.S.: Acesulfame-K (e.g. Sunett); Aspartame (e.g. Nutrasweet; Equal); Neotame; Saccharin (e.g Sweet'N Low); and Sucralose (e.g. Splenda, which is heat-stable and can be used in baking or cooking). Rebaudioside A (e.g. Stevia; Truvia), a "highly purified product" from the stevia plant (and hundreds of times sweeter than sugar according to the American Diabetes Association website) is also being used and is considered a "food additive" by the FDA.

Saccharin was the first non-nutritive sweetener on the market; original packaging for saccharin, from the Zucker Museum, Berlin. Originally thought to cause bladder cancer in rodents, but no longer.

Photographer: User: FA2010. Released to Public Domain by user/Wikipedia Commons.

These sweeteners are called "artificial" which, of course, means "man-made," but ironically, "artificial" can also mean "contrived or fabricated for a particular purpose, especially for deception." *(Oxford English Dictionary)* They deceive us into thinking they are an energy source, for example, because evolutionarily, we have an innate preference for sweets, and we are programmed to associate sweet taste with sources of energy. Are nonnutritive sweeteners deceiving us in other ways as well?

At this point, there is no evidence that these substances are toxic or carcinogenic. Reports have been surfacing for years, though, that these sugar substitutes have been associated with metabolic abnormalities, such as type II diabetes or even weight gain. Studies on humans have produced inconsistent results; some have even found that they are effective for people in controlling weight. Other researchers have suggested that the sweet taste associated with lower calorie foods may paradoxically increase appetite and lead people to overeat. There is even the suggestion that these substances may change the flora in our intestinal tracts and interfere with the metabolism of many medications. Recently, as well, researchers have discovered that our gut has "sweet taste receptors" just as we have in our mouth, and it is possible that nonnutritive sweeteners may affect the release of hormones involved in satiety and the regulation of glucose metabolism, though there is to date, "no solid evidence" yet in humans.

Pepino and Bourne, writing in the journal *Current Opinion in Clinical Nutrition and Metabolic Care* (2011), make several important points regarding nonnutritive sweeteners:

- Researchers still don't know whether nonnutritive sweeteners are metabolically as inert as originally thought.

Sweet'N Low (saccharin) sugar substitute. One of several non-nutritive sweeteners on the market.

Raysonho@Open Grid Scheduler/Grid Engine. Creative Commons CC0 1.0 Universal Public Domain Dedication. Wikipedia.

Campaign by the US Food Administration to curtail sugared drinks. Commercial color lithograph, 1917. Artist: Ernest Fuhr, Carey Printing Company. Metropolitan Museum of Art, NYC.

Gift of William C. Moore, 1972, Public Domain.

- they may not necessarily promote "diet healthfulness"
- they may interfere with physiological responses that are responsible for metabolic homeostasis by "dissociating sweetness from calories"
- they may affect glucose control by interacting with newly discovered so-called sweet-taste receptors in our gut
- future research must acknowledge that sensitivity to sweet taste can vary considerably among different species (e.g even rats and mice differ in their sensitivity to sucralose)

Bottom line: The jury is still out on these nonnutritive sweeteners. For some of you, they may play a role in watching your overall caloric consumption. For others (probably those of you somewhat genetically challenged), they may make you hungrier and want to eat more, and they may even affect the reward systems in your brain. As every scientific article states, "Further research is needed..." ■

GUT REACTION: CAN G.I. BACTERIA CAUSE WEIGHT GAIN OR WEIGHT LOSS?

Changing the flora of our intestinal tract to regulate weight

34—Posted February 19, 2013

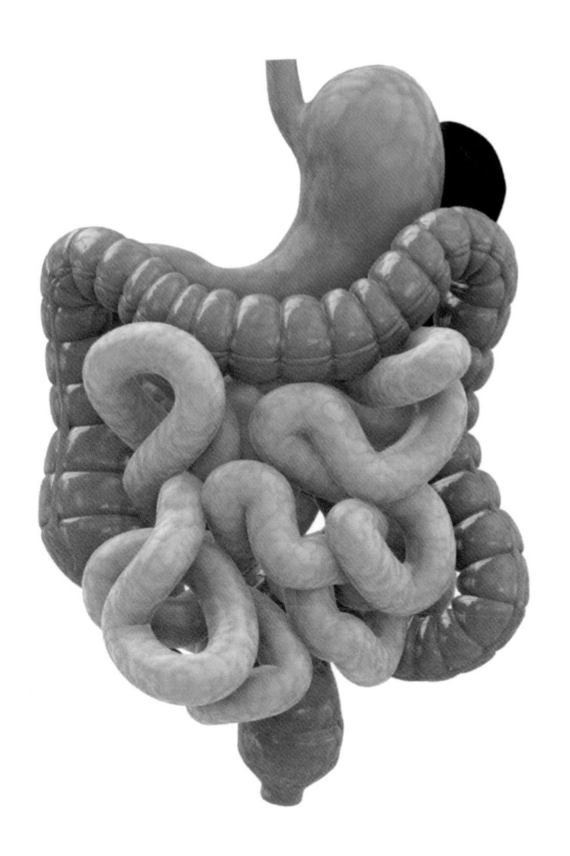

Our G.I. tract contains trillions of bacteria.

Used with permission, istockphoto.com. Credit: Eraxion.

No one really knows why overweight and obesity rates have skyrocketed to epidemic proportions in most of the world over the past thirty or so years. What we do know is that excess fat accumulation results from an energy imbalance or disequilibrium (increased caloric intake and/or decreased caloric expenditure) and is the result of an extraordinarily complex interaction of genetic, neurochemical, metabolic, behavioral, psychosocial, and other environmental factors. There are many theories regarding the burgeoning increase in the prevalence of obesity in the past thirty or so years. One of my previous blogs (12/17/12) highlighted top ten contributors to the growing obesity epidemic. Included in that list is the possibility that viruses or even bacteria may play a role. Here I focus on the role of bacteria in our gut. Could there be a connection between the flora in our intestinal tract and human obesity?

According to researchers Christina A. Tennyson and Gerald Friedman, writing several years ago in the journal *Current Opinion in Endocrinology, Diabetes, and Obesity*, the intestinal tract of an adult human may contain as many as 100 trillion micro-organisms (with from 15,000 to 36,000 bacterial species represented), called our "microbiota." Though they are found throughout our small and large intestines, they are most predominant in our colon, and they perform many beneficial functions, including aiding metabolic functioning and our immune system.

Since human fetuses are sterile while *in utero*, how do we acquire all these microorganisms? Infants first gain exposure to their mother's bacteria by vaginal delivery, and the more prolonged the delivery, the greater the exposure. Those born by Caesarian section can acquire them through general maternal contact such as breastfeeding or being kissed and even by contact with nursing staff and other infants. It is ultimately the job of our immune system to differentiate harmful bacteria from the majority of those that are beneficial to us.

Million and Raoult report, in a recent (2013) issue of the journal *Current Infectious Disease Reports*, that it has been a common practice in agriculture in industrialized countries for years to change the gut flora in pigs, calves, and chickens, by administering "growth promoters"—low-dose antibiotics or probiotics— for the purpose of inducing weight gain in farm animals. Probiotics are live micro-organisms that change intestinal flora and may lead to health benefits when given in sufficient amounts either to animals or humans. Probiotics can be purchased in health food stores as dietary supplements and taken as capsules, but they (e.g. *Lactobacillus* and *Bifidobacterium*) are also increasingly added to yogurts and other fermented products.

What researchers are increasingly finding is that both animals and humans may either gain or lose weight when their gut microbiota are manipulated either by exposure to antibiotics, probiotics, or even prebiotics. Tennyson and Friedman report that prebiotics, such as bran or psyllium (e.g. Metamucil), are "dietary components that stimulate the growth and metabolism of these beneficial organisms"

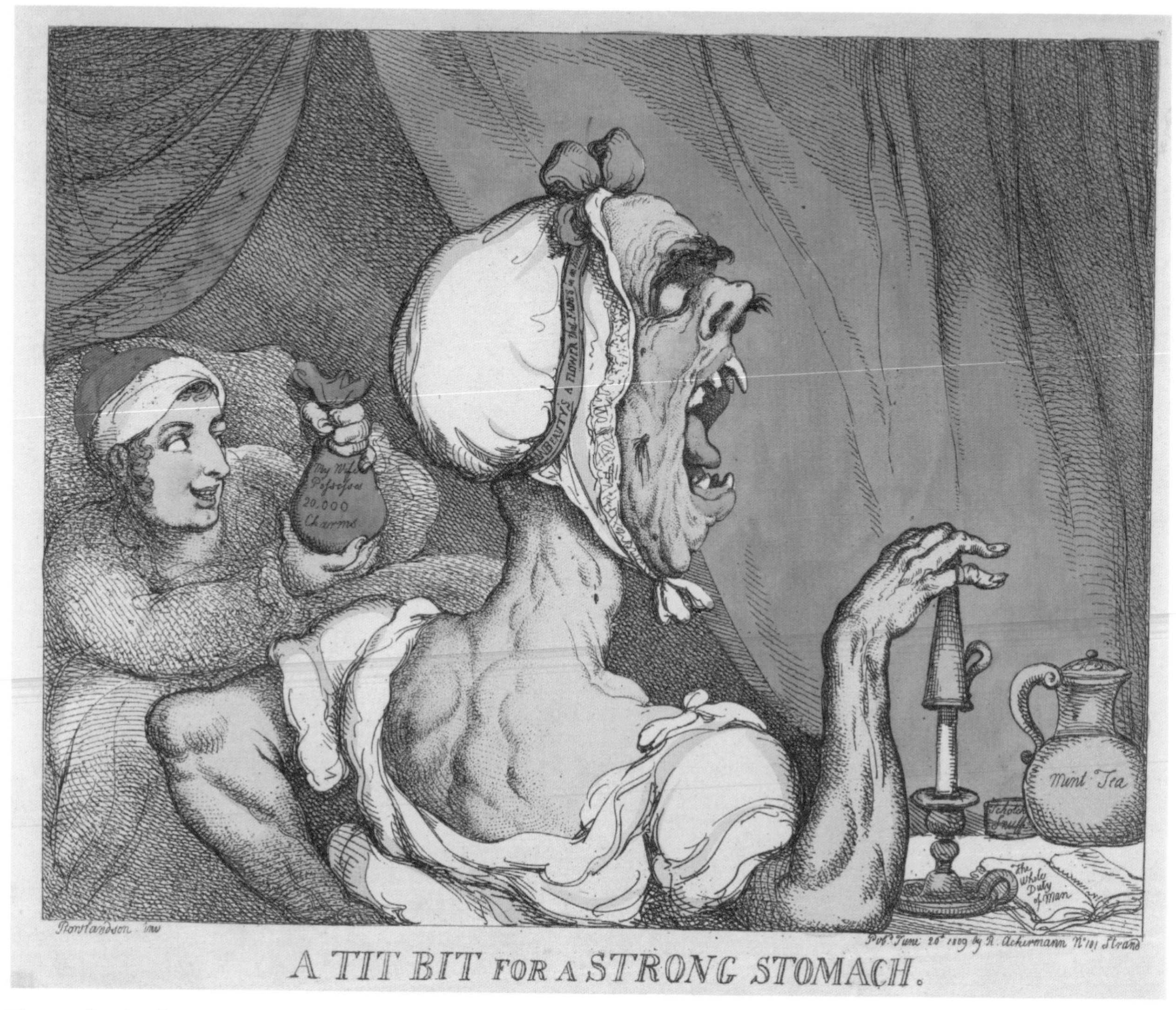

Thomas Rowlandson, *A Tit Bit for a Strong Stomach*. Publisher: Rudolph Ackermann, London, June 20, 1809. Metropolitan Museum of Art, NYC.

The Elisha Whittelsey Collection, The Elisha Whittelsey Fund, 1959. Public Domain.

◀ Buddhist Monk Budai, Qing Dynasty, 17th–18th century, Chinese. Some researchers believe that our microbiome—the millions of bacteria in our body—may lead to obesity. Metropolitan Museum of Art, NYC.

The Friedsam Collection, Bequest of Michael Friedsam, 1931, Public Domain.

and hence change the ratios of some bacteria to others. It has been found that lean people are more apt to have a significantly greater proportion of *Bacteroidetes* relative to *Firmicutes*, the two most prevalent *phyla* in our intestinal tract, and when obese people lose weight, they reverse that ratio so that they have more *Bacteroidetes*. Even obese mice have a greater proportion of *Firmicutes* than lean mice, and these obese mice have "extracted calories more efficiently." In other words, it is as if obese people (and mice), because of the percentage of certain flora in their G.I. tract, are eating more calories from the same food! There is likely a genetic component as well.

Could certain probiotics, though, also cause a decrease in weight, and even be used as anti-obesity agents? There have been studies in humans to suggest that probiotics may affect glucose and fat metabolism and even result in weight loss. The underlying mechanisms, however, are not entirely known and may be complex and multi-determined. There is clearly a need, as well, for more research on the safety of probiotics. Million and Raoult, for example, note that research on probiotics is subsidized or sponsored by the food industry, and unlike the pharmaceutical industry, the food industry, to date, does not have to reveal any financial conflicts of interest. As a result, perhaps we should interpret some of this research on the positive effects of probiotics cautiously.

At this point, it remains to be seen whether probiotics have a place in the treatment of overweight and obesity. Whether there is an infectious component to weight control is an intriguing idea but many may still have an initial gut reaction of skepticism. ■

▲ Some yogurt contain probiotics that change the intestinal flora of our gastrointestinal tract. Many cuisines around the world feature yogurt.

Used with permission, istockphoto.com, AlfPhoto.

THE HARE AND THE TORTOISE: AESOP'S FABLE AND WEIGHT LOSS

Does rapid weight loss necessarily lead to weight regain?

35 — Posted March 11, 2013

MOST EVERYONE KNOWS Aesop's fable of *The Hare and the Tortoise*: a hare makes fun of a tortoise for moving so slowly. The tortoise challenges him to a race. Feeling quite sure of himself and seeing how far ahead of the tortoise he is initially, the hare decides he has time for a rest. When the hare finally awakens from his nap, it is too late and the tortoise has already won the race. The moral: "Slow and steady wins the race."

What about with weight loss? Is there an optimal rate at which we should lose weight? Is "slow and steady" the secret to avoiding weight regain? Weight regain, after all, is a serious problem. Even when dieters lose significant amounts of weight, they typically regain one-third to one-half of their lost weight in the year following treatment. (Perri et al, 2008, *Archives of Internal Medicine*). Eventually, over the next several years, most people regain all their lost weight and then some.

In his excellent article, "Myths, Presumptions, and Facts about Obesity," published recently in *The New England Journal of Medicine* (January 31, 2013), Dr. David B. Allison, Quetelet Endowed Professor of Public Health at the University of Alabama (Birmingham), and his colleagues found that it is actually a myth that "large, rapid weight loss is associated with poorer long term weight loss outcomes, when compared with slow gradual weight loss." Using an evidenced-based model, Allison et al found seven myths about obesity that are "false and scientifically unsupported beliefs" but remain "pervasive in both the scientific literature and the popular press."

French painter Gustave Courbet, *After the Hunt*, circa 1859. Metropolitan Museum of Art, NYC.

H.O. Havemeyer Collection, Bequest of Mrs. H.O. Havemeyer, 1929, Public Domain.

This particular false and persistent belief most likely arose from studies that involved dangerous, very low calorie and "nutritionally insufficient" diets (VLCD) of fewer than 800 calories a day. Years ago, those diets, particularly when not supervised by a physician or nutritionist, had been associated with severe medical (e.g. cardiac) complications and even death. Characteristically, when patients do lose weight rapidly by severe caloric restriction, they are likely initially losing water and even muscle mass, particularly if they are not getting sufficient amounts of protein in their diets. Typically, most textbooks recommend 1 to 2 pounds a week as a safe amount to lose. This is often based on the fact that roughly, there are 3500 calories to a pound: to lose a pound, you have to eliminate 500 calories a day from your diet and to lose two pounds a week, 1000 calories a day. (Of course, as you lose weight, your body compensates and requires fewer calories for maintenance.) For most people, cutting all those calories is difficult enough, particularly in our obesogenic environment, without cutting even more.

Steady weight loss is not necessarily less likely to prevent weight regain for everyone. Astrup and Rössner, for example, in *Obesity Reviews* (2000), found that a greater initial weight loss, as long as it was the first step in a weight management program, could result in "improved sustained weight maintenance." As a result of their survey of the literature, these authors concluded that obese patients should not necessarily be encouraged to set lower weight-losing goals. In fact, losing weight more rapidly can provide considerable psychological reinforcement for the dieter. What tends to prevent weight regain, though, is adherence to a long-term program—what's called a "continuous care" model since obesity (and weight control) are chronic problems for those so genetically challenged. For most, long-term follow-up—whether face-to-face or even by telephone or computer—provides the much needed accountability, i.e., consistent monitoring of what we eat, how much we exercise, and what we weigh as measured on a scale. Furthermore, what also is more likely to lead to maintenance of any weight loss (i.e. avoiding weight regain) is continuing to practice lifestyle interventions such as daily exercise and watching our diet (including monitoring of calories) indefinitely. More recently, Nackers et al (2010) also found that those who lost more weight initially in the first six months were more successful long-term and were not more susceptible to regaining their lost weight over 18 months of follow-up. In this study, those who lost weight most rapidly were more likely to have greater adherence to the program (e.g. attended more follow-up sessions, completed more food records, eaten fewer calories, and exercised more) than those who lost weight the most slowly. And they agree that losing at a slow rate, may be "less reinforcing to participants than losing at a moderate or fast initial rate."

Bottom line: The recommendation to lose weight more slowly works for some but may actually interfere with long-term goals and successful weight maintenance for others. It sometimed depends on how you lose the weight and there are substantial genetic differences among people. We don't always know why some people lose weight more slowly than others. Allison and his colleagues, though, note "heritability is not destiny!" Environmental factors can be significant contributors to weight control as well. ■

Diego Velázquez, Portrait of Aesop, circa 1638.
Prado Museum, Madrid.

Wikipedia Commons/Public Domain.

Tile with two rabbits, two snakes, and a tortoise. Illustration for Zakariya al-Qazwini's Book, *Marvels of Things Created and Miraculous Aspects of Things Existing* (13th century.) Earthenware, Iran, 19th century. Louvre Museum.

Jastrow, 2006, Wikipedia Commons/Public Domain.

THE MEDICALIZATION OF WEIGHT: ARE WE "DISEASE MONGERING?"

From the moral to the medical: changing views on overweight and obesity

36—Posted March 29, 2013

"THERE'S A LOT OF MONEY to be made from telling healthy people they are sick. Some forms of medicalizing ordinary life may now be better described as 'disease mongering': widening the boundaries of treatable illness in order to expand markets for those who sell and deliver treatments." So say Ray Moynihan and his colleagues in a 2002 article in the *British Medical Journal.* Do overweight and obesity fit into this category of "disease mongering?"

What we do know is that overweight and obesity can be risk factors for other diseases. Overweight and obesity are defined by an excess accumulation of fat, as measured fairly imprecisely by body mass index, BMI, (the weight in kilograms divided by the height in meters squared.) This equation, originated by Adolphe Quetelet in the 19th Century, did not come into use as a measure of overweight and obesity until the 1970s—and the norms we use now were not even established until the late 1990s by the World Health Organization. (Skin calipers are an even more inaccurate way to measure fat.) BMI in and of itself, though suggestive of health (the higher the BMI, the more likely health-related issues), does not actually indicate the health of a particular person. Other than an excess of fat, however, there are no other clinical signs or symptoms that are seen in everyone who is obese or overweight. Heshka and Allison, in a 2001 article in in the *International Journal of Obesity*, acknowledge that physicians cannot even predict who will necessarily develop an overweight or obesity-related health problem, of which, of course, there are many (e.g. cardiovascular disease, type II diabetes, various metabolic disturbances, hypertension, certain forms of cancer, osteoarthritis, sleep apnea). These authors note that, even though obesity is a public health problem, there are, in fact, overweight and obese people "who will live long lives free of any of the morbidities known to be influenced by obesity." Heshka and Allison add, "We are therefore placed in the conceptually awkward position of declaring a disease which, for some of its victims, entails no affliction."

There have been significantly more references over the years to overweight in both the lay press and the medical literature, and according to Jutel, writing in the journal *Social Sciences and Medicine* (2006), the references have shifted from sign or symptom to disease entity. Jutel makes the point that two important factors have contributed to making overweight into a diagnosis: the importance of measurability in establishing health and disease and a strong emphasis in Western society on normative appearance. For example, scales for weighing did not even enter the physician's office until the end of the 19th century. Jutel believes that overweight is not necessarily a disease "any more than slenderness is an indication of health." In other words, it is merely "a description of physical appearance."

Chang and Christakis, in reviewing how one classical medical textbook wrote about obesity in subsequent editions throughout the years of the 20th century, noted that though the basic model of obesity was always seen as a result of excess calories over expenditure, there were different causal factors superimposed. Over time, obese individuals were progressively held less responsible for their condition in subsequent editions of the textbook—from fat as a personal or even moral failing to the medicalization of fat as a sickness. Remember that gluttony and sloth are two of the "Seven Deadly Sins."

An obese man consulting his physician. Coloured etching. Wellcome Library, London.

Abigail Saguy, in her book, *What's Wrong with Fat?* explains that the word "obesity" implies a medical frame, which, in turn, "implies that fat bodies are pathological." Saguy continues, by "framing fatness as a matter of health raises the stakes. No longer is it a question of appearance, fatness becomes a matter of life and death." And Dr. David L Katz, Director of the Yale University Prevention Research Center and Editor-in-Chief of the journal *Childhood Obesity*, wrote in a recent editorial (February 2013), "... weight per se was never what mattered....what makes shape and size problematic is they are often harbingers of ill health...." "What matters here is health. Everything else is fashion." ■

Claude Monet, *Dr. Leclenché*, 1864.

Gift of Mr. and Mrs. Edwin C. Vogel, 1951.Metropolitan Museum of Art, NYC. Public Domain.

Vincent van Gogh, *Dr. Paul Gachet*, 1890. Musée d'Orsay.

Google Art Project. Wikimedia Commons/Public Domain.

Hieronymus Bosch, *The Seven Deadly Sins*, between 1505–1510, Prado Museum, Madrid. Obesity used to be considered the result of two of the deadly sins—sloth and gluttony, as depicted by Bosch. We now know that obesity is a chronic disease of complex etiology.

Wikipedia Commons/Public Domain.

HEAVY: UNEASY LIES THE FAT THAT WEARS A CROWN

The "crown-like structures" of dying white fat cells that 'flesh is heir to'

37—Posted April 27, 2013

Overweight or obesity is an excess accumulation of fat. The greater the amount of fat, the worse the level of obesity. Other than an excess accumulation of fat, as I have said in my last blog entry, there is no other sign or symptom that is present in everyone who is obese or even overweight. So let's wax a bit Shakespearean (e.g *Henry IV, Part II* and *Hamlet)* and ask what do we really know about fat?

Fat, or more technically, adipose tissue, was long thought to be an inert substance whose sole purpose was to cushion and support our other organs, as well as provide insulation against the cold. Our body has white fat, browned white fat (also known as "beige fat"), and brown fat. For those interested in beige and brown fat, please see blog 21, *Special Delivery: What Can (Brown) Fat Do for You?*

The primary function of white adipose tissue is energy storage. We now now that white fat is quite a remarkable substance and far from inert: it is a highly metabolically active endocrine organ that secretes about 100 substances, including the hormones leptin and adiponectin, and dangerous "pro-inflammatory" substances such as tumor necrosis factor alpha and interleukin-6, as well as many so-called *adipokines* whose functions are still unknown. What is fascinating is that white fat is an organ found in multiple places all over our body and is constantly undergoing remodeling.

White fat and the pattern of its distribution can have a "profound influence" on our health and the risk for disease, according to researchers Lee and colleagues, writing in the journal *Molecular Aspects of Medicine* (2013). White fat that is found in the areas of the upper body around the abdomen just below the skin (i.e., subcutaneous) and particularly, viscerally (i.e., abdominal fat that encases our internal organs) results in a large waist and the "apple" shape appearance more commonly seen in men. Fat in this central location is potentially the most dangerous kind and is more likely associated with metabolic abnormalities such as insulin resistance, glucose intolerance, abnormal levels of trigylcerides and cholesterol, hypertension, and ultimately cardiovascular disease. Subcutaneous fat found in the gluteo-femoral areas (i.e,. the "pear" body shape of fat located predominantly on the hips) may even be somewhat protective of metabolic disturbances and is more commonly seen in women. Where our fat accumulates when we gain weight is most likely genetically based. With obesity, adipose tissue can also accumulate in other organs, forming so-called "ectopic" fat deposits, such as in the liver (e.g. fatty liver as a precursor to cirrhosis), skeletal muscles, heart, and blood vessel walls and cause substantial damage to these organs.

"Crown-like" structures of macrophages form of top of dead fat cells. Ritual Crown with the Five Transcendent Buddhas, lates 14th-early 15th century, Tibet.

Gift of Jeffrey Kossak, The Kronos Collections, 1985. Metropolitan Museum of Art, NYC, Public Domain.

Polish researchers, Wronska and Kmiec, in the journal *Acta Physiologica* in 2012 note that white fat is composed predominantly of spherical adipocytes that are lipid (i.e., trigylceride)-filled cells, but it also contains precursor cells, or pre-adipocytes, that do not contain lipid but have that potential to become lipid-filled, as well as endothelial cells of blood vessels and lymph tissue, nerve fibers, and macrophages that are cells involved in inflammation. White fat tissue also contains stem cells that can differentiate into other kinds of cells, including neurons and even liver cells. Adipose tissue depends on a rich network of blood vessels to transport oxygen and other substances to it and provide a route away for its many secretory adipokines. When we gain weight initially, we get "enhanced angiogenesis" (i.e., more blood vessels are created) and when we lose weight, these blood vessels regress. It is "is probably the most highly vascularized tissue in the body." (Lemoine et al, *Thrombosis and Haemostasis*, 2013) There is speculation that when fat

▲ Flemish artist Jacob Jordaens, *The King Drinks,* first half of 17th century. Chronic overindulgence of food and drink can lead to inflammation, crown-like abnormalities around fat cells, and the development of obesity.

The Yorck Project, Wikimedia Commons, Public Domain.

◀ Wall painting in the Heiliger-Geist-Hospital Holy Spirit Hospital, Luebeck, Schleswig-Holstein, Germany.

Image/BROKER/Alamy Stock photo, used with permission.

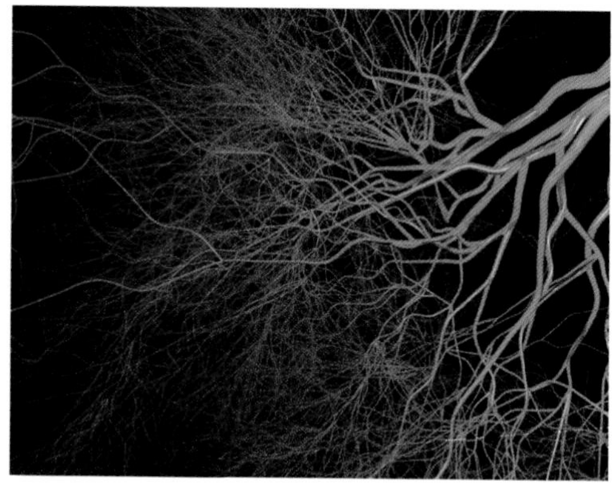

▲ Adipose tissue needs a rich supply of blood vessels.

Used with permission/istockphoto.com. Credit: Inok.

◀ Alexander Benois, *Design for the Costume of Carabosse (Wicked Fairy Godmother)*, 1927, for the *Ballet La Belle au Bois Dormant (Sleeping Beauty)*. Metropolitan Museum of Art, NYC.

Bequest of Sallie Blumenthal, 2015. Image copyright, The Metropolitan Museum of Art. Image source: Art Resource, NY. Copyright 2018 Artists Rights Society (ARS), New York/ADAGP, Paris. Used with permission.

cells become too massively enlarged, their blood supply becomes insufficient, hypoxia (lack of oxygen) develops, and there occurs an infiltration of macrophages and subsequent inflammation. With obesity, when white fat cells are dying or die, they become surrounded by macrophages that create the so-called "crown-like structures" around the top of the fat cells. These crown-like structures around dying or dead (necrotic) fat cells are more common in visceral fat than subcutaneous fat and are thought related to the development of the metabolic disturbances. In fact, obesity has been described as a chronic low-grade inflammatory and metabolic disease. (Suganami and Ogawa, *Journal of Leukocyte Biology*, 2010) Adipose tissue enlarges primarily by hypertrophy (i.e., increased size of an individual fat cell) or, if the obesity is severe, by hyperplasia (i.e., increasing the actual number of fat cells.) There is significant turnover of fat cells, but unfortunately, even when we lose considerable weight by diet or bariatric surgery, our fat cell numbers remain the same; they just shrink in volume.

It is possible that eventually interfering with the blood supply of fat tissue will be a treatment for some forms of obesity. ■

HUMAN BONDAGE: AMERICA'S NOT SO MAGNIFICENT FOOD OBSESSION

Four books on their authors' personal struggles with our obesogenic environment

38—Posted May 8, 2013

This week marks the publication of *Obsessed: America's Food Addiction—And My Own* by Mika Brzezinski, author and co-host of the television program *Morning Joe.* This is one of several excellent books that have appeared in recent years that explore their authors' personal struggles with food—particularly the highly addictive processed foods laden with sugar, fat, and salt—as well as their authors' concern for a nation whose children and adults have continued to grow more overweight and obese over the past thirty years.

What makes her book different and particularly appealing is that, for all the public to see, although she has been a size 2 and in seemingly perfect control of her eating, Brzezinski has been quietly preoccupied with food for most of her life. Her book grew out of a confrontation she had with one of her best friends, Diane Smith, an Emmy-award winning journalist and author, who had a much more visible weight problem and whose weight had ballooned to the point she had more than 75 pounds to lose. Brzezinski challenged Smith to lose weight and write the book with her.

Brzezinski received considerable positive reinforcement for her size 2 body. No one knew how much control (and what was required) on a minute-to-minute basis to maintain her weight at that level. Says Brzezinski, "Most people assume all food addicts are fat, but I'm here to tell you they are not. Just because I have a healthy body weight doesn't necessarily mean I have a healthy relationship with food." Brzezinski became fascinated with the concept of the body's set point. Set point, though, is a theoretical concept—no researcher has yet to locate one area in the brain or anywhere else in the body that regulates the body's weight around a particular range. Nevertheless the concept is an appealing one and helped Brzezinski appreciate that her ideal weight was about ten pounds heavier than she had been struggling to maintain.

Obsessed describes research corroborating what Brzezinski has suspected all along from her own experience, namely there is such a thing as food addiction (with tolerance and even withdrawal symptoms). Some researchers, though, question whether any food can be addictive. After all, we can never be without food; we can speak of food use and food abuse, but, unlike the language of drug abuse, we can never speak of a non-user. Nora Volkow, MD, Director of the National Institute for Drug Abuse (NIDA), though, has been writing for years on the addictive quality of certain foods that illuminate the reward centers in our brain in exactly the same way as cocaine or other drugs of abuse.

The juxtaposition of two public figures—one thin and the other fat— but each similarly preoccupied with her weight and when to get her next fix of junk food, provides the back story for America's growing addiction to processed foods, made abundantly clear in Michael Moss's recent book, *Salt Sugar Fat.* Moss describes how our large food companies are all striving for "stomach share," i.e., "the largest share of what people eat," where the "cardinal rule of processed food, (is) when in doubt, add more sugar." Moss describes how these companies search for the bliss point, the "defining facet of consumer craving," or the "precise amount of sweetness—no more, no less—that makes food and drink most enjoyable." David Kessler, former head of the FDA, in his book *The End of Overeating,* also confronted the food industry and described our "conditioned hypereating" brought on by the "hyperpalatable" foods in our environment. Kessler, too, struggles with his

Auguste Rodin, *Study for Obsession,* c. 1896.

Metropolitan Museum of Art, NYC., Public Domain. Gift of Auguste Rodin, 1912

People with an "addiction" to food may feel metaphorically enslaved. American artist John Steuart Curry, *John Brown*, 1939.

Arthur Hoppock Hearn Fund, 1950. Metropolitan Museum of Art, NYC. Public Domain.

own food demons, particularly chocolate chip cookies and the "enticement of a Cinnabon."

Born Round, written by Frank Bruni, *New York Times* columnist (and former restaurant critic for the *Times*) was published a few years ago and is an extremely funny (but painful) story of Bruni's own life-long preoccupation with food that started early in childhood. Says Bruni, "A third burger isn't good mothering. A third burger is child abuse." His mother, incidentally, always "believed that somewhere out there was a holy grail of weight loss...." Unfortunately, many share that belief. David B. Katz, MD, Director of the Yale Prevention Center, Editor-on-Chief of the journal *Childhood Obesity*, and one of the experts interviewed by Brzezinski, has spoken of the "prevailing gullibility" of a public "beguiled by a belief in weight-loss magic" such that "any weight loss claim is accepted at face value." Those who work in the field and those who struggle with weight control know there is no magic to weight control.

We live in an obesogenic environment. The word "obesogenic" came into the scientific literature around the year 2000. One of the earlier references appeared in a 2001 paper in the journal *Obesity Reviews* by Drs. Cynthia M. Bulik and David B. Allison, both major researchers in the field of obesity, who raised the provocative question, "Why are certain people not obese, given the environment in which we live?" The answer lies in the fact that weight control is the result of a highly complex interaction of genetic, behavioral, and environmental factors. Food both disturbs our body's homeostasis and is required for its homeostasis. Some people are more genetically inclined to be predisposed to food preoccupation (and even addiction) and overweight and obesity in our current environment. While we cannot do much about our genetics, we can work on our environment. Brzezinski offers suggestions for ways to change the way we think about weight and confront our environment, including publicizing the costs of obesity and funding more research, as well as "celebrating a healthy thin in the media." Her book is one of several that explain why it is so important to do so. ■

GREAT OR NOT SUCH GREAT EXPECTATIONS: WEIGHT LOSS GOALS

The goalkeepers: does it matter if your goals are ambitious or more realistic?

39—Posted May 26, 2013

EVEN THOUGH MOST researchers assert that a weight loss of 5% to 10% of body weight can lead to substantially beneficial health effects, most patients set expectations for themselves of far more ambitious weight loss goals, regardless of the method they use to lose weight. For example, Heinberg et al, in a 2010 article in *Obesity Surgery*, report that patients undergoing bariatric surgery (the most dramatic way to lose weight) will lose, depending on the surgical procedure used, up to 80% of their *excess* body weight, but many of these patients will lose considerably less and may have unrealistic expectations to lose close to 100% of their presurgical *excess* body weight! These researchers even caution that surgical informed consents should make explicit to patients the amount of weight loss actually expected from surgery. Typical weight loss for the gastric lap band surgery (recently undergone by Governor Chris Christie of *New* Jersey), for example, is only about 20 to 25% of general body weight, according to Sjöström *et al* (The New *England Journal of Medicine*, 2007) Results from lifestyle changes (e.g. diet and exercise) and even psychopharmacological treatment generally yield far lower percentages.

American sculptor Robert Tait McKenzie, *The Competitor*, bronze, 1906. Credit: Rogers Fund, 1909.

Metropolitan Museum of Art, NYC, Public Domain.

Dieters can be somewhat persistent and not so malleable in their wish to change their expectations. For example, Wadden *et al*, in a 2003 article cleverly called *Great Expectations: 'I'm losing 25% of my weight no matter what you say'* (and the inspiration, other than Charles Dickens, for this blog's title) found that even when patients are "informed repeatedly" that their weight loss goals were unrealistic, they still expected to lose twice as much as they had been advised. These researchers found that some obese patients, though, may actually "be surprised to find they are generally satisfied by the 10% weight loss now recommended by health professionals."

So does it matter whether you set so-called "realistic goals" for weight loss? This has become common parlance among many clinicians who discuss weight control with their overweight and obese patients. After all, so the argument goes, setting goals that are too ambitious may lead dieters to give up and experience considerable disappointment and a sense of failure. Apparently, though, there is no scientific evidence (i.e., "no statistically significant relationship") to support this widely held belief and it is actually one of several weight loss myths uncovered by David B. Allison, PhD, Quetelet Endowed Professor of Public Health, and his group at the University of Alabama at Birmingham.

For this particular myth, Dr. Allison *et al* did a thorough search of the literature from 1998 to 2012 and conducted the first meta-analysis of randomized controlled studies (11 met their inclusion criteria) that investigated the relationship between goals set by patients for weight loss and their actual weight loss. Publishing this year in the journal *Obesity Reviews*, they found "current evidence does not demonstrate that setting realistic goals leads to more favourable weight loss outcomes," apparently regardless of the weight loss intervention (e.g. cognitive behavioral therapy, dietary modification, use of medication or even bariatric surgery.) Even when patients' goals became more realistic, they did not necessarily lose more weight

American artist Thomas Eakins, *The Champion Single Sculls (Max Schmitt in a Single Scull)*, 1871.

Purchase, The Alfred N. Punnett Endowment Fund and George D. Pratt Gift, 1934, Metropolitan Museum of Art, NYC, Public Domain

or even have better psychological outcomes as measured by less evidence of depression or greater self-esteem. The researchers found, though, that there was not a consistency among studies in the terminology used in their description of "weight loss goals." For example, some studies assessed "maximum acceptable weight loss goals" while others assessed "dream/ideal weight loss goals." One limitation of their meta-analysis, they note, is that they focused on goals for the "active phase" of weight loss rather than on goals during the important phase of maintenance of the weight lost, and they acknowledge that these phases (and the behaviors involved) may be quite different. The researchers conclude, "While the assertion that unrealistic goals lead to disappointment and discontinuation of weight loss efforts makes intuitive sense, the empirical evidence does not support this conclusion."

What may be more important in achieving ambitious weight loss goals, for example, may be an individual dieter's motivation, autonomy ("ownership over newly adopted behavioral patterns") and self-determination, i.e., "a sense of choice and volition," rather than feeling pressure to comply or feeling controlled, according to Teixeira *et al* (*International Journal of Behavioral Nutrition and Physical Activity*, 2012). These researchers even question the standard "continuous care" model for treating excess weight in that they believe it may interfere with some dieters' own sense of self-efficacy. For Teixeira *et al*, general self-efficacy (i.e. an "overall sense of assurance or self-confidence") may be more predictive of successful weight loss than self-efficacy for specific behaviors (e.g. eating) or even goals patients set.

For a discussion of other weight control myths found by Dr. Allison and his group, please refer to their article, *Myths, Presumptions, and Facts about Obesity*, published this past year in the January 31st issue of *The New England Journal of Medicine*, as well as refer to blog 35, *The Hare and the Tortoise: Aesop's Fable and Weight Loss* on another of their myths concerning the spurious relationship of rapid weight loss leading necessarily to weight regain. ■

'WHEN THE BOUGH BREAKS': EXCESSIVE WEIGHT BEFORE PREGNANCY

Guidelines on how much weight a pregnant woman should gain

40 — Posted July 5, 2013

Most people know the lyrics to the old nursery rhyme, "Rock-a-Bye Baby:" "Rock-a-bye baby in the treetop. When the wind blows, the cradle will rock. When the bough breaks, the cradle will fall. And down will come baby, cradle and all." Just as, in retrospect, it may seem like poor judgment to place a baby-laden cradle on a treetop bough, it may be just as poor judgment to begin a pregnancy with excessive weight. Our nursery rhyme—the heavy bough broken by the wind— is a metaphor for all the complications that can arise when a woman begins her pregnancy already either overweight or obese.

Back in Ancient Greece, Hippocrates had called attention to the fact that "unnaturally fat women" may have irregular menstrual periods, cycles without ovulation, and even difficulty conceiving as "fat presses against the womb." As obesity has increased worldwide, though, so have the numbers of pregnant women who begin their pregnancy already overweight or obese. In fact, Buschur and Kim, writing in the *International Journal of Gynecology and Obstetrics* (2012) refer to "pre-conception (i.e., pre-pregnancy) obesity" as a "growing epidemic."

Vincent van Gogh, *La Berceuse (Woman Rocking a Cradle; Augustine-Alix Pellicot Roulin*, 1889.

The Walter H. and Leonore Annenberg Collection, Gift of Walter H. and Leonore Annenberg, 1996, Bequest of Walter H. Annenberg, 2002, Metropolitan Museum of Art, NYC, Public Domain.

What are some of the problems associated with pre-pregnancy obesity, as defined as a body mass index (BMI) of 30 kg/m^2. (weight in kilograms divided by the height in meters squared)? The range of complications is extensive. In fact, Galliano and Bellver, writing in *Gynecological Endocrinology* (2013), note that obstetrical complications in obese women can occur in all three trimesters of pregnancy, as well as during labor and delivery. Obese women are more likely to suffer from diabetes that develops during pregnancy (i.e., gestational diabetes mellitus, a significant predictor of future type 2 diabetes in the mother), hypertension (including preeclampsia, a potentially deadly hypertensive condition during pregnancy for which there is no treatment except premature delivery), dysfunctional labor, delivery by Caesarian section, and even pre-term birth. Maternal pre-pregnancy obesity has also been linked to spontaneous abortion, stillbirths, neural tube defects, and macrosomia (babies large for gestational age, or greater than the 90th percentile for birth weight,) as well as wound, genital tract, and urinary tract infections and even post-partum hemorrhaging in the mother. Furthermore, pre-pregnant overweight and obesity were also associated with fetal distress, longer hospital stays, and even need for referral to a neonatal intensive care unit. In fact, Sujatha et al (2012), writing in the journal of *Clinical Gynecology and Obstetrics*, believe that with all these increased risks associated with a pregnancy complicated by initial excessive weight, these women should be considered in the "high risk" pregnancy category. Say Galliano and Bellver, "In short, women should attempt to conceive at a normal weight if they wish for better obstetric outcomes."

Despite overnutrition, obese pregnant women may actually have nutritional deficiencies. Thornburgh (2011) in the journal *Seminars in Perinatology*, notes that despite excessive caloric intake, obese women may have deficiencies in iron, folic acid, and B12. Dao and colleagues, from

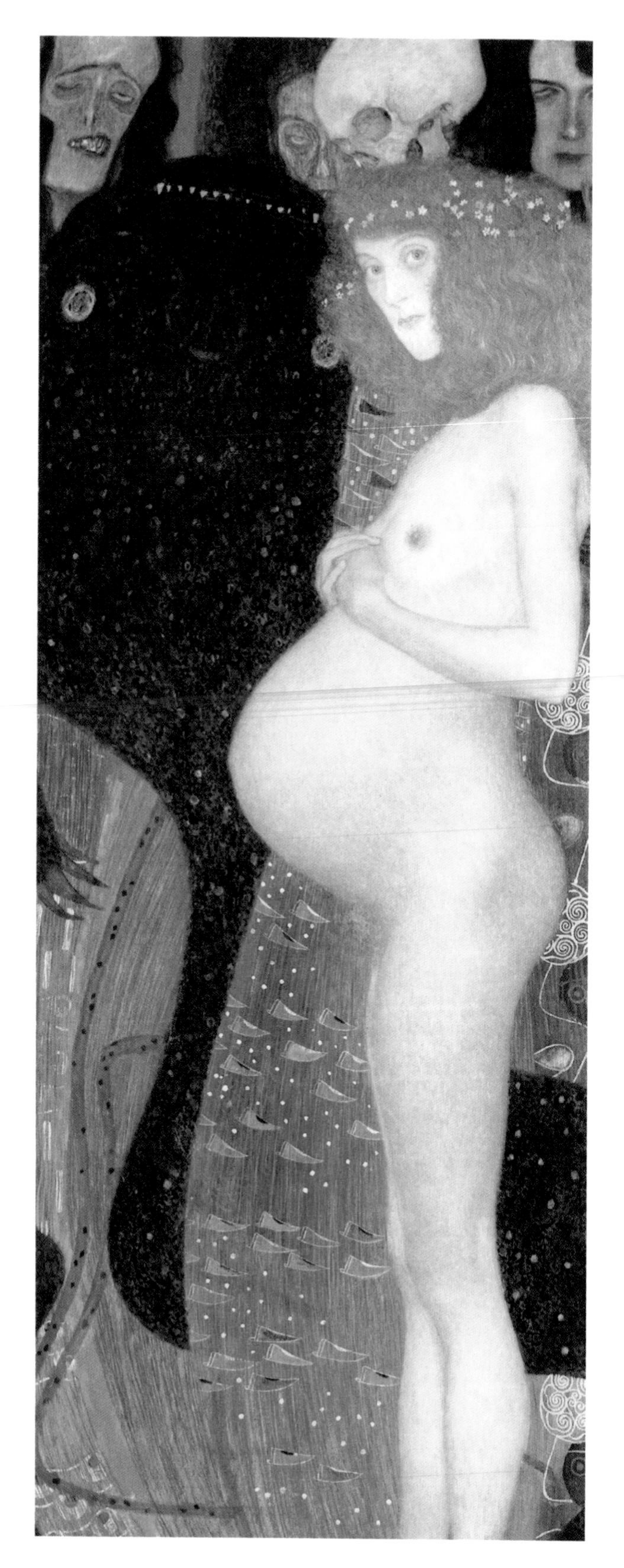

▲ Madonna del Parto, 15th century unknown master, Accademia of Venice.

Wikimedia Commons/Public Domain.

◀ Gustav Klimt, *Die Hoffnung (Hope I)*, 1903, National Gallery of Canada, Ottawa.

Wikipedia Commons/Public Domain (Hope I).

Nikolai Argunov, Portrait of Praskovia Kovalyova, 1803, Kuskovo Museum, Moscow.
Wikimedia Commons/Public Domain.)

Attributed to Marcus Gheeraerts the Younger, Portrait of an Unknown Lady, circa 1595. Tate Britain, London, 2011.
Wikipedia Commons/Public Domain.

Boston, for example, have reported in a 2013 article in the *Journal of Perinatology*, that obesity may lead to impaired iron stores in the fetus and newborn, through a process of chronic low-grade inflammation and consequently, this iron deficiency may have a major impact on the developing nervous system and lead to "lifelong and irreversible effects on neurodevelopment." Babies of obese women are themselves also more likely to develop obesity, type 2 diabetes, cardiovascular disease, and even cancer. (Galliano and Bellver, 2013) This is the concept of "fetal programming" or the importance of the environment of the womb on subsequent development. In other words, through epigenetic changes, there may be a "metabolic memory" for inflammation, insulin resistance, etc. that leads to possible increased risks for metabolic disorders and even certain cancers, etc.

In 2009, the prestigious Institute of Medicine (IOM), in response to the increasing age of women who become pregnant as well as increasing gestational weight gain, issued general guidelines for weight gain based on pre-pregnancy BMIs. Those who begin their pregnancy in the normal range of BMI should gain between 25 to 35 pounds; those who are overweight initially should gain only 15 to 25 pounds; and those who are obese initially should gain only 11 to 20 pounds. Significantly, though, Gaillard et al, in a large study in the journal *Obesity* (2013), report that pre-pregnancy overweight and obesity were associated with more adverse pregnancy outcomes compared with excessive weight gain that occurred just during pregnancy. To maintain these recommended values, pregnant women should exercise daily, monitor their food intake by watching calories, and weigh themselves frequently. It is crucial, as well, that those women with excessive initial weight have regular follow-up after delivery, with concern for their weight control. Thornburg reports that obese women, particularly those who gain excessive weight during their pregnancy, often fail to lose their weight after giving birth and this complicates subsequent pregnancies. They are also less likely to breastfeed and if they do begin breastfeeding, are more likely to discontinue it by six months. ■

IT'S NOT EXACTLY BETTER LIVING THROUGH CHEMISTRY

Can environmental chemical pollutants lead to weight gain?

41 — Posted August 9, 2013

17th century copy of the Quentin Metsys painting, Presumed Portrait of the Physician Paracelsus, 15th century.

Bequest of A.F. de La Coste Vivier, 1907, Louvre Museum, Paris.Wikipedia Commons/ Public Domain.

THE SLOGAN, "Better Living Through Chemistry," was a popular variant of an advertising slogan by the DuPont Company that was used from the mid 1930s until the early 1980s. (Apparently now it is also the name of a 2013 film, directed by Moore and Posamentier, and featuring, among others, Jane Fonda and Ray Liotta.) For our purposes here, we'll use it to call attention to the complex relationship of environmental pollutants to weight. We are talking about endocrine disrupting chemicals, i.e., potentially toxic synthetic compounds that may interfere with the regulation of our body's complex endocrinological (i.e. hormonal) systems or even possibly our circadian rhythms. They include dioxins, polychlorinated biphenyls (PCBs), brominated flame retardants, organochlorine pesticides, phthalates used to manufacture shampoos, cosmetics, and nail polish as well as bisphenol A used in the manufacture of plastic food and beverage containers, among others. These are chemicals that are found pervasively throughout our environment, including potentially in our water and food supply and now, even in a mother's breast milk! There is some indication these chemicals may play a role in the burgeoning worldwide obesity epidemic.

A series of articles in this June's issue of the prestigious journal *Obesity* has brought the controversy front and center. An editorial by David B. Allison, PhD, Quetelet Endowed Professor, and his colleague Julia Gohlke, both at the University of Alabama at Birmingham, discusses some of the complex issues involved. They quote from Paracelsus, a 16th physician and all-around Renaissance man, who had said, "All things are poison, and nothing is without poison; only the dose makes a thing not be poison." In summarizing the research presented in the Journal, Gohlke and Allison explain that different scientists can look at the same data and have "divergent interpretations." For example, Heindel and Schug, believe that environmental chemicals can be considered obesogens (i.e. they cause weight gain over time on exposure), particularly when exposure is during the "critically sensitive" "plastic" phase of early development and in those who are already overweight or obese because these chemicals are stored in fat tissue (i.e., the more fat, the more exposure.) They note that there are "now nearly 20 chemicals shown to cause long-term weight gain based on exposures during critical periods of development due to their ability to disrupt normal hormone and neuronal signaling pathways." Others, such as Sharpe and Drake, refute such a conclusion. They believe that the association of human obesity to so-called environmental "obesogenic chemicals" may be circumstantial. They believe that our fast food diets may have more of a contributory role. Still others are not sure and want more evidence.

Compounding the difficulties is that randomized controlled studies, the *sine qua non* of scientific investigation, may be unrealistic because of ethical issues of knowingly exposing a population to a potentially toxic chemical—weight-related or not. An example would be determining the effects of breastfeeding on exposure. Furthermore, causality is extremely difficult to prove: there are so many confounding, uncontrolled variables, such as contributions from diet (e.g. fat intake may be a factor in amount of exposure), length of exposure required for an effect, and even when that exposure takes place. There are also male and female differences in the effects of exposure so that any research must take gender differences into account. Furthermore, we are exposed to a toxic soup of these compounds in the environment so it may be difficult to separate out how exposure to particular ones affects someone.

Researchers, though, are trying. For example, Juliette Legler, in this issue of *Obesity*, reports on the OBELIX Study, which stands for "Obesogenic endocrine disrupting chemicals: linking prenatal exposure to the development of obesity in later life" that is being conducted in four European countries. The strength of that investigation is that it is studying "the same chemicals in an integrated toxicological and epidemiological approach," that includes research involving mothers and their children followed over eight years, as well as animal experiments with mice exposed to endocrine disrupting chemicals. Legler and her colleagues have found that there is an inverse relationship between a marker for PCB in infant cord blood and birth weight, not unlike the low birth weights and later life obesity found in some of the Dutch population exposed to starvation during the years of World War II.

Bottom line: we really do not completely understand why obesity rates have reached epidemic proportions in the past thirty years throughout the world. We do know that overweight and obesity result from a complex interaction of genetic, environmental, behavioral, and metabolic factors, among many others. How much a role endocrine disrupting chemicals have in weight gain may still be open to question. Say Gohlke and Allison, "Despite the divergent interpretations of individual studies and the differences in overall conclusions drawn…it is clear we have imperfect knowledge." ■

Plastic bottles have endocrine disruptors in them. The chemicals not only ruin the environment but may affect our body weight internally.

Used with permission, istockphoto.com. Credit: EuToch.

Bust of Paracelsus, who said "All things are poison…only a dose makes a thing not poison."

Used with permission, istockphoto.com. Credit: sagasan.

COLLEGE WEIGHT GAIN: DEBUNKING THE MYTH OF THE 'FRESHMAN 15'

How much weight do college students really gain in their freshman year?

42—Posted September 10, 2013

Meeting of Doctors at the University of Paris. French manuscript, 1537. Étienne Colaud, manuscript illustrator.
Wikipedia Commons/Public Domain.

There are many presumptions—"beliefs that persist in the absence of supporting scientific evidence"—and even myths—"beliefs that persist despite contradicting evidence"—about weight control. Seven of these myths, including about setting realistic goals for weight loss, the rapidity of weight loss, weight loss readiness, and breast feeding, were recently debunked in a comprehensive review by David B. Allison, PhD and his colleagues that was published this past year in the *New England Journal of Medicine.*

What about the so-called "Freshman 15'? This is the season when students return to school and many thousands are leaving home to begin their freshman year at college. Do these newly matriculating students really gain fifteen pounds when they go off to college? The answer, despite articles in the popular media to the contrary, is no. Studies to date have shown that freshman college students do tend to gain weight but far less than the reported fifteen pounds for most freshmen—and over time, may even tend to lose much of any weight they gain.

Freshman weight gain was apparently first noted in the literature in the mid-1980s in an article by Hovell and colleagues. These researchers compared women living on campus with a comparable sample of freshman women living in the community. They found those freshman living on campus were almost three times as likely as the community sample to gain weight but by their junior year they were almost back at baseline levels. Reasons for the initial weight gain during freshman year were attributed to cafeteria food and dormitory living, as well as to the psychological stress of living away from home, family, and friends. The subsequent weight loss over the next few years of college was attributed to a move from the dorms and away from cafeteria-style eating. According to Cecelia Brown, in a review article published in 2008, the term "Freshman 15" was coined and first appeared in a 1989 article in the magazine *Seventeen.* For the years 1985 through 2006, Brown surveyed peer-reviewed journals, magazines, and newspapers, including university newspapers, and found considerable "misinformation" in the media articles that did not reflect research findings: half of the *popular press publications* claimed a 15-pound weight gain while the 14 research studies in the peer-reviewed journals indicated a typical weight gain was less than five pounds. Brown's conclusion was that the "Freshman 15" was "one of college and university students' most dreaded fears" and nothing other than an "urban myth." Furthermore, she noted that freshman students' preoccupation with gaining so much weight might even lead to eating disorders such as bulimia and anorexia nervosa, whose incidences peak during these late adolescent years.

More recently, researchers Smith-Jackson and Reel, in a 2012 article surveyed over 200 freshman women, all of whom had heard of the dreaded "Freshman 15," and who reported "intense fears about gaining weight." Many perceived weight gain as "inevitable" and even "a self-fulfilling prophesy." Vending machines, availability of fast food (and limited healthy choices), increased alcohol use, buffet-style cafeterias, and "food independence" (i.e. "increased choice over one's food intake and preparation") were all cited as potential factors that can lead to initial weight gain in freshman year transition. Communal eating (temptation to eat more when in a group) and less physical exercise have also been noted in the literature as contributors to this tendency to gain weight.

Research published in 2013 by Dr. Brian Wansink and his group at the Dyson School of Applied Economics and Management at Cornell University, Ithaca, found that Cornell students tended to eat more unhealthy snacks (e.g. potato chips, fries, chicken fingers) rather than healthy snacks (e.g. yoghurt, fresh fruit, and granola bars) in the final two weeks of the fall semester around the time of final exams and term paper deadlines. The researchers recommended that health professionals and those who advise students caution them about this possibility, and they also suggested that the food service provide more healthy convenient alternatives for the students.

Bottom line: Potential weight gain is understandably a realistic concern for freshmen college students, often more for women than men, as they make the transition from home to school. An unrealistic fear of the "Freshman 15" myth, however, may exacerbate anxiety and even lead to eating disorders in particularly vulnerable students with body image preoccupations. Students should be aware of changes in their eating habits, even fairly minor ones, that may lead to weight gain. Accountability, i.e., monitoring one's behavior, whether by having a diet buddy friend or weighing oneself on a regular basis, may help avoid creeping and excessive weight gain. ■

University life is conducive to weight gain, though it is a myth that freshman gain 15 pounds. Claudia Cohen Hall (formerly Logan Hall), University of Pennsylvania.

Author: MatthewMarcucci, English Wikipedia. Wikipedia Commons/Public Domain.

A.D. White Reading Room, Uris Library, Cornell University, Ithaca. Photo by eflon, 2008.

Creative Commons Attribution 2.0 Generic. Wikipedia Commons/Public Domain.

THE PROVERBIAL SAYING, "The proof of the pudding is in the eating" dates back several centuries and has been attributed to Cervantes in his *Don Quixote*, though some believe it was a mistranslation of the original Spanish. "Proof" in this case means "test" and what it signifies is that we have to put something to a test or trial to know its value. You will find David L. Katz's new book, *Disease Proof*, a worthy contender.

Proof of the Pudding is in the Eating, 1938, color lithograph, English, Private Collection.

Bridgeman Images, used with permission.

WHEN THE "PROOF OF THE PUDDING" IS NOT IN THE EATING

Waterproof, weatherproof, bulletproof, shatterproof, and now "Disease Proof"

43 — Posted October 15, 2013

David L. Katz, MD, MPH is the founding director of Yale University's Yale Griffin Prevention Research Center, editor-in-chief of the journal *Childhood Obesity*, and author or coauthor of over 100 professional papers and fifteen books. Incidentally, for those with a special interest in nutrition, there is no better or more comprehensive book than Dr. Katz's *Nutrition in Clinical Practice* (2008).

This time, with his co-writer, Stacey Colino, Dr. Katz has written a very readable, psychologically-minded, practical approach to maintaining health in this age of epidemic obesity, type II diabetes, and heart disease. By the year 2050, for example, according to the Centers for Disease Control (CDC), says Dr. Katz, unless we reverse the current trends, one in three (or about 100 million people) will have diabetes in the U.S. alone. Katz is sensitive to the plight of those with serious weight issues: "The weight of shame, blame, and self-recrimination is far heavier than any number of pounds."

Reversing disease and maintaining health (not just losing weight), though, are no easy tasks in this age of overwhelmingly poor choices and enormous portions. Dr. Katz tells how his wife Catherine, who has her PhD in neuroscience from Princeton, could understandably not figure out which of four French breads—one with too much sodium, another multi-grain with the least amount of fiber, still another with partially hydrogenated (and potentially dangerous) fat, and the fourth with high fructose corn syrup—was the healthiest. Dr. Katz has devised a nutrition guidance program, called the Nu-Val system, (now used in thousands of markets) to help consumers choose among different options. He reminds us, "The longer the shelf life of the product, the shorter the shelf life of the person who consumes it regularly." Reading labels can help us discern when a pasta sauce, for example, has more sugar than a chocolate sauce, says Dr. Katz.

Disease Proof focuses on the importance of sleep, exercise, and the people in one's life, as well as proper diet. Katz

The Plumb-Pudding in Danger;—or State Epicures Taking "un Petit Souper." British artist James Gillray. Published by Hannah Humphrey, London. Hand-colored etching, 1805.

Gift of Philip van Ingen, 1942. Metropolitan Museum of Art, NYC. Public Domain.

even includes recipes and exercises, and discusses the questionable value of vitamin supplements that may contribute to "nutritional noise." The book takes a cognitive behavioral approach: Katz emphasizes the importance of and need for self-efficacy, i.e., the conviction that a person can succeed at a task. Katz suggests that dieters create a "social contract" with other people, i.e., make a public announcement of their intentions as a kind of way of being accountable. He also recommends what he calls "impediment profiling"'—assessing barriers that interfere with achieving goals, and he believes that when someone is not able to lose weight on a diet, it is not evidence of failure—rather, that person did not have the right skill set. He recommends constructing a "decision balance analysis" where people list advantages and disadvantages to changing behavior.

Bottom line: Disease Proof is a welcome addition to any library as it can give people some tools to help take control of their health. ■

THE MYTH OF "NUTRITIONAL PRECISION: "WHAT DO WE REALLY KNOW?

Reading between the lines: Twinkies and food quality illiteracy

44—Posted November 13, 2013

THERE ARE MANY MYTHS in the field of weight control. In previous blogs, I have recommended Dr. David B. Allison's and his colleagues' 2013 article published in the *New England Journal of Medicine* on some of the most common myths about weight, and I have explored myths about the "freshman 15" and setting weight loss goals.

A new book *Nutritionism: The Science and Politics of Dietary Advice* by Dr. Gyorgy Scrinis, from the University of Melbourne in Australia, explores still another myth, the myth of *nutritional precision*, i.e., an "exaggerated" belief held by many people that scientists understand more about the complex interactions of nutrients, food, and their effects on the body than they really do.

For example, when health professionals recommend a diet of 30% fat intake, they are giving a *veneer of scientific precision*, rather than dispensing solid knowledge necessarily based on double-blind controlled clinical studies. When we label foods as "good" or "bad," we have a "one dimensional' simplistic understanding of health, rather than appreciating that nutrients can be beneficial or harmful depending on the company they keep. When we hear about the dangers of trans fats in our foods, says Scrinis, we are not focusing on the fact that the health consequences of removing these deadly fats will ultimately depend on what chemical processes are involved and what compounds are substituted instead.

Even a magnifying glass will not necessarily help identifying the food processing involved.

Used with permission, istockphoto.com. Credit: yosmanor.

Scrinis says that we talk about calories "as if they had an independent existence," rather than that they are a fairly "crude measure" of energy. After all, a calorie is the amount of energy required to raise one kilogram of water one degree on the Celsius scale. A calorie is a "highly abstract concept," not a concrete entity. To focus on calorie counting, while important as a general way of measuring food, obscures the extraordinary complexity of metabolism in the body, as, for example, how foods eaten together may interact with each other in unpredictable ways. For Scrinis, this "singular calorie focus" is an example of *nutritional reductionism*.

According to Scrinis, we have gone through three historical periods. The first was the era of *quantifying nutrition* where the focus was on consuming adequate nutrients to avoid diseases (e.g. vitamin C to prevent scurvy.) The second period was the era of *good and bad nutritionalism*, where the focus was on preventing chronic diseases (e.g. lowering cholesterol to prevent heart disease.) Since the mid 1990s, we have entered the third period, that of *functional nutritionalism*, where the focus has been on emphasizing the beneficial aspects of nutrients and creating engineered "functional" or "superfoods" that claim actual health benefits, though the term is "so poorly and broadly defined that virtually any food with added nutrients...seems to qualify" (e.g. probiotic yogurt drinks; orange juice with added calcium.)

Scrinis acknowledges that most foods, except those eaten raw, have some form of processing, including peeling, cooking, or chopping, but clearly the most healthy for us are those that have minimal processing. The second level of processing involves extraction, refining, and concentration of "food components." This level may involve chemical, temperature, and pressure manipulations that enable these foods to be transported over distances and have a considerably longer shelf life. Examples of this level of processing include production of white flour, cane sugar, fruit juices, or even high fructose corn syrup. The third level, and most problematic, level of processing involves reconstituting food, often with chemical preservatives, and added refined flour, sugar, salt, and fat. These are sometimes called "fabricated" or "synthetic" foods. Examples of this level are chicken nuggets and Twinkies, as well as those ammonia-treated, heated beef trimmings used as a filler for hamburger meat.

Scrinis believes we should base our dietary guidelines on how food is produced and the level of processing involved. Even our current nutritional labeling "fails to distinguish between nutrients intrinsic to the ingredients in food and those added during processing." He suggests a "food quality label" that would not only list all the ingredients, but specify the types of processing involved.

Bottom line: We all need to become considerably more "food quality literate" and continue to question not only how our food is processed but demand healthier food choices from the food industry. ■

Snack foods are examples of highly processed foods laden with sugar, fat, salt, and chemicals.

Photographer: AlejandroLinaresGarcia, 2009. Wikipedia Commons/licensed under GNU Free Documentation License, Creative Commons Attribution Share-Alike 4.0 International.

Finely textured beef or boneless beef trimmings sometimes have been added to ground beef. US Department of Agriculture, inspection of meat plant.

Photo taken by government employee. Wikipedia Commons/Public Domain.

HEALTHY OBESITY: AN OXYMORON?

Are those who are healthy and obese really "patients-in-waiting"?

45 — Posted December 22, 2013

Édouard Manet, *Before the Mirror*, 1876. People with obesity may be "patients-in-waiting." Many researchers do not believe that someone can be obese and remain healthy over time. Solomon R. Guggenheim Museum, NYC.

Gift of Justin K. Thannhauser, 1978. Wikimedia Commons/Public Domain.

BACK IN THE 1950S, Jean Vague observed that central, also called abdominal or visceral obesity (i.e., the so-called "apple"-shaped body, as distinguished from the "pear"-shaped body of non-central or subcutaneous obesity), is more likely associated with metabolic abnormalities. But even that observation, while generally true, has been called into question. Young and Gelskey, publishing in *JAMA* almost 20 years ago, for example, concluded that "non-central obesity," even in a cross-sectional sample of almost 2800 Canadian adults, was most definitely not benign compared to the non obese, even though their metabolic profiles were not as severe as those with central obesity.

For years, researchers and clinicians, though, have also documented the existence of a subgroup (some reports as high as 1/3 of obese people), as measured by elevated body mass index (BMI), including those with abdominal obesity, who have no metabolic abnormalities. In other words, these individuals have been considered metabolically healthy, with no evidence of hypertension, abnormal lipid or glucose levels, insulin resistance (or even overt type II diabetes), as well as other markers of inflammation (e.g. increased blood levels of C-reactive protein) that are typically found in obese people.

There are those, furthermore, who believe in "health at any size" and that cardiovascular fitness, as measured by exercise on a treadmill, for example, is far more important than percentage of body fat or even BMI in terms of mortality or even morbidity. This was the conclusion of a large 2013 study, published in the *European Heart Journal* by Ortega et al. The study population, though, was not a typical one: they were overwhelmingly Caucasian, well-educated, held professional or executive positions and those who were obese were only in the Class I obesity range as assessed by BMI. Ortega and colleagues acknowledged that one of the problems with the concept of benign obesity is that there is no clear, generally accepted and standardized definition of metabolic health. For example, is it acceptable to have one metabolic abnormality or two (and if so, which one is the most significant) to still count as healthy? It depends on the study, and so it is even difficult to know its true prevalence among those with excessive weight.

So is there really such a thing as "healthy" or "benign" obesity" or is that an oxymoron, i.e., a contradiction in terms? A recent study, published in the *Annals of Internal Medicine* by Caroline K. Kramer, MD, PhD and her colleagues, calls into question the concept of healthy obesity. These researchers report on 12 studies and conducted a meta-analysis on 8 longitudinal studies that investigated mortality and/or cardiovascular events in over 61,000 normal weight, overweight, and obese adults. They found that even when obese people had no metabolic abnormalities, if followed longitudinally for at least 10 years, there is "no healthy pattern of increased weight" and these people are "at increased risk for adverse long-term outcomes." This is not a completely new finding—for years, others have called into question the concept of metabolically benign obesity. For example, John McEvoy and his colleagues called it "a

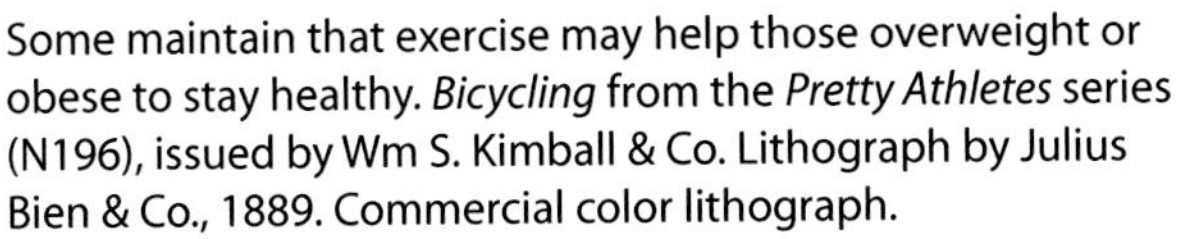
Some maintain that exercise may help those overweight or obese to stay healthy. *Bicycling* from the *Pretty Athletes* series (N196), issued by Wm S. Kimball & Co. Lithograph by Julius Bien & Co., 1889. Commercial color lithograph.

The Jefferson R. Burdick Collection, Gift of Jefferson R. Burdick. Metropolitan Museum of Art, NYC. Used with permission from Permissions Department.

There maybe a fine line, as if walking a tightrope, between health and disease when obese or overweight. *Tightrope Walking,* from the *Pretty Athletes* series (N196), issued by Wm S. Kimball & Co. Lithography by Julius Bien & Co, 1889. Commerical color lithograph.

The Jefferson R. Burdick Collection, Gift of Jefferson R. Burdick. Metropolitan Museum of Art, NYC. Used with permission from Permissions Department.

wolf in sheep's clothing" in a 2011 article in the journal *Athlerosclerosis* and suggested that "nomenclature in the field of obesity should forever abstain from all use of the word "benign." The Kramer study, though, was the first systematic review and meta-analysis that focused on long-term follow-up. If followed over time, are these really so-called "patients-in-waiting?"

The first reference I could find to the concept of "patients-in-waiting" was in a 2004 article by Parry and colleagues on the "limbo" feeling patients may experience prior to having their diagnosis of type II diabetes confirmed. Much more recently, Timmermans and Buchbinder, in their 2010 article in the *Journal of Health and Social Behavior*, describe "patients-in-waiting" as "an umbrella concept" for patients who, as a result of genetic screenings, may find themselves potentially living "under medical surveillance between health and disease." In other words, these patients may learn that they have a more likely genetic predisposition to contract a disease (e.g. Alzheimer's or breast cancer) eventually, even though they have no evidence of that specific disease at the time of the genetic screening. It was this kind of information, for example, that prompted

David Levine, *English Working Class,* also called *A Fat Man Confronting a Thin Man,* 1964.

Gift of Caryl Horwitz, 2011. Copyright, Matthew and Eve Levine. Used with generous permission. Image copyright, The Metropolitan Museum of Art. Image source: Art Resource, NY.

actress Angelina Jolie, though apparently asymptomatic at the time but with a strong family history of breast cancer, to opt for a preventive bilateral mastectomy.

Bottom line: Whether benign obesity does exist, at least in a small percentage of people, may still be open to question, but these studies have significant clinical implications. Those who are overweight or obese, regardless of where their fat accumulates and regardless of being presently asymptomatic, should not be complacent and should not assume they will necessarily remain healthy, especially if their BMI increases over time. Their metabolic health may, in fact, be transient. Over time, they may likely develop the typical complications seen most commonly in those with central obesity. In other words, they should see themselves as potentially "patients-in-waiting" and consider the importance of lifestyle modifications such as exercise and diet. ■

English poet John Milton by unknown artist, circa 1629. Milton is the author of the play/poem *Comus*, from which my title comes.

"WHAT HATH NIGHT TO DO WITH SLEEP?" NIGHT EATING SYNDROME

The controversies about making a diagnosis of an eating disorder

46—Posted January 25, 2014

THE 17TH CENTURY English poet John Milton (famed for *Paradise Lost*) wrote a play/poem popularly called *Comus*, after Komos, the Greek mythological god of nighttime festivities and excesses of all kinds. In his work, technically called a "mask," (an elaborate musical performance) Milton tells the story of three siblings traveling through a forest at night. The sister, who becomes separated from her brothers, comes across a disguised and treacherous Comus (son of Bacchus and Circe) who tries to seduce her with an intoxicating potion and tempting food—"lickerish baits"... "all manner of deliciousness...tables spread with all dainties" as he claims he can reunite her with her brothers. Says Comus, "What hath night to do with sleep?"

Those English scholars among you appreciate that Milton's poem of self-control and self-indulgence is about much more than nighttime eating and revelry. (There has been the suggestion that Milton's work was loosely based on an actual legal case of rape.) Comus's line, though, provides a metaphor for those who are tempted by Milton's description of "swinish gluttony" to eat most of their food at night—those who have the so-called night eating syndrome—and for some, have an increased risk of obesity.

Originally described in the mid-1950s by psychiatrist and major researcher in the field, Albert J. Stunkard, MD, the night eating syndrome consisted of three distinct characteristics: nocturnal hyperphagia, insomnia, and morning anorexia (i.e., negligible interest in eating breakfast.) Stunkard initially believed that the clinical syndrome developed in the context of psychological stress and often led to "untoward reactions to dieting," such as depression and anxiety, as well as obesity.

Over the years since Stunkard's initial description, studies in the literature have used somewhat different and inconsistent criteria. There was apparently considerable discussion and controversy during preparation of the fifth edition of the *Diagnostic and Statistical Manual of Mental Diseases (DSM-5)* on whether to include the night eating syndrome as a psychiatric diagnosis for this very reason. While binge-eating disorder has now been included, ultimately, the panel for Eating Disorders for DSM-5 opted not to include the night eating syndrome in this edition of the *Manual*. For those interested in a more in depth discussion, see the review article by psychologist Ruth Striegel-Moore, a member of the Panel, and her colleagues, including the options they had considered for inclusion in DSM-5 in their 2009 article in the *International Journal of Eating Disorders*. Their review found "significant limitations and gaps in the scientific literature of the night eating syndrome." They also noted that earlier reports focused on overeating while more recent reports have focused on the timing (i.e. nighttime) of the eating. Furthermore, Striegel and her colleagues could not find consistent support for the notion that night eating (rather than excessive caloric intake) necessarily leads to weight gain or obesity.

Lorenzo Costa, *The Reign of Comus,* early 16th century.
Wikipedia Commons/Public Domain.

Stunkard and his group, including Dr. Kelly C. Allison, at the University of Pennsylvania School of Medicine, have been at odds with Striegel-Moore and her colleagues. They had recommended that the night eating syndrome be included as an eating disorder in DSM-5 and have proposed the following constellation of symptoms: daily pattern of eating demonstrates at least 25% of food intake is consumed after the evening meal for at least two or more days a week; awareness and recall of the eating episodes (as different, for example, from reports of eating that are *not recalled* when taking the sleeping medication Ambien); and a clinical picture characterized by three of the following: negligible interest in eating breakfast (i.e., early morning anorexia) at least four or more mornings a week; presence of a strong urge to eat after dinner and during the night; presence of a belief that one must eat in order to return to sleep; and depressed mood that is worse in the evening. Furthermore, this disordered eating pattern is associated with significant distress, has been maintained for at least three months, and is not secondary to a medical condition

Artist: Thomas Rowlandson, *Good Night,* After George Moutard Woodward, British. Publisher, Rudolph Ackermann, 1799, hand-colored etching.
The Elisha Whittelsey Collection, The Elisha Whittelsey Fund, 1959. Metropolitan Museum of Art, NYC, Public Domain.

Paul Klee, *The Hour Before One Night,* 1940.

The Berggruen Klee Collection, 1984. Metropolitan Museum of Art, NYC, Public Domain.

or substance use disorder or other psychiatric illness. For a more complete discussion of their criteria, see the article by KC Allison et al in the 2010 *International Journal of Eating Disorders*.

Vander Wal, in a critical review of the literature, in a 2012 article in the journal *Clinical Psychology Review*, believes that the night eating syndrome is associated with obesity. Vander Wal also noted that it appears to run in families, may involve the neurotransmitter serotonin, as well as involve a circadian rhythm "dysregulation" of certain hormones such as leptin and melatonin. Medications such as the selective serotonin reuptake inhibitors (SSRIs), (e.g., sertraline-Zoloft) have been used with some effect, as has bright light therapy, and cognitive behavioral therapy. Along those lines, Kucukgoncu et al, in a cross-sectional study at Yale University's Department of Psychiatry, report in the recent 2014 article in *European Eating Disorders Review,* that the night eating syndrome was seen in over 21% of 155 patients with major depression, as diagnosed by the criteria proposed in that 2010 article by KC Allison and her colleagues. Their conclusion was that the night eating syndrome was common and was associated with increased weight (significantly higher body mass index—BMI) (and increased food intake) as well as with smoking and should be "carefully evaluated" by clinicians for their depressed patients.

Bottom line: The night eating syndrome has suffered from a lack of consistent criteria until more recently and as a result, unfortunately was not included as an eating disorder in DSM-5. Nevertheless, clinicians should be aware that their patients may have its symptoms and specifically inquire about it. ■

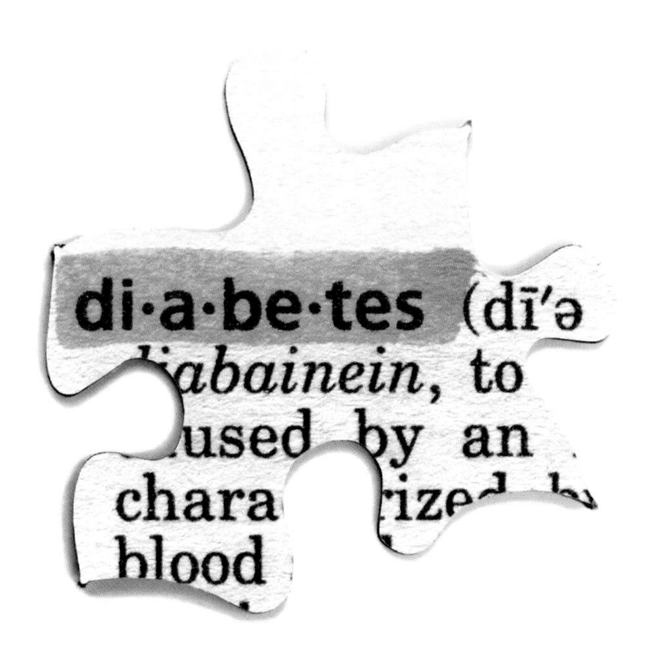

THE PUZZLING CONDITION OF PRE-DIABETES

Dual citizenship in the "kingdom of the well" and in the "kingdom of the sick"

47 — Posted February 26, 2014

SUSAN SONTAG, in her essay, *Illness as Metaphor*, writes that we all hold "dual citizenship in the kingdom of the well and in the kingdom of the sick." Eventually, says Sontag, we will, "sooner or later," experience passage from one to the other. We can use Sontag's "dual citizenship" metaphor to call attention to type 2 diabetes and more specifically, that intermediate state of so-called "pre-diabetes."

Approximately 6.2 million people in the U.S. alone have undiagnosed type 2 diabetes (Cohen et al, *Journal of Clinical Endocrinology and Metabolism*, 2010.) Apparently, by 2030 about 472 million people worldwide will have pre-diabetes. (Gosmanov and Wan, 2014). Pre-diabetes is a state of intermediate glucose metabolism not quite meeting the criteria for full-blown diabetes. Shaw, in his 2011 article in the *Medical Clinics of North America* writes that pre-diabetes is not a disease itself, but rather a "risk state." Eventually everyone who develops type 2 diabetes passes first through that state of pre-diabetes. One of the main reasons for diagnosing pre-diabetes is to identify those who may be at increased risk for developing not only diabetes but its many complications, such as cardiovascular disease, diabetic retinopathy, kidney disease, and neuropathy. Many studies suggest that proper blood glucose control has been associated with fewer long-term medical complications. Once identified, these patients can be better managed with the implementation of lifestyle changes and possible pharmacological interventions. Typically type 2 diabetes is diagnosed by measuring fasting blood glucose levels or assessing glucose levels by an oral glucose tolerance test.

It was Dr. Samuel Rahbar, in the late 1960s, who discovered, quite by accident, that those with diabetes had elevated levels of hemoglobin A1c (HbA1c), i.e., glycolated hemoglobin (glucose irreversibly attached to hemoglobin) in their blood. (Gebel, *Diabetes Care*, 2012) Hemoglobin A is the most abundant form of hemoglobin, the substance that carries oxygen in human red blood cells from the lungs to the tissues, and there are five minor components of hemoglobin A, including HbA1c. The lifespan of a red blood cell is about 120 days; the HbA1c reflects the average blood glucose level for the previous 8 to 12 weeks. Hemoglobin A1c accounts for about 3 percent of hemoglobin in normal adult red cells. In diabetes, though, this level may be two to three times higher than normal. Rahbar's discovery was not initially appreciated, but eventually has become an important clinical marker for those with both type 1 (insulin deficient, auto-immune disorder, more often seen in childhood) and type 2 diabetes (most closely associated with overweight or obesity and most often seen in adults, but with the spread of obesity, now seen in children and adolescents.) In 2010, the American Diabetes Association recommended using HbA1c to diagnose both pre-diabetes and diabetes. Currently someone with a percentage of HbA1c of 6 to 6.5% has a high risk for developing diabetes, while those with a percentage of 5.7 percent to 6.4 percent have "pre-diabetes."

Most recently, though, Gosmanov and Wan, in a 2014 *American Journal of the Medical Sciences*, have called into

The puzzling condition of pre-diabetes.
Used with permission/istockphoto.com. Credit: Lanier.

Life of Saint Bernard of Clairvaux: Saint Bernard exhorts the sick in body to enter his church with their spirits, in doing so returning to find their infirmities healed. Unfortunately, the condition of pre-diabetes may need more than prayer to avoid the development of overt type 2 diabetes. Design attributed to Master of Saint Severin, after 1535, German stained glass.

Gift of Stanley Mortimer, 1959, Metropolitan Museum of Art, NYC, Public Domain.

Tappy-on-the-Window-Pane, "This'll scare em sick," from the *Terrors of America Set* (N136), by Duke Sons & Co. Lithography by Knapp & Company, 1888-89. Often receiving the diagnosis of pre-diabetes does scare someone enough to change unhealthy eating habits.

The Jefferson R. Burdick Collection, Gift of Jefferson R. Burdick. Metropolitan Museum of Art, NYC. Public Domain

question the value of using HbA1c to diagnose pre-diabetes. Their prospective study of 66 patients found that the predictive value of HbA1c for the diagnosis of pre-diabetes is low and patients with values of HbA1c of 5.7 percent to 6.4 percent should undergo confirmation by an oral glucose tolerance test, the "gold standard" for diagnosing diabetes. In their study, those in this mid-range missed the diagnosis of diabetes, as assessed by an oral glucose tolerance test, in 12 percent of patients and only 39 percent had pre-diabetes based on the oral glucose test. In other words, the HbA1c can be misleading and over-diagnose pre-diabetes in more than half of the subjects. The HbA1c test, though, is convenient, does not require fasting or a two-hour time period, has international standardization, and less day-to day variability, though it is more expensive than a fasting blood glucose and may not be as accurate. Furthermore, race, age, and even medications may affect values, and there is just more "inter-individual physiological variability" than initially appreciated. Cohen et al, as a result recommends that the mid-level HbA1c range be called "impaired HbA1c" (rather than pre-diabetes), and they also recommend confirmation by a fasting blood glucose or oral glucose tolerance test.

For a systematic review of 16 prospective studies on HbA1c, involving over 44,000 patients, see the Zhang et al article in *Diabetes Care* (2010). Their review found that HbA1c values between 5.5 and 6.5 percent were associated with a "substantially increased risk" for developing diabetes but also concluded that errors could be reduced when done in combination with fasting blood glucose levels and an oral glucose tolerance test.

Bottom line: HbA1c is a screening blood test that reflects glucose levels over the past 8 to 12 weeks. It is most useful for identifying patients with a normal glucose tolerance if HbA1c is below 5.7 percent and has a high predictive value if the HbA1c is above 6.4 percent for diabetes. If your levels fall within that mid-range (i.e., neither completely normal nor overtly diabetic), consider asking your physician for an oral glucose tolerance test and/or a fasting blood glucose level to ensure a more accurate assessment. ■

FAT SHAMING AND STIGMATIZATION: HOW FAR IS TOO FAR?

A debate rages over the use of shame in public health.

48 — Posted March 30, 2014

BACK IN THE LATE 1990S, Dr. Jerome P. Kassirer and Dr. Marcia Angell, both former editors-in-chief of *The New England Journal of Medicine*, wrote an editorial in the journal about the "dark side to this national preoccupation" with weight loss, with failed attempts often leaving dieters blaming themselves for being undisciplined and self-indulgent, and feeling guilt and self-hatred.

In their column, Kassirer and Angell acknowledged that being overweight is indeed correlated with substantial medical morbidity but they also noted that obese people were often "criticized with impunity" by those "critics merely trying to help them." They added: "Some doctors take part in this blurring of prejudice and altruism when they overstate the dangers of obesity and the redemptive powers of weight loss."

Eve hides her head in shame. Rodin, *Eve After the Fall*, 1881. Museum Boijmans Van Beuingen, Rotterdam, The Netherlands.

MicheleLovesArt. Licensed under the Creative Commons Attribution Share Alike 2.0 Generic. Wikipedia Commons.

They also speculated about whether being overweight was actually "a direct cause of those illnesses associated with it"—an idea as controversial then as now—but they added, "few would claim that becoming obese is consistent with optimal health."

The World Health Organization calls obesity a "global epidemic." About 1.5 billion adults worldwide are overweight and at least 500 million are obese as defined by body mass index (BMI), according to its most recent statistics. Some researchers predict that these numbers will increase substantially in the next 15 years. Ironically, even as our population has grown increasingly overweight, there is still considerable prejudice—and overt discrimination—against those who are weight-challenged, even among the professionals who work with this population. (See blog 13, *We Hold Those Truths to be Self-Evident...*)

Herein lies the conundrum: Given the prejudice against the overweight and obese, what measures can or should be taken to stem the increasing obesity rates? In the name of public health, is it ever appropriate to stigmatize and shame people into losing weight?

Bioethicist Daniel Callahan, writing in the *Hastings Center Report* (2013), suggests that there is a place for so-called "fat shaming" to attempt to overcome this epidemic. He suggests three major strategies:

- "Strong and somewhat coercive public health measures" (taxing sugared drinks, banning unhealthy food advertising to children, posting calorie information in restaurants, and reducing the costs of healthy foods through government subsidies);
- Childhood prevention programs (working through lunch programs, providing exercise opportunities in school and working through parents to discourage sedentary activities such as TV watching at home);
- Most controversially, "social pressure on the overweight." Callahan believes that "whether they recognize

their own role or not, (the public) need to understand that obesity is a national problem, one that causes lethal disease." His solution is social pressure "that does not lead to outright discrimination," or what he calls "stigmatization lite." He suggests a series of questions that could be asked to "nudge" people in the right direction: "If you are overweight or obese, are you pleased with the way you look?" for example, or, "Fair or not, do you know that many people look down upon those excessively overweight or obese?"

In a recent article in the journal *Biothethics* (2014), medical ethicist Christopher Mayes takes issue with Callahan. He explains that Callahan sees obesity not just as a clinical or personal issue but frames it as an "ethical issue with social and political consequences," in which obese people not only harm themselves but others as well, because of their increased economic costs to society. The problem is that obesity is far more complicated than mere individual choice—there are social, cultural, environmental, and biological variables to consider as well.

In general, coercing individuals toward healthy behaviors is mostly ineffective and potentially harmful in that it may increase stigmatization.

Although sociologist Erving Goffman prominently wrote of stigma in the 1960s, there is still no widely accepted definition of the term. It is a cultural phenomenon that involves an us-vs.-them mentality, in which people distinguish and differentiate themselves from others seen as having undesirable characteristics. Stigma can be a potent source of social control that can result in both the loss of status in a community, as well as overt discrimination.

A posture of shame. Francisco de Goya, *For Being Born Somewhere Else,* between 1810–1811. Those who are obese are often stigmatized and made to feel they don't fit anywhere. Prado Museum, Madrid.

Wikipedia Commons/Public Domain.

Edwin Roscoe Mullins, *Cain or My Punishment is Greater than I Can Bear,* (from Genesis 4:13) circa 1899. The shame and possible ostracism of being obese may feel as terrible metaphorically as Cain's shame. Glasgow Botanic Gardens, Kibble Palace.

Photographer: Daniel Naczk. Licensed under Creative Commons Attribution-Share Alike 4.0 International. Wikipedia Commons.

Saint Francis of Assisi Receiving the Stigmata, attributed to Jan van Eyck, between 1430–1432. Stigmata are believed to be the marks or pain sensations that corresponded to Christ's crucifixion wounds. Because obesity is not something they can hide, some obese people may feel stigmatized as if they are metaphorically bearing external marks of their own pain and suffering.

Philadelphia Museum of Art, Wikimedia Commons/Public Domain.

Those stigmatized often resort to attempts at concealment. This may happen with a disease that is not always evident like HIV/AIDS or epilepsy. Concealment, though, is not an option for the overweight and obese.

Law professor Scott Burris, writing in the *Journal of Law, Medicine, and Ethics* (2002), raised the provocative question of whether there can ever be "good stigma," as in the ongoing campaign against smoking, in which the activity was deliberately stigmatized and transformed "from being a glamorous activity" into "antisocial self-destruction." The dangers of smoking were emphasized and smokers were stigmatized as the habit became socially unacceptable and even restricted in most public places.

Burris, however, writes that stigmatizing a person because of an addiction or disease is an "offensive" form of "social warfare" that does not belong in campaigns for public health. This was even addressed in the 1962 Eighth Amendment ("cruel and unusual punishment") decision by the Supreme Court in *Robinson vs. California* regarding alcoholism. The Court found it "barbarous" to allow "sickness to be made a crime and [to permit] sick people to be punished for being sick." But Burris does distinguish between actually stigmatizing people and labeling

Rarely are there "judgment-free" zones for the obese.

Used with permission/istockphoto.com. Credit: VladSt.

behaviors, such as smoking, unsafe sex, and overeating as "bad." "Criticism and negative attitudes," he says, "are not stigma."

Bottom line: There is no straightforward answer, in the name of public health, to the question of how paternalistic a society should be in its attempts to "protect" citizens from unhealthy behavior. But certainly shaming, prejudice, and discrimination have no place. ■

"WHAT POTIONS HAVE I DRUNK?" CONCERNS ABOUT DIET SUPPLEMENTS

Are some so-called remedies for weight control worse than the disease?

49—Posted April 26, 2014

Apothecary, 15th century—Plate IV of XII. Portrayed by Swiss illustrator, Warja Honegger-Lavater.

Wikipedia Commons/Public Domain.

THE DISTINGUISHED PHYSICIAN and medical philosopher William Osler once said, "Man has an inborn craving for medicine…the desire to take medicine is one feature which distinguishes man, the animal, from his fellow creatures." Osler also said, "One of the first duties of the physician is to educate the masses not to take medicines." This particular aphorism has relevance to our perpetual quest for supplements that promise health benefits including appetite suppression and rapid weight loss.

In 2012 alone, Americans apparently spent $32.5 billion on dietary supplements despite concerns and confusion about their effectiveness and safety, according to Garcia-Cazarin et al in an article (2014) in the *Journal of Nutrition*. Supplements, which may include vitamins, botanicals and herbs, and sports nutrition products, are not classified as either medication or food and hence not regulated by the Food and Drug Administration (FDA.) They are "not intended to prevent, diagnose, treat, mitigate, or cure diseases," even though many people may think of them as having medicinal powers. As a result, the products themselves and the product labeling for these substances are not well-controlled: different manufacturing companies may produce vastly different substances (e.g. containing impurities and dosage irregularities) under the same name, and labels themselves may be misleading, inaccurate, overtly false, and even fail to mention safety concerns. Owens et al, in a 2014 article in *The American Journal of Medicine*, notes that the widespread use of the Internet makes the inadequate labeling all the more concerning. These researchers found there were 1300 English language sites that contained information on herbal products. Roughly half of these were retail sites, and fewer than 1 of 10 of these sites include information on their products' potential adverse effects or ability to interact with other medications. They also found that almost 14% of these retail websites included information that clearly violated FDA regulations by making overt medical claims. Further, website testimonials by clients may be misinterpreted and confusing. Not all sites, of course, are misleading: non-retail sites are more likely to contain appropriate and authoritative information on safety concerns and more likely to suggest consulting a physician before using a dietary supplement.

In a 2014 article in *The New England Journal of Medicine*, Harvard physician Pieter A. Cohen noted "our woefully inadequate system for monitoring supplement safety" because "unlike prescription medications, supplements do not require pre-marketing approval" before they are offered for sale. Furthermore, though the FDA is supposed to identify and remove supplements that are dangerous, it may never receive the information. Clinicians can certainly voluntarily report adverse events, but many supplements are sold directly to consumers through the Internet. Cohen reports that between 2008 and 2010 there were over 1000 reports made to poison centers (but not to the FDA)

Unknown artist, Italian School. *Early Italian Pharmacy,* 17th century. Science History Institute, Philadelphia, PA.

Gift of Fisher Scientific International, Science History Institute Collections. Wikipedia Commons/Public Domain.

on problems with adverse reactions from supplements. Many of these supplements are particularly toxic to the liver. Abdualmjid and Sergi, reporting in a 2013 article in the *Journal of Pharmacology and Pharmaceutical Sciences,* reviewed over 250 studies involving dietary supplements and found substantial evidence for mild to severe liver damage and even deaths attributed to some of these compounds in susceptible patients. These authors, as well as Cohen, recommend a nationwide data base that could be maintained by a multidisciplinary team involving the FDA and poison centers, as well as "rigorous safety testing" for all supplements before they are marketed and sold.

Safety is certainly a major issue but what about efficacy? Astell et al, in a 2013 article in *Complementary Therapies in Medicine*, systematically reviewed double blind randomized controlled clinical studies of botanical compounds purported to aid in weight control. Of the over 5200 studies found, only 326 were randomized controlled studies and of these, another 216 had to be eliminated due to methodological weakness (e.g. small sample size, studies of short duration). Ultimately, only 14 studies of plant extracts met their inclusion criteria! Among the compounds they reviewed were green tea extract and Garcinia cambogia (both of which were touted recently for weight control on the *Dr. Oz* program.) According to their systematic review, they did not find convincing evidence that plant extracts used as appetite suppressants for weight loss in the treatment of obesity are either particularly effective

or necessarily safe. Their conclusion was that though some plant extracts "show promising results in the short term, there is need for longer duration clinical trials to verify" their claims of suppressing appetite. They recommend further studies to ascertain optimal dose, mechanism of action, adverse reactions, and long term safety considerations. This incidentally, was the conclusion back in 2001 in an article by David B. Allison, PhD and colleagues in the journal *Critical Reviews in Food Science and Nutrition.* They added, "The lack of rigorous research is regrettable because it leaves health professionals without a sound basis for making recommendations to patients..." They also called attention to "economic reasons for this state of affairs," namely that most of these products cannot be patented and hence there is "no assurance of a return on investment in research and development." More recently, though, the National Institutes of Health (NIH) is committed to funding research for dietary supplements. (Garcia-Cazarin et al, 2014)

Bottom line: Buyer beware. Dietary supplements are not necessarily effective and for some, may be overtly dangerous. Search reputable, non-retail web sites for information and safety concerns before you purchase any products and consult your physician before starting any supplement as some may interact with medications you may be taking. Be sure to report any adverse reactions to your physician.

The Shakespearean scholars among you will know my title comes from *Sonnet 119.* ■

DOWN THE RABBIT HOLE: WHEN MEDICATION LEADS TO WEIGHT GAIN

A particularly vicious circle when prescription medication makes you fat

50—Posted May 21, 2014

Many medications, including antidepressants, antipsychotics, mood stabilizers, corticosteroids, beta-blockers, hormonal contraceptives, insulin, and even medications for allergy such as diphenhydramine (Benadryl), cause weight gain—even considerable weight gain—in susceptible patients. Far more medications result in weight gain than in weight loss. Initially there were only anecdotal reports of weight gain with prescription medications but the extent of the problem was delineated when Allison and his colleagues conducted comprehensive literature searches (Cheskin et al, 1999; Allison et al, 1999) almost 15 years ago and found medication-related weight gain was "under-recognized" by clinicians and sometimes resulted in patients' noncompliance with treatment.

How much weight is someone willing to gain when he or she is on a medication? That question was posed by Sansone and colleagues about ten years to a sample population of over 200 Midwestern, suburban (and primarily women) in a primary care practice. For either a medical or psychiatric non-life threatening condition, this sample would accept a weight gain of about 5 1/2 pounds. If the medical or psychiatric condition involved a life-threatening condition, people were able to tolerate a weight gain of 13 pounds or higher. Of note, though, in this particular sample, more than 5% were unwilling to gain any weight. In other words, for some, any weight gain is intolerable, regardless of the efficacy of the medication prescribed. For others, though, it is not just an issue of aesthetics: weight gain produced by medication may lead to serious metabolic abnormalities such as insulin resistance, hypertension, abnormal blood lipid levels, and even overt type 2 diabetes in those genetically vulnerable. This is particularly common in the so-called second generation antipsychotics.

Medications can cause weight gain in the short term (within the first 8 to 12 weeks) and the long term (several months to a year), according to Hasnain and Vieweg, in the journal *Postgraduate Medicine* (2013.) There is a suggestion that those who gain weight in the first several weeks of treatment are more likely to continue to gain, though some medications like the selective serotonin reuptake inhibitors (SSRIs) result in some weight loss initially but ultimately weight gain over the year.

Why do some medications cause weight gain? There are several factors, and the more mechanisms involved, the more likely weight gain will occur. For example, some medications may cause an increase in appetite specifically by receptor blockade. Wysokiński and Kloszewska in a recent article in the *Journal of Advanced Clinical Pharmacology* (2014) reviewed the complex hormonal system involved in short-term satiety and long-term energy storage. These authors note that histamine H1 blockade and serotonin 5-HT2C receptor antagonism are responsible for the weight gain that is seen with antipsychotics such as clozapine (Clozaril), olanzapine (Zyprexa), quetiapine (Seroquel), and risperidone (Risperdal), as well as antidepressants such as some of the SSRIs, most notably with paroxetine (Paxil).

*Used with permission/istockphoto.com. Credit: Vlada_*

Henri de Toulouse-Lautrec, *Woman before a Mirror*, 1897. Standards of beauty have changed over the years.

Metropolitan Museum of Art, NYC. Public Domain. The Walter H. and Leonore Annenberg Collection. Bequest of Walter H. Annenberg, 2002.

Advertisement, 1895. Many medications unfortunately lead to weight gain. Ironically, this advertisement, from the 19th century, is offering a product to encourage weight gain. Baker Art Gallery, Columbus, Ohio.

The Gribler Bank Note Co. Wikipedia Commons/Public Domain.

Because a medication such as aripiprazole (Abilify), used primarily to treat psychosis but also now marketed (and highly advertised on TV) as an adjunct for treatment of depression, is a partial agonist, rather than an antagonist, it is generally thought of as weight neutral and may sometimes be substituted for those that cause the most weight gain such as clozapine and olanzapine. H1 receptor blockade is also responsible for weight gain with antidepressants, such as mirtazapine (Remeron) and trazodone (Desyrel), or the antihistamine such as hydroxamine (Vistaril.)

Other medications increase appetite by a direct effect on the many hormones involved in appetite regulation, including leptin, ghrelin, and insulin. For example, some antipsychotics (e.g. clozapine and olanzepine) also block the action of leptin, resulting in increased, but ineffective levels of this hormone (leptin resistance) and accumulation of fat tissue. Both antipsychotics and antidepressants can also affect the levels of insulin, creating a state of insulin resistance and even an increased risk of type 2 diabetes. Wysokiński and Kloszewska caution, however, that

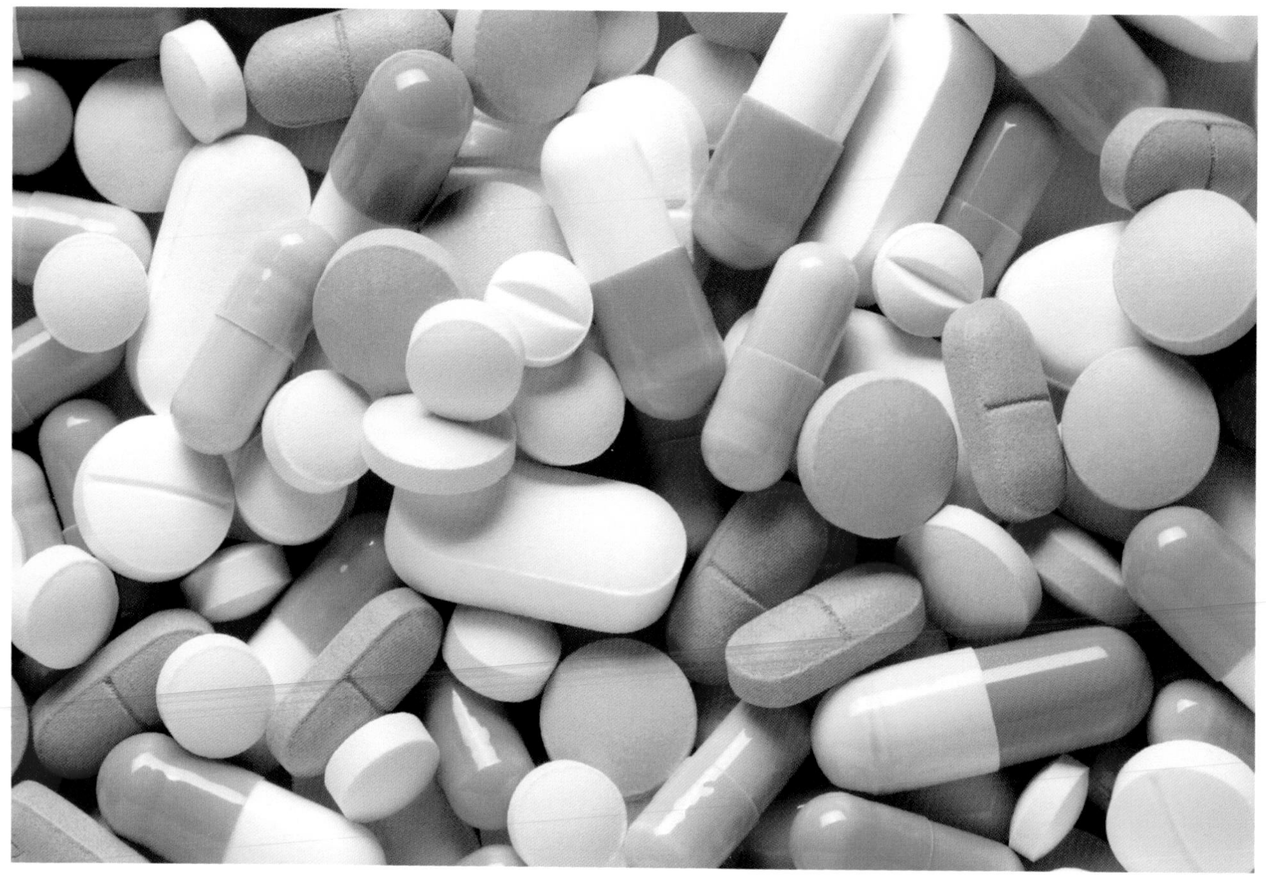

Many medications cause weight gain.
Used with permission/istockphoto.com. Credit : DNY59.

changes in these hormones may be secondary to weight gain rather than the cause of weight gain.

Sometimes, medications don't affect appetite but rather can change (i.e. decrease) a person's resting metabolic rate and hence cause weight gain. This has been seen with the older tricyclic antidepressants such as impramine (Tofranil). Further, tumor necrosis factor alpha (TNF-α) is a cytokine that can also lead to weight gain with some of the antipsychotics such as clozapine and olanzapine, but also with lithium, amitriptyline (Elavil), and mirtazapine. Wysokiński and Kloszewska report that activation of this TNF-α system seems to occur early in treatment and might eventually become a sensitive marker that weight gain will occur. Other mechanisms resulting in weight gain include drinking highly caloric beverages because of dry mouth that may accompany medication or even increased sleeping time due to the sedating effects of medication and hence less energy expenditure. Sometimes patients are taking several medications at once, and concomitant medications may interact in a way to increase weight gain. Further, ethnicity, gender, and age also contribute to differences in medications' effects on weight. For example, some studies report that weight gain is more common in women and more likely to occur in those predisposed to excessive weight in general.

Many of these mechanisms involve mutations in specific genes and eventually genomic studies will lead to more specific individual recommendations for patients. For example, some patients are "poor metabolizers" and some are "ultra-rapid metabolizers," according to Altar et al, writing in the *International Review of Psychiatry* (2013.)

Bottom line: Weight gain can occur in both the short and long term and may interfere with treatment compliance. Clinicians should monitor patients carefully for weight-related and metabolic changes, as well as educate patients regarding healthy lifestyle choices of diet and exercise. It is often possible to switch to a more weight neutral medication or be able to lower the dose of the offending medication. Eventually, there will be more widely available genetic screenings that will lead to individualized recommendations. ■

FROM THE FRYING PAN INTO THE FIRE? SATURATED FAT AND HEALTH

Have we been misled about the evils of eating saturated fat for all these years?

51 — Posted June 14, 2014

"Jack Sprat could eat no fat. His wife could eat no lean. And so betwixt the two of them, they licked the platter clean." So goes the children's nursery rhyme. Poor Jack Sprat. He may have had it wrong all along. This is the conclusion of Nina Teicholz in her provocative and extremely well-researched new book, *The Big Fat Surprise: Why Butter, Meat, and Cheese Belong in a Healthy Diet.*

Teicholz, who is a journalist and not a scientist, explains she can bring a fresh perspective to the field of nutrition, without any axe to grind. Her thesis is that we have all been seriously misled for the past fifty years about the so-called dangers of saturated fat (e.g. the fat found in red meat, whole milk, cheese, eggs, butter, lard, etc) in causing heart disease. Her story has both heroes and villains but her chief villain who started and perpetuated this myth was Ancel Keys, one of the leaders in the field of nutrition. Keys is known for several important contributions, including K rations in the military ("K" standing for Keys); his semi-starvation experiments on conscientious objectors during World War II that produced a two-volume tome; and his first using the term body mass index (BMI) to popularize the formula (weight in kilograms over height in meters squared) devised by the statistician Quetelet centuries earlier that we use today as a rough and imprecise but standardized measure of overweight and obesity.

It was Keys, though, who initiated the first international epidemiological *Seven Countries* study that purported to find that diets rich in animal fat more likely led to heart disease. Teicholz went back to the original Keys data and found that of the over 12,000 participants, Keys had evaluated food consumption in less than 4%. She also found there was no consistency among the countries studied in how the data were actually collected. Furthermore, Teicholz found that some of the data collection (e.g. in Crete) occurred during the 48 day period of Lent when most of the population would be consuming considerably less animal meat. She noted that Keys seemed to choose only those countries that seemed to fit his hypothesis that consumption of animal fats led to heart disease and a diet low in saturated fat could prevent it. Apparently, as well, Keys seemed to ignore the fact that even within certain countries, such as those people living in Eastern Finland, died of heart disease at rates triple that of those living in the west of Finland, despite that their "lifestyles and diets, according to Keys' data, were virtually identical."

Studies involving food consumption and diet, though, are extraordinary difficult to conduct, especially over time. Participants may be reluctant to tell the investigators the truth of what they eat, may tell them what they think the investigators want to hear, or even change their diet over time, particularly if they think one diet may be healthier than another. Furthermore, diet recall is subject to distortions in memory. And people who adhere to diets may be different in other ways: they may be more health-conscious, more likely to exercise, less likely to smoke, etc.

Over the years, some, including Rockefeller University researcher E.H. Ahrens, who studied fat and set up the first gas-liquid chromatography lab in the U.S, questioned Keys' diet-heart hypothesis. Ahrens suggested that it was carbohydrate consumption, rather than fat, that more likely led to obesity and heart disease. (Ahrens' view in the late 1970s was prescient as we have now come to believe that excessive carbohydrate intake, particularly in the form of refined sugars and white flour, leads to obesity, type 2 diabetes, and other metabolic abnormalities in those predisposed, a view more recently popularized by science journalist Gary Taubes.) Keys, though, was apparently such an intellectual

Melting butter, an example of saturated fat.

Jessica Merz, licensed under Creative Commons Attribution 2.0 Generic. Wikipedia Commons/Public Domain.

Most researchers believe that saturated fat is strongly associated with cardiac disease. William P.W. Dana, *Heart's Ease* 1863. Gift of S. Howland Russell,1890.

Metropolitan Museum of Art, NYC. Public Domain.

German painter, Jan Spanjaert, *Butter Made in a Barn.* Butter and other products made from animals have saturated fat. Circa 1664.

Wikipedia Commons/Public Domain.

Bacon, an animal product from the pig, is another example of saturated fat.

Used with permission/istockphoto.com. Credit: Drbouz.

bully that his influential views managed to hold sway over the scientific and eventually political landscape for decades.

Unfortunately, says Teicholz, Americans, with the support of the American Heart Association and other scientific organizations, came to demonize saturated fat: the gospel became that if we want to remain heart-healthy, that we need to exchange it for polyunsaturated vegetable oils, either in liquid form (e.g. safflower, cottonseed, soybean, peanut, corn, and canola oil) or even as hardened fat such as margarine or Crisco. It was in the hardened form that these fats contained the artificially created (by a process of partial hydrogenation) trans fats that make cakes and cookies moist and increase shelf life. At one point, says Teicholz, "there were partially hydrogenated oils in some 42,720 packaged food products." We now know, though, that these are particularly dangerous to health. And now that we have eliminated trans fats, we may be using even more harmful substitutes, such as vegetable oils that oxidize when heated and create toxic chemical compounds that remain in our food and body. Teicholz's

Copper frying pan, 5th to 4th centuries B.C., Greek, Thessaloniki.

Gts-tg, Licensed under Creative Commons Attribution Alike 4.0 International. Wikipedia Commons.

solution: return to beef tallow and butter and other saturated fats that are stable when heated and do not oxidize.

Teicholz also addressed the questionable science behind the Mediterranean Diet (e.g. olive oil, fish, vegetables, fruits, grains, nuts, red wine), also popularized by Keys and recommended so commonly today. She asks, "Did any single Mediterranean Diet even truly exist? There was so much variation in eating patterns across countries and even within countries that it seemed nearly impossible to define any kind of overarching dietary pattern with any specificity." And, she adds, "What is 'a little meat,' and 'a lot' of vegetables?" Teicholz noted that Keys focused only on certain Mediterranean countries and excluded African and Middle-Eastern countries that also bordered the Mediterranean Sea. Apparently Keys loved Italy (and even purchased a home there). Says, Teicholz, "One has to wonder whether we would know more about the diets of other long-lived peoples, such as the Mongolians or Siberians, if researchers were equally drawn to the landlocked countries with desert steppes and long, freezing winters?" It is also possible, says Teicholz that the "Mediterranean Diet is associated with good health because it is low in sugar."

Bottom line: Teicholz's book is well worth reading. It is an eye-opening dissection of some of the long-held nutrition myths we have accepted as fact. Her conclusions are so "counterintuitive" that many people may find them hard to digest, even with the evidence she provides. In other words, just as the character says in Woody Allen's *Sleeper,* "Everything we thought to be unhealthy is precisely the opposite..." Time will tell how these new and controversial ideas will hold up. ■

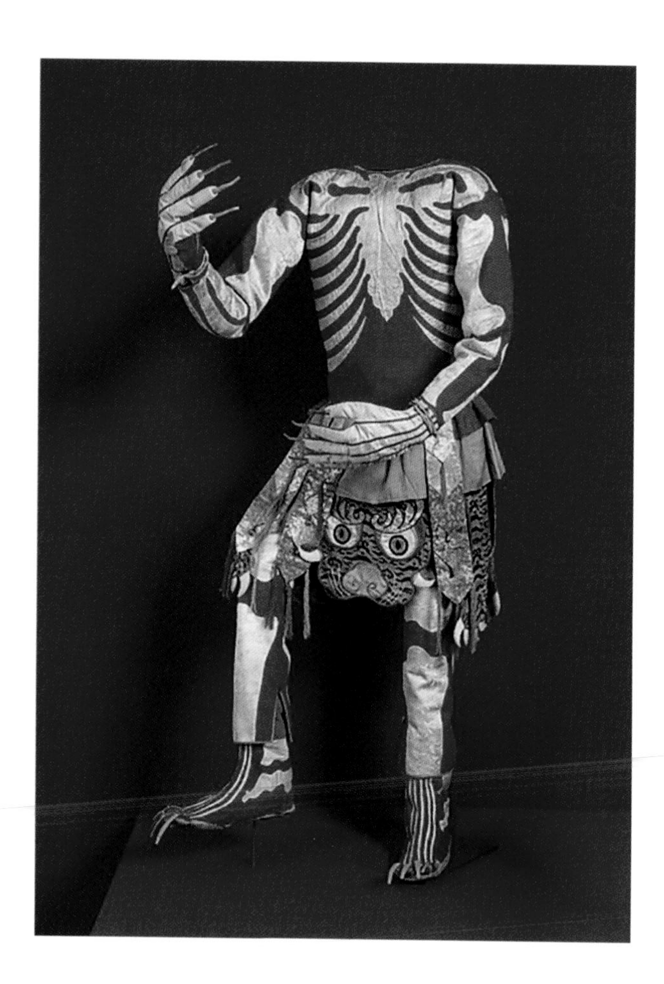

Skeleton Dance Costume, late 19th, early 20th century, Tibet.

Gift of Mrs. Edward A. Nis, 1934. Metropolitan Museum of Art, NYC. Public Domain.

A BONE OF CONTENTION: OSTEOPOROSIS AND WEIGHT

"Down to the bone" may have some new connotations: the bone-fat connection

52—Posted Juyl 21, 2014

"THY BONES ARE MARROWLESS..." (Act III, Scene IV) says Shakespeare's frightened Macbeth as he tries to reassure himself there is no danger upon seeing the ghost of Banquo, the Scottish general he had ordered to be killed. How does bone marrow, though, relate to body weight and even to osteoporosis? "Is osteoporosis the obesity of bone?" That is the provocative question that researchers Rosen and Bouxsein, in an article appearing several years ago in the journal *Nature Clinical Practice (Rheumatology)*, asked.

Osteoporosis is a serious condition that affects both men and women and can occur at any age. Rosen and Bouxsein note that signs and symptoms include back pain, fractures (most commonly at the spine, hip, distal radius, and proximal humerus) without accompanying significant trauma, and low bone mineral density, a measure of bone mass. A bone's quality is a function of bone geometry and bone strength, as well as its mineral density. When bone mineral density is low, though, bones are fragile and prone to fracture. One of the best ways to measure it is by dual-energy X-ray absorptiometry (DXA), particularly of the spine, hip, and wrist. DXA, incidentally, is also one of the most accurate ways to measure body composition and specifically our percentage of our body fat. Osteoporosis typically occurs in the elderly (and especially women after menopause). It can also occur in the context of nutritional deprivation (i.e. severe food restriction) as seen in those with anorexia nervosa, as well as secondary to organ transplantation, chronic liver or kidney disease, Cushing's Disease (with its increased production of glucocorticoids), rheumatoid arthritis, lymphoma, and types 1 and II diabetes. Further, both excessive alcohol consumption and cigarette smoking have been associated with increased risks of osteoporosis.

Rosen and Bouxsein note that both obesity and osteoporosis have some features in common: both are "disorders of body composition that are growing in prevalence;" both may have a genetic basis as well as influences from the environment; both diseases tend to develop over time and are associated with "significant morbidity and mortality;" and perhaps most importantly, both "can be traced to dysregulation of a common precursor cell," i.e., both fat cells (adipocytes) and bone cells (osteoblasts) derive from the same embryonic mesenchymal cells. These researchers describe how what are called pluripotential bone-marrow mesenchymal stem cells can differentiate into osteoblasts (cells that form bone) or adipocytes (fat cells), depending on a complex process involving "switches" within the cells "suggesting significant plasticity" between the two cell types. Shapses and Sukumar, writing in the *Annual Review of Nutrition* (2012) note one important obvious difference: osteoporosis is often considered a "silent disease" (i.e. its first sign can be a fracture) in contrast to obesity which has "high visibility."

Originally it was thought that those with excessive weight (i.e. who have an "increased mechanical load" due to their weight) were less likely to develop osteoporosis. Sharma et al in a recent article (2014) in the *Journal of Midlife Health*, report that large population-based studies now call into question the notion that increased weight is protective of bone health. The situation, though, is complex: remarkably, both a lowered body weight (and even a recent weight loss of only 5%) and excessive weight can be risk factors for increased bone loss and increased risk of fractures. Further, obese patients who are followed longitudinally after gastric bypass surgery and who have subsequently lost considerable weight, can lose significant bone mineral density, according to Rosen and Bouxsein, but a patient's age, sex, ethnicity, and lean body mass are contributing factors for the development of osteoporosis.

As we age, there tends to be a fatty infiltration of bone marrow, confirmed by MRIs, and this is associated with a greater tendency for bones to be fragile. According to studies reported by Lecka-Czernik and Stechschulte (2014) in the *Archives of Biochemistry and Biophysics*, adipose tissue accumulates in our long bones and vertebrae. Postmenopausal women can have twice the fat in their bone marrow as pre-menopausal women.

Kawai et al, writing in the *Journal of Internal Medicine* (2012) describe how there has been a "paradigm shift" in thinking of adipose tissue as an inert energy storage substance to a focus on it as a major "endocrine modulator of satiety, energy balance, and pubertal development." These adipocytes in the marrow, just like adipocytes elsewhere in the body, can secrete inflammatory substances (cytokines) that may lead to bone resorption. Furthermore, there is another

Jan Bruegel the Elder, *Triumph of Death,* late 16th century.
Museum Joanneum, Austria.
Wikipedia Commons/Public Domain.

Frans Francken the Younger, *Death Playing the Violin,* 17th century.
Wikipedia Commons/Public Domain.

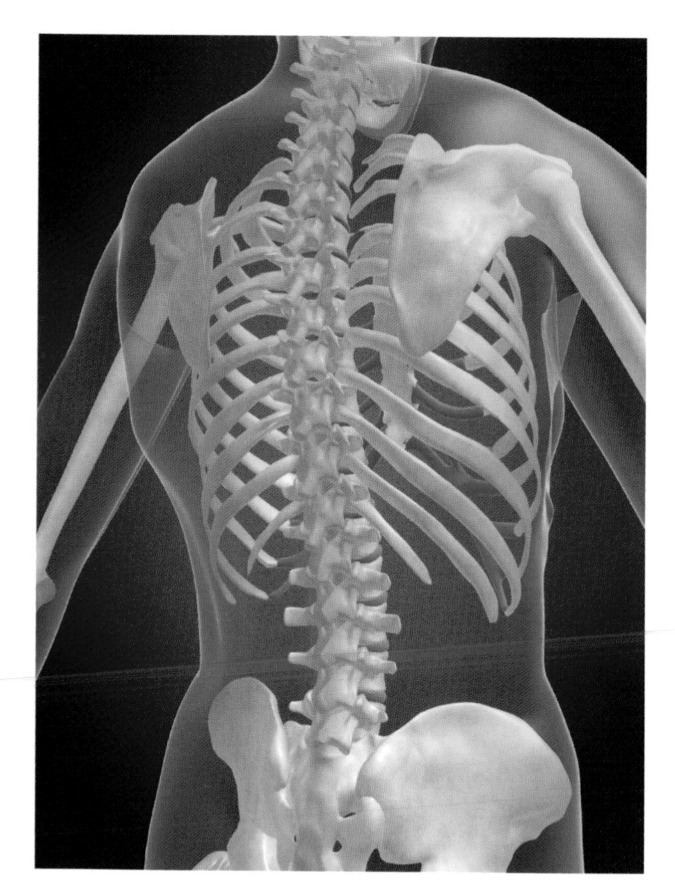

Used with permission/istockphoto.com. Credit : Eraxion.

connection between fat and bone because the fat hormone leptin, which regulates energy balance among its many other functions, can also influence bone mass. Rosen and Bouxsein note, though, that the function of fat in bone marrow is not completely known and may be either protective or harmful. Lecka-Czernik and Stechschulte raise the question whether adipocytes in bone marrow actually have a negative effect on bone mass or whether it is the low bone mass that stimulates the accumulation of adipocytes. Furthermore, both brown and beige adipocytes have been found in bone marrow and may also be involved in regulating bone mass. For example, Kawai et al speculate that these may even create a "favorable microenvironment" for bone formation by functioning as its energy source or even as a regulator of temperature.

Bottom line: The relationship between fat and bone is an extraordinarily complex and poorly understood one. If there are any beneficial effects of fat accumulation in bone marrow, according to Lecka-Czernik and Stechsculte, they may ultimately lead to new therapeutic possibilities for treatment of both osteoporosis and obesity. ■

HOLDING A MIRROR UP TO "WHITE HAT BIAS" IN RESEARCH

Mirror neurons, righteous zeal, and weight control

53 — Posted August 28, 2014

"...HUMAN UNDERSTANDING is like a false mirror, which, receiving rays irregularly, distorts and discolors the nature of things by mingling its own nature with it." Francis Bacon, *Novum Organum*, 1620

The Lone Ranger may have worn a black mask but his other characteristic, must-have accessory was his large-brimmed white hat, the mark of a hero in the old cowboy movies. It was this symbol of the classical Western hero of yore that led researchers Drs. Mark B. Cope and David B. Allison of the University of Alabama at Birmingham to label a specific kind of bias, "White Hat Bias," that they first identified in their review of the literature on obesity.

There are, of course, many potential sources of bias in scientific research, defined as any systematic error, as opposed to an error by chance—that can affect the design or implementation of a study. In fact, clinical epidemiologist and biostatistician Dr. David L. Sackett had identified over fifty different kinds, further subcategorized by the stage of research (e.g. conducting a literature review, selecting a sample population, measuring of exposures and outcomes, publishing of results, etc.), in his classic 1979 paper. For Sackett, bias was anything that "systematically deviates from the truth."

Cope and Allison define "White Hat Bias" as "bias leading to distortion of information in the service of what might be perceived to be righteous ends." In their 2010 papers in the *International Journal of Obesity* (London) and *Acta Paediatrica*, these researchers explain this kind of bias can manifest itself in several ways, including misleadingly and inaccurately reporting data from scientific studies by "exaggerating the strength of the evidence." It can also present in media press releases that distort, misrepresent, or even fail to present the facts of the actual research, particularly exaggerating claims of significance or application and failing to report any caveats or limitations. In their own articles, Cope and Allison focused on two examples they found in the obesity literature: the complexities and misrepresentation of research on the relationship of breast feeding to subsequent development of obesity in children and the role of sugar-sweetened beverages in contributing to the obesity epidemic. They note that 'White Hat Bias' can be either intentional or unintentional and can 'demonize' or 'sanctify' but regardless which way, presents a bias "sufficient to misguide readers."

I was reminded of Cope's and Allison's papers after reading Dr. Gregory Hickok's fascinating new book, *The Myth of Mirror Neurons*. Mirror neurons were originally discovered by a group of Italian researchers in the 1990s in a specific area (F5) of the brains of macaque monkeys. Their special property was that they were active (i.e., fired) not only when a monkey performed an activity but when a monkey saw the examiner performing an activity. From these original animal experiments sprang an entire avalanche of speculation about the potential importance of mirror neurons for humans. They captured the imagination not only of researchers (e.g. called by one "the neurons that shaped civilization") but also of the media and were even dubbed at one point in *The New York Times* as "cells that could read minds." Over subsequent years, they were misleadingly and inaccurately presented as the neurons that are responsible for what makes us human, including our ability to empathize with others. Ironically, until recently, their existence in humans was not even established but this did not stop the media and even some in the scientific

The Lone Ranger and his horse Trigger stamp: This television character always wore his white hat.

Used with permission/shutterstock.com.

Thomas Eakins, *Cowboy Singing*, circa 1892. Notice that even Eakins' cowboy is wearing the signature white hat.
Fletcher Fund, 1925. Metropolitan Museum of Art, NYC. Public Domain.

Snow White Mirror, an illustration from page 30 of 1852 Icelandic translation of the Grimm-version fairytale. Image proved by Landsbókasafn Íslands, Project Gutenberg.
Wikipedia Commons/Public Domain.

community to link mirror neurons and their dysfunction with autism (i.e., the so-called "broken mirror theory"), schizophrenia, and even the complex relationship between patient and therapist in psychotherapy. The hype that surrounded mirror neurons, particularly the exaggerated claims of significance and application in both the scientific literature as well as in the media press releases, without hard data to support their claims, but with potentially righteous intentions, seemed a clear example of Cope's and Allison's "White Hat Bias."

Eventually, though, researchers began to appreciate that mirror neurons and brain functioning are far more complex. In his new book, Hickok systematically and quite even-handedly delineates eight enormous problems—"anomalies," including evidence from neurological disorders, as well as noting that we have the ability to understand actions, such as playing a sport or a musical instrument, that we cannot necessarily perform ourselves—with the theory and its application. The hype, incidentally, led to mirror neurons being associated with everything from

Statue of Sir Francis Bacon, author of *Novum Organum,* 1620.
Used with permission/istockphoto.com. Credit: vailatese.

Frederick Carl Frieseke, *Woman with a Mirror,* 1911. Metropolitan Museum of Art, NYC.

Gift of Rodman Wanamaker, 1912. Public Domain.

love, smoking, aesthetic response to music, spectator sport appreciation, and yes, even to obesity! For example, Hickok cites a paper by Deborah A. Cohen (2008) from the journal *Diabetes* in which Cohen describes "neurophysiological pathways to obesity" and states in her *Abstract*, among ten pathways, that mirror neurons "*cause* (my emphasis) people to imitate the eating behavior of others without awareness." In the body of the paper, she further notes that mirror neurons "could be the mechanism through which "obesity is contagious in social networks." Cohen adds, "Although the existence of mirror neurons is not new, in the current environment, they can serve as a mechanism to amplify increases in energy consumption…" The point here is Cohen's language is unfortunately misleading and in her presumably righteous zeal to explain at least one contribution to the burgeoning obesity epidemic, expands her deductions far beyond what the data from the research can support.

Bottom line: As Francis Bacon said, human understanding is like a false mirror. So let the reader of obesity literature beware! Be sure to hold that mirror up to scientific scrutiny! ■

WHEN HEALTHY EATING TURNS UNHEALTHY: ORTHOREXIA NERVOSA

Excessive preoccupation with food quality & a judgmental attitude toward others

54—Posted September 27, 2014

Orthorexia nervosa, as originally defined, is a condition manifested by "an unhealthy obsession with eating healthy food." The word "orthorexia" was coined by physician Steven Bratman in the 1990s from a combination of the ancient Greek for "straight, correct, or right" and "appetite" or more literally "desire"—and so becomes what might be called "righteous eating." Bratman chose the word as a "parallel" to anorexia nervosa and first described his clinical observations in the *Yoga Journal.* Subsequently, he published a book *Health Food Junkies* (2000). Since then, there has appeared a growing number of articles worldwide in the scientific literature from countries such as Italy, Hungary, Turkey, India, and Korea, but many of these reports are individual clinical case vignettes and not evidenced-based studies. Furthermore, some of this research is not available in English translation. The term, though, has caught on in the media and now even appears in the *Oxford English Dictionary.*

Most recently, Moroze and his colleagues from the University of Colorado published a discussion of the syndrome in the journal *Psychosomatics* (2014, online ahead of print). According to Moroze et al, this extreme condition is often associated with dietary restrictions that lead to "unbalanced and insufficient diets" significant enough to lead to weight loss and medical conditions related to malnutrition (e.g. low sodium and potassium levels, metabolic acidosis.) Those with orthorexia will spend "inordinate amounts of time" each day thinking about the ingredients in their food, and they are cautious and vigilant about their food preparation. Unlike patients with bulimia or anorexia nervosa who wish to be thin, have body image distortions, and are concerned with the *quantity* of food they eat, those with orthorexia are generally more preoccupied with the *quality* of the food they ingest.

Chaki et al, in the *Journal of Human Sport and Exercise* (2013), note, though, that being preoccupied with the quality may not distinguish them from those with other eating disorders. They also note "There is a very thin margin between selectivity about the type and quality of food consumed and developing a psychological obsession about the diet…" They describe how the patients they have seen in their population become extremely selective about their food purity and will avoid any foods with artificial ingredients, such as artificial colors, flavors, or preservatives, and will avoid genetically modified ingredients and those that might contain pesticide residues. They are often not concerned with their weight per se, and there seems to be no specific relationship to the condition and body mass index. The condition may evolve slowly over time and begin innocently enough as a wish to eat a healthier diet (e.g. incorporate protein shakes into their diet) or improve a medical condition. Eventually, though the obsessive preoccupations and compulsive behaviors predominate with their self-imposed regimen, and there may develop a condescending, judgmental "sense of moral superiority" toward others who are not so preoccupied with the purity of their food, according to a review of the "evidence and gaps in the literature" by Varga et al in the journal *Eating and Weight Disorders* (2013). Social isolation may result.

Currently, the diagnosis is made by self-report. Several questionnaires have evolved from the original questions proposed by Bratman, but these are often not particularly specific and of questionable validity. As a result, information on the actual prevalence and incidence of orthorexia

Healthy green smoothie. Sometimes, healthy eating can be taken to the extreme.

Used with permission/istockphoto.com. Credit: tanjichica7.

Basket of Fruit and Vegetables, Italian, Wax and wicker. 18th century Italian, Naples. Gift of Loretta Hines Howard, 1964.
Metropolitan Museum of Art, NYC, Public Domain.

Vegetables sold in the market in Bohol, Philippines, 2006.
Jasper Greek Golangco. Wikipedia Commons. Unrestricted use.

are not known. Some studies have suggested the condition is more common in dietitians and students of nutrition, but methods of assessment have varied considerably. Orthorexia is not technically recognized as an official psychiatric diagnosis in the latest edition (2013) of the Diagnostic and Statistical Manual (DSM-5). It is an example of what I would call "disordered eating." If categorized as an eating disorder, though, it would now be classified under the wastebasket category of "unspecified feeding or eating disorder." Moroze et al have suggested the condition could be classified as a subgroup of the "avoidant/restrictive food intake disorder," though that is a disorder typically beginning in childhood. They note that there is "precious little empirical research" for this condition and standardized, validated diagnostic criteria have not yet been established. These researchers have proposed their own diagnostic criteria for orthorexia nervosa, including the person's obsessional food preoccupations impair either their physical health due to nutritional imbalances or impair their social, academic, or vocational functioning. They have also noted those suffering may experience guilt and worries if they "transgress" from their rules of healthy eating and consume "impure foods," may be particularly intolerant to others who do not share their beliefs, and may spend excessive amounts of money on food they believe to be of higher quality. Moroze et al have also included the caveats that the behavior is not related to the observance of organized orthodox religious food rituals or to specialized food requirements subsequent to allergies or other medically diagnosed conditions.

Bottom line: Whether orthorexia nervosa will ever achieve the status of a psychiatric disorder remains to be determined. Remarkably, it has taken sixty years for binge-eating disorder, first described by Dr. Albert Stunkard in the 1950s, to be recognized officially as a diagnosis. Clearly, well-designed clinical studies, with more sophisticated instruments for assessment, are warranted. But clinicians should be suspicious when patients seem unduly and obsessively preoccupied with the quality of their food to the exclusion of almost everything else. ■

WHY OUT OF SIGHT REALLY IS OUT OF MIND

Research supports an underrated path to achieving your goals.

55 — Posted October 28, 2014

AT THE MAD TEA PARTY in Lewis Carroll's *Alice's Adventures in Wonderland*, the Hatter says indignantly to Alice, "You might just as well say that 'I see what I eat' is the same thing as 'I eat what I see.'"

Clearly the Hatter has not read Brian Wansink's astute observations on human behavior. Wansink is the John Dyson Professor of Consumer Behavior at Cornell University in Ithaca, NY, and director of its Food and Brand Lab. For many years, he and his colleagues have designed ingenious studies exploring the connection between human nature and our eating environment. He is best known for his 2006 best-selling book *Mindless Eating: Why We Eat More Than We Think*.

Wansink's team has discovered that people will eat fewer chicken wings if the plates of half-eaten wings are left piled on the table rather than removed by a waitress. He also designed the "bottomless soup bowl" experiment, in which he found that people seem to keep eating, regardless of how full they may be or how much they have actually eaten, if the soup bowl, for example, never empties. In other words, many people still take quite literally the "clean your plate" dictum from childhood. Wansink has seen that people will eat more M and M candies if they are labeled "fat-free," even though there really is no such product, and they will eat *fewer* candies if the pieces are placed farther away, preferably out of sight in a drawer (and even covered by opaque foil wrapping or a lid.) We do very clearly tend to eat what is conveniently available—and what we can see. Further, because portions tend to be so much larger today than in the past—even the recent edition of *The Joy of Cooking* supersized its portion sizes—we all tend to suffer from "portion distortion" and lose track of what are called "consumption norms," i.e., what is an appropriate serving size.

Keeping candy wrapped and out of sight is a way to minimize the temptation.

Used with permission/istockphoto.com. Credit: CreativeBrainStorming.

Wansink has now written the intelligent and eminently practical manual, *Slim by Design: Mindless Eating Solutions for Everyday Life*. Wansink's thesis is that *everyone* is capable of so-called "mindless eating," given the right environment. Since we may make hundreds of food choices every day, including whether to finish a particular dish, take soup or salad, or have dessert, he believes we need to work *with* human nature, not *against* it.

"Becoming slim by design," he writes, "works better than trying to become slim by willpower." In other words, "It's easier to change your eating environment than to change your mind." Systematically, Wansink focuses on the many often subliminal eating decisions we all confront—in our own kitchens, at work, at restaurants and at supermarkets—plus the schools that our children attend. Wansink has been called the "Sherlock Holmes of food." He and his colleagues, often doing their detective work under cover, keenly observe how people behave in their natural habitats and, particularly, how slim people behave differently from those who are not.

For example, Wansink found that slim people approach an "all you can eat" buffet by "scouting out" what is available—"getting the lay of the land," as it were—before they grab their plates and pile on food. They are also more like to sit facing away from, and to choose a table farther away from a buffet; more likely to choose small plates; and, if eating Chinese food, eat with chopsticks. "None of these behaviors has anything to do with counting calories or choosing bean sprouts over Peking duck," he writes.

Wansink has been a consultant to numerous companies, organizations—even the Pentagon—and restaurants

Brian Wansink has found that people eat fewer chicken wings if the half-eaten wings are not cleared away from the table.

Used with permission/istockphoto.com/ Credit: colematt.

At the mad tea party in *Alice in Wonderland*, the Hatter says, "You might as well say 'I see what I eat is the same as I eat what I see.'" (A coloured plate. 'What day of the month is it?'), 1907.

Lewis Carroll and Charles Robinson. London: Cassell & Company, opp. P. 94. Wikipedia Commons/Public Domain.

nationwide, and notes that supermarkets, grocery stores, and restaurants don't have the goal of making us fat—they have the goal of making money. He offers many practical suggestions, such as how healthy items (rather than candy and chips) can be placed on a check-out line to avoid impulsive junk food purchases or can be enticingly described on a menu (e.g. crisp summer salad with shrimp, pineapple and avocado) that not only get us to choose more wholesome options but leads to increased revenue for these businesses. Further, he suggests that we write to our favorite haunts to get them to change their practices to make it easier to choose healthier alternatives by default. He also offers recommendations for reorganizing our own kitchens—he found that the size of our plates, the color of both our plates and our walls, and even what foods we have set out on our counters could all have an impact on our weight.

Bottom line: Unfortunately, most of us have difficulty in exerting our willpower consistently, particularly when confronted with the sight of hundreds of appealing food items, whether at home, at work, in school, or at restaurants or supermarkets. Rather than changing human nature, we have a better chance of succeeding at controlling what we eat by modifying our food *environments*. Wansink's *Slim by Design* offers up hundreds of empirical strategies to alter our behavior.

Note: Since the publication of my book's first edition, some of Dr. Wansink's research has been called into question, and Dr. Wansink, after an investigation, has been accused of scientific misconduct. Nevertheless, many of Dr. Wansink's conclusions about human behavior make sense, and he offers some practical suggestions in his books for avoiding overeating. I have chosen to keep this blog, first posted October 28th, 2014, within this collection. I will let the reader decide for him or herself. ■

OF EPIDEMIC PROPORTIONS: THE PRIMARY COLORS OF OBESITY

A metaphor by any other name: the obesity epidemic

56—Posted November 23, 2014

THERE ARE VIVID, graphic descriptions, both historical and literary, of epidemics spreading throughout an entire population. In Fifth century B.C. Athens, for example, the great historian of the Peloponnesian War, Thucydides, described a highly contagious disease that decimated much of the Athenian population. Over the span of about five years, this disease killed as many as 100,000 people or about 25% of the city's population, including the famous statesman Pericles. Despite Thucydides' meticulous description of victims' symptoms, medical historians still cannot agree on whether the disease was smallpox, typhus or even, as one researcher had suggested back in 1996, Ebola! One of the most famous literary descriptions of an epidemic is found in Albert Camus's gripping novel *The Plague*, where the seemingly minor occurrence of Dr. Rieux's inadvertently stepping on something squishy—one "dead rat lying in the middle of the landing"—signaled the epidemic's inauspicious beginning.

Recent hysteria about the potentially highly contagious spread of deadly Ebola from localized regions in Africa to the U.S. made me reflect on the concept of "epidemic" and its applicability in describing the burgeoning rates of obesity. The word "epidemic" comes from the Greek "upon" and "people." According to Porta's *Dictionary of Epidemiology*, an epidemic is the "occurrence in a community or region of cases of an illness, specific health-related behavior, or other health-related events clearly in excess of normal expectancy."

Epidemics, such as the recent Ebola epidemic in Africa, can create overwhelming fear.

Used with permission/istockphoto.com. Credit: DieterMeyrl.

The number of cases that indicate the presence of an epidemic can vary, depending on such factors as the vector involved, as well as the size and type of population exposed, and the time and place where the epidemic occurs. A "pandemic," another word used to describe the increasing prevalence of obesity, indicates an epidemic that has spread worldwide.

In his 1992 book *Explaining Epidemics*, Charles Rosenberg notes that we use the word "epidemic" in many ways that are mostly metaphorical and often to connote the "emotional urgency" and "dramatic intensity" typically characteristic of an epidemic. For Rosenberg, a "true epidemic" has a "unity of time and place." It is "an event, not a trend" and is a "social phenomenon" calling for an immediate response. Further, out of overwhelming anxiety, people often want to explain the susceptibility of others in order to reassure themselves they are immune: they look for behaviors in themselves and others that are seen as under individual control and then often "blame the victim" morally and socially. This was certainly the case in the initial phase of the AIDS epidemic. In fact, says Rosenberg, "Epidemics have always provided occasion for retrospective moral judgment." He further explains that the public only acknowledges an epidemic's existence when its presence becomes unavoidable. Once there is public acknowledgment, though, there is often pressure for the community to act to control the epidemic. With obesity, we are now somewhere between "blaming the victim" and insisting the community does something to control the situation. An epidemic, though, often ends quietly, with a whimper, as it were, rather than a bang, as incidence returns to previous levels in the population.

Isabel Fletcher, in a recent 2014 article in *Sociology of Health and Illness*, suggests that it was the adoption between the 1970s and 1990s of Body Mass Index (BMI) as a standard that defined and measured obesity and overweight that "was crucial in the framing of obesity as an epidemic." Fletcher notes that this standardization created easily obtainable

One dead rat in Camus' *The Plague* marked the beginning of the epidemic.

Used with permission/istockphoto.com. Credit: epantha.

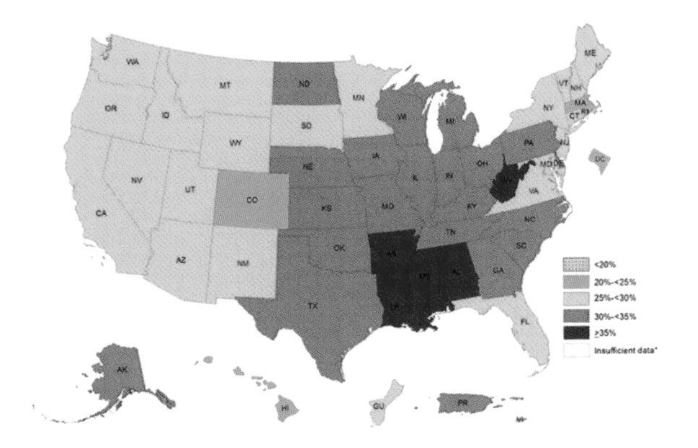

Prevalence of Self-Reported Obesity among U.S. Adults by State and Territory, 2016.

Public Domain, Centers for Disease Control, Behavioral Risk Factor Surveillance System.

"large data sets" that could now track population-level changes in body weight. In other words, researchers could now tally prevalence rates. Several researchers, such as J.E. Oliver, have suggested that the Centers for Disease Control (CDC) contributed to labeling obesity an epidemic when it created its now infamous Powerpoint presentation in the 1990s. The slides graphically demonstrated the increasing prevalence of obesity in the U.S., as defined by BMI levels, by designating those states with the highest levels of obesity in the inflammatory primary colors of red and yellow, as if there were a hot zone of spreading infection. Most recent rates from the CDC's website for 2013 indicate 23 states with a prevalence of obesity between 25 and 30% and 18 states with a prevalence of between 30 and 35% of the population. Two states—Mississippi and West Virginia—have a prevalence greater than 35%.

Several years ago, researcher Katherine Flegal of the CDC, in the *International Journal of Epidemiology*, wrote about the concept of the obesity "epidemic." She noted that population samples in U.S. were first compiled in the early 1960s, i.e., the National Health and Nutrition Examination Survey (NHANES). Researchers found that from this first sample to 1980, when the second NHANES was conducted, there was little change, but by the third survey in the late 1980s to early 1990s, there were "unanticipated increases" in BMI in the U.S. population that was "difficult to explain."

Marble bust of the ancient Greek general and historian Thucydides, who wrote about a major epidemic that swept through Athens in the Fifth Century B.C. Royal Ontario Museum.

Photograph Captmondo, 2005, Wikipedia Commons/Public Domain.

Arnold Böcklin, *The Plague,* 1898. The increased prevalence of obesity over the past thirty years has been compared to an epidemic. The higher the body mass index, the more likely an earlier mortality. Kunstmuseum Basel, Switzerland.

Wikipedia Commons/Public Domain. Credit: Gottfried Keller-Stiftung, Montana State University.

These rates have generally continued to increase over the years though there may be some recent stabilization in some segments of the population. The increases in prevalence made Flegal question whether obesity qualifies as an epidemic: two aspects of the concept, though—a high prevalence and a rapid spread—"clearly fit the general definition." Flegal emphasized, though, that it was not a totally new observation to find a high prevalence of overweight in the U.S.: even in that first national survey, for example, in the early 60s, 45% of the population was overweight. Flegal's conclusion is that the word "epidemic" has "some drawbacks as a descriptor" since there is no specific quantitative definition. She also clarifies that the word "masks" some aspects of obesity, such as the endemic (i.e., constant presence within a population) nature of overweight in the population in general, the sustained upward trends in weight that is associated with economic development, the controversies regarding long-term deleterious health effects with different levels of weight, and even the difficulties in defining the end to an epidemic with regard to obesity. Nevertheless, she says, obesity does have some characteristics of an epidemic in its "surprising and unexpected increases." In a recent 2014 article in the journal *Circulation*, Dr. David Katz has called rates of obesity "more correctly" "hyperendemic."

Bottom line: Whether obesity fits the definition of a classic epidemic remains a matter of dispute, but the hype surrounding the use of the word does keep it importantly in the public's attention.

Final note: I have avoided any discussion of the disease nature of obesity. For more on the controversy of obesity as a disease, please see blog 36 on disease mongering and the medicalization of weight. ■

"SUFFICIENT UNTO THE DAY:" THE COMPLEXITIES OF SATIETY

"Give us this day our day bread"— and a sense of fullness as well

57—Posted December 30, 2014

MANY YEARS AGO, Rockefeller University researcher Jules Hirsch noted that over a lifetime, a person might consume about 70 million calories or 14 tons of food that the body must process, and considering all the variables, does so remarkably well. There are many complex physiological measures involved in how the body prepares itself for the ingestion, digestion, and metabolism of our food. These anticipatory responses are called cephalic-phase responses and they can be physical (e.g. motility of our gastrointestinal system), secretory (e.g. release of enzymes and hormones), or even metabolic (e.g. process of thermogenesis.) The process of eating, as Power and Schulkin note in their 2009 book *The Evolution of Obesity*, is ironically both required to maintain homeostasis and simultaneously a threat to homeostasis. Initiating a meal is "largely opportunistic," according to Chambers et al in a 2013 issue of the journal *Current Biology* and involves cognitive assessments, such as food availability, time of day, palatability, and learning. What determines, though, how the body prepares itself to stop eating? Determining the end of a meal, in fact, involves literally a "cascade" of physical and biochemical processes, including the release of multiple hormones that enable us to stop eating. The intestinal hormone CCK, released preferentially in response to a fatty meal, for example, delays the emptying from the stomach and reduces food intake and meal size and hence is one factor that leads to satiety. Other hormones that inhibit food intake are the intestinal GLP-1 (glucagon-like peptide 1), and the pancreatic hormones Peptide YY (PYY) and amylin. Many of these hormones are mediated through the vagus nerve.

William Kentridge, *Johannesburg, 2nd Greatest City after Paris; Mine; Sobriety; Obesity and Growing Old*, 1989–91, in his Portfolio/Series: *Drawings for Projection.*

There are separate physiological processes involved in short-term *satiation* (i.e., fullness and decrease in hunger during a meal) that leads to termination of that particular eating episode and long-term *satiety* (i.e., involving food intake over the entire day or longer and food intake frequency from one meal to another.) There is also *sensory specific satiety* that occurs after eating a specific food, such that we lose our taste for or interest in it. Even Shakespeare knew that when he said, "A surfeit of the sweetest things /The deepest loathing to the stomach brings." (*A Midsummer-Night's Dream*, ii, 2, 137). The more variety of food choices to which we are exposed (e.g. a buffet), the more we tend to eat.

The Russian scientist Ivan Pavlov, at the turn of last century, performed some of the early experiments with dogs with what he called "fictional feeding"—or "sham feeding," in which ingested food did not reach the stomach but drained out of an artificially created fistula. These experiments with dogs and later ones by others with rats demonstrated that when food does not reach the stomach, animals eat more. In other words, food in the mouth is not enough to elicit satiation or satiety. Gastric distention from food in the stomach—i.e., that sense of fullness we get when we eat, seems to be required to stop eating. Chambers et al note that there is "no question that gastric volume is a rate-limiting factor in meal size." Incidentally, the prophet Mohammed seemed to know that intuitively: it is reported that when he was hungry, he tied a stone to his belly to ward off the feelings of hunger. In recent years, we use horizontal or vertical lap band bariatric surgery to make the stomach capacity literally much smaller.

Sometimes, though, when the mechanical and biochemical safety brakes fail, people eat beyond that sense of fullness to the point of discomfort in what we call excessive eating (i.e. gluttony.) Gluttony, described by Pope Gregory in the Sixth century, was designated one of *The*

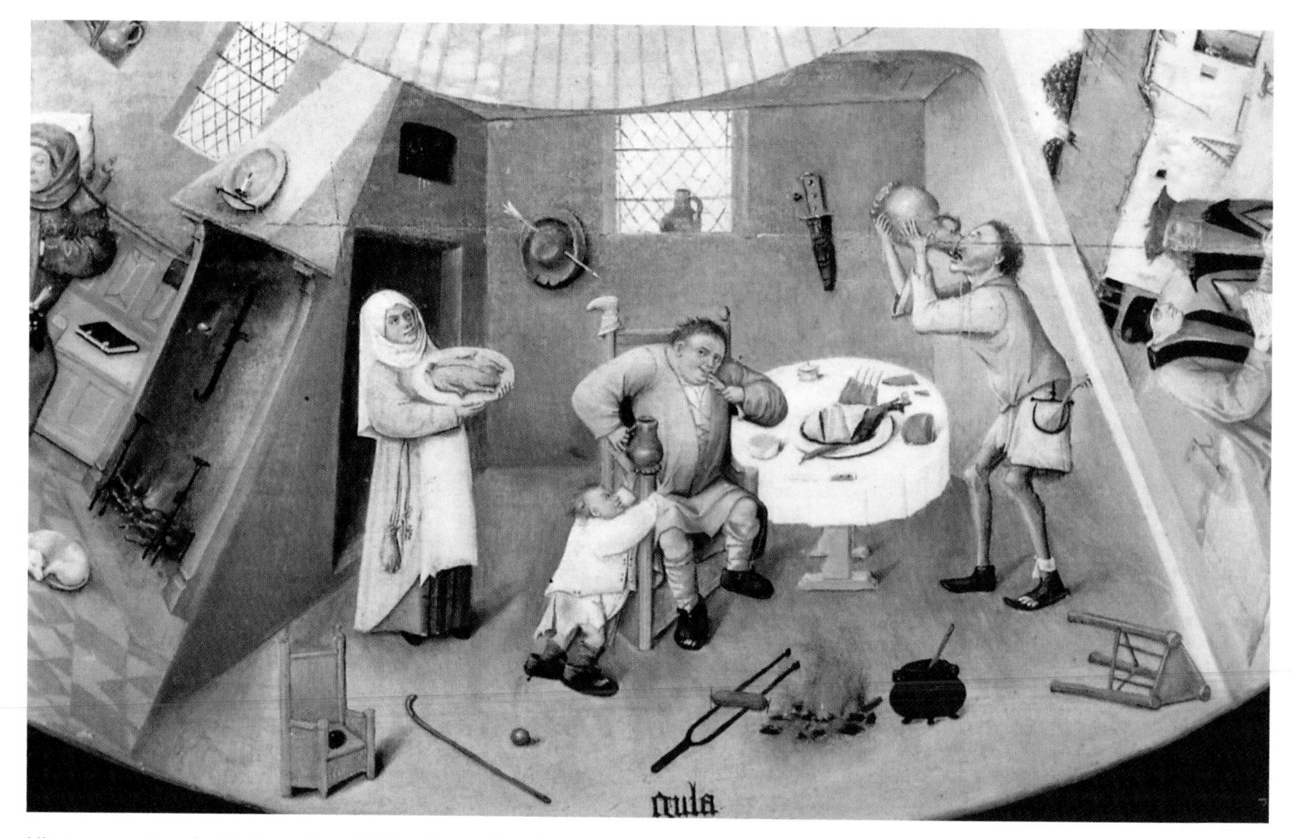

Hieronymus Bosch, *Gluttony*, from *Table of Seven Deadly Sins* (1505–1510), Prado Museum, Madrid.
Wikipedia Commons/Public Domain.

Seven Deadly Sins in early Catholic teaching. Saint Thomas Aquinas, philosopher and theologian in the 13th century, elaborated on the sin of gluttony and described several ways that people commit this sin: eating excessively; eating too expensively or luxuriously; eating too eagerly; eating too daintily or too elaborately; and eating at an inappropriate time.

Jules Hirsch reported on experiments whereby lesions of the lateral hypothalamus produced decrease eating (hypophagia), profound weight loss, and even death by starvation, whereas lesions of the ventromedial hypothalamus produced hyperphagia, massive obesity, and even examples when hyperphagia led to stomach rupture in rats. The ventromedial hypothalamus then became known as the satiety center. In the psychological thriller *Se7en*, a serial killer uses *The Seven Deadly Sins* as a template for his murdering spree. The glutton dies by being forced to eat himself to death, ostensibly rupturing his stomach by his excessive consumption.

Williams, writing in the journal *Physiology & Behavior* (2014) notes that for years, homeostatic eating (i.e. for maintenance of energy balance and providing negative feedback on eating) was seen as distinct and as operating in opposition to hedonic eating (i.e. eating for pleasure and the system that can override the homeostatic system and lead to overeating.) For Williams, these two systems overlap and it is "inaccurate" to continue to think of them as two separate systems. Williams believes that the hormone orexin A, found predominantly in the lateral hypothalamus, is involved in increasing our motivation to obtain and continue to eat highly rewarding food at the expense of our sense of satiation.

Satiety, of course, is particularly important to dieters. Rebello et al, writing in a 2013 issue of *Advances in Food and Nutrition Research*, emphasize that success of a weight-loss dietary regimen is closely related to compliance, which, in turn, is "largely dependent on hunger, appetite, and satiety." For example, certain, but not all, dietary protein is the most satiating of the food groups for some people. Not only does it take more energy to metabolize protein, but there is also speculation that satiety due to the ingestion of protein is related to increases in two hormones that decrease eating, GLP-1 and PYY, and to a decrease in ghrelin, the hormone that leads to an increase in eating.

Flemish artist Pieter Coecke van Aelst, *Gluttony,* circa 1550–60. Tapestry. Metropolitan Museum of Art, NYC.

Gift of Mrs. Frederic R. Coudert Jr., in memory of Mr. and Mrs. Hugh A. Murray, 1957. Public Domain.

Bottom line: At one time, we thought that satiety was related to specific regions of the hypothalamus. We now know that many complex endocrine, cognitive, and neural systems are also involved, with multiple and "redundant" protective mechanisms in place. There are many biological factors involved in the control of how much we eat that have yet to be elucidated. For example, Allen S. Levine and colleagues, (Olszewski et al, *Current Opinion in Endocrinology, Diabetes, and Obesity,* 2017) recently have been studying the anorexigenic effects of the hormone oxytocin and believe oxytocin may eventually have a "therapeutic potential" for inducing satiety by controlling appetite and reducing food intake. With all of these systems, we may, unfortunately, have much less conscious control over eating than we think we have. ■

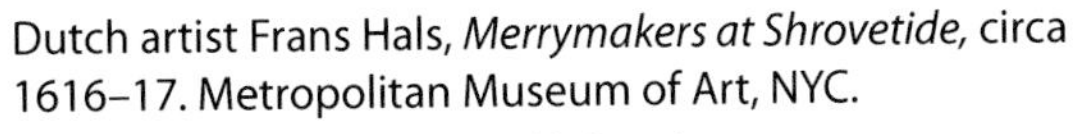

Dutch artist Frans Hals, *Merrymakers at Shrovetide,* circa 1616–17. Metropolitan Museum of Art, NYC.

Bequest of Benjamin Altman, 1913. Public Domain.

FOOD CRAVINGS: THOSE "TORMENTS OF EXPECTATION"

"Subdue your appetites, my dears and you've conquered human nature."

58 — Posted January 24, 2015

In chapter five of Charles Dickens' *The Life and Adventures of Nicholas Nickleby,* Nicholas observes the brutal schoolmaster Wackford Squeers: "Smacking his lips" in obvious enjoyment, Squeers "calmly" breakfasts on milk, bread and butter, and beef while he cruelly tantalizes his five young charges who watched him eat and "remained with strained eyes in torments of expectation." "Speaking with his mouth quite full of toast and beef," Squeers says to these starving little boys, "Subdue your appetites, my dears and you've conquered human nature." This vivid depiction is about abject hunger and the unfortunate mistreatment of unwanted children, but Dickens' phrase "torments of expectation" made me think of the intense desire or yearning people commonly experience with food cravings.

Charles Dickens, the author of *Nicholas Nickleby.* Dickens' character Wackford Squeers, the brutal schoolmaster, tells his starving boys, "Subdue your appetites..." Ary Scheffer, portrait of Charles Dickens, 1855.

What differentiates food cravings from ordinary hunger, though, is their particular specificity and intensity (Meule and Kubler, *Eating Behaviors,* 2012). The word "craving" comes from Old English *etymological roots,* "to demand" or "require." It denotes an urgent desire or longing and even may connote "to force or exact," according to the *Oxford English Dictionary (OED.)* The *biological roots* of cravings, though, are less well understood.

Food cravings, from an evolutionary perspective, can be seen as beneficial as they may lead to an interest in and search for a variety of foods (i.e., food-seeking) and hence to a greater tendency to meet our body's nutritional requirements, particularly in times when resources were scarce. Carbohydrate craving, in particular, "appears to be part of a biologically conducive system...to sustain life," say Ventura et al, in a comprehensive review article (*Nutrition,* 2014) that explores five theories for the neurobiological basis of carbohydrate cravings. These researchers note that there is speculation that the wanting and liking of food evolved separately. The "like" aspect is reflective of a food's palatability and the hedonic or pleasurable aspect of eating that stems from the opioid system. The "want" aspect reflects a motivational desire to obtain food and is dopamine-driven. Mela (*Appetite,* 2006) distinguishes not only liking a food from desiring (craving) it now or in the near future, but also from preference, i.e., comparison and selection of food from alternative choices.

Hormes and Rozin, in a 2010 issue of *Addictive Behaviors,* note that not all languages have a concept of "craving." In our culture, though, almost everyone has cravings for certain foods now and then. Cravings are more common in women generally and may exhibit a cyclic pattern during the days prior to menstruation or during certain stressful times such as pregnancy. Studies by Hormes and her colleagues (2014, *Appetite*) have found that chocolate (which typically has fat and sugar as well as chocolate in various proportions of cocoa itself) is the most craved substance especially among women in North America, but not

Vik Muniz, *Action Photo, after Hans Namuth*. Hans Namuth had been a photographer famous for his portraits of artists, including Jackson Pollock. Mr. Muniz uses the image of dripping chocolate to mimic the iconic photograph, taken by Namuth, of Pollock's style of painting. Though it may not subdue cravings, this creative use of chocolate is far less caloric.

necessarily worldwide and conclude that chocolate craving may be a "culture-bound syndrome." These researchers have found that U.S. women are more likely to think of chocolate as both "pleasurable and forbidden" simultaneously, and they found that there were differences between men and women in their study: women were more likely to have more frequent and intense cravings and have heightened responsiveness to the food environment than men. Men reportedly are more apt to crave savory foods. Hormes et al conclude, "...although physiological or biochemical hypotheses regarding the reasons for craving are appealing, individual and contextual factors appear to play a more significant role." Restricting intake can lead to increases in cravings, and cravings can be conditioned by particular cues of sight or smell. Advertisers (and restaurants) use this to our disadvantage when they use overt or covert images of delicious-looking food to instill cravings in their customers.

Food cravings are typically considered benign, particularly when contrasted with cravings for alcohol, drugs of abuse, or cigarettes, though some people may experience guilt when they succumb to their cravings. Furthermore, cravings can become out of control and have been connected to less dietary restraint in general, disordered

Italian artist Pietro Longhi, *The Morning Chocolate*, 1775–80. Ca' Rezzonico, Venice.

Wikipedia Commons/Public Domain.

In many countries, but not all, chocolate is the most craved food.

Used with permission/istockphoto.com. Credit: loooby.

patterns of eating, and more specifically, binge eating disorder and even obesity. They have also been associated with those who report higher levels of so-called "food addiction," a controversial concept developed and measured by the *Yale Food Addiction Scale* of Gearhardt and her colleagues (2009, *Appetite).*

In their 2014 article in the journal *Frontiers in Psychiatry*, Potenza and Grilo note that though often thought of in terms of drugs or alcohol, as well as food, the concept of craving entered psychiatric nomenclature as a criterion for addiction (i.e., the uncontrolled, compulsive seeking and use of a substance despite negative health and social consequences) only in our most recent *Diagnostic and Statistical Manual* (*DSM*-5.) DiLeone et al (2012, *Nature Neuroscience*) describe drug addiction as 'hijacking' the reward pathways in the brain. They acknowledge that there are similarities in compulsive food seeking and drug addiction, but "important pieces of the story are still missing" and "we have a greater understanding of the detailed neural and behavioral basis of drug intake and seeking than we do of food intake and seeking." There are also clear differences: after all, we cannot be completely "food abstinent." Craving has been studied by neuroimaging, and the anterior cingulate cortex(ACC), an area related to reward and cognitive control seems one (of many areas) to be implicated, but as Wilson and Sayette explain in a 2015 issue of *Addiction*, "intensity of the urges matters." Brain responses during MRIs require further study as they are often done during only mild states of desire rather than the "overpowering desire" such as when a person cannot think of anything else.

Allen S. Levine and colleagues (Olszewski et al, *Physiology & Behavior*, 2011) emphasize that reward itself is "a dynamic and individual-specific state," that can be considerably modified by many factors, including physiological status, memory and learning, and the characteristics of the food. The opioid system is involved and can enhance "pleasure related to intake of any food--palatable, bland, or even potentially dangerous..."

Bottom line: The concept of craving, whether for particular foods or substances of abuse, is a complex one that needs "further refinement." (Wilson and Sayette) "Subduing your appetites," as Dickens wrote, may lead to conquering human nature. There are, however, neurophysiological variables as well as those involving measurement, definition, and even culture to understand before we can do so. ■

SMOKING AND WEIGHT: THOSE "BURNT-OUT ENDS OF SMOKY DAYS"

The burden of weight gain after smoking cessation

59—Posted February 20, 2015

"A CUSTOM LOATHSOME to the eye, hateful to the nose, harmful to the brain, dangerous to the lungs," and with its "stinking fumes," it resembles the bottomless pit of the underworld's River Styx. So wrote King James 1st of England (*aka* King James VI of Scotland and after whom the King James version of the Bible was named) in his 1604 treatise *A Counterblast to Tobacco*. Ironically, though, King James noted that the "unsavory" and "filthy custom" of smoking "refreshes a weary man," but also "makes a man hungry." Many people, though, (and particularly women) have used smoking to control their appetite and report that one major reason to continue smoking, despite serious medical morbidity and mortality, is the fear of gaining considerable weight when they stop.

My title phrase "burnt-out ends of smoky days" comes from T.S. Eliot's poem *Preludes*, written just about 100 years ago. Though Eliot, according to some critics, evokes the gritty alienation of urban life, he furnishes us with a poetic image for smoking cessation.

Édouard Manet, *Gypsy With Cigarette*, date unknown, possibly 1862, Princeton University Art Museum.
Wikipedia Commons/Public Domain.

Despite dire warnings by the Surgeon General, over 18% or more than 42 million U.S. adults continue to smoke, according to statistics from the Centers for Disease Control (CDC). Cigarette smoking and exposure to second-hand smoke kill more than 480,000 Americans a year.

Researchers have found that not only does smoking impact cardiac and pulmonary functioning specifically, but it also has multiple effects on endocrine and metabolic functioning and has been associated with impaired adrenocortical functioning (e.g. decreased cortisol levels), thyroid disease (e.g. Graves disease), reduced fertility, insulin resistance, osteoporosis in post-menopausal women, and even type 2 diabetes. (Berlin, 2009, *Current Medical Research and Opinions.*)

For years, researchers have known that smokers typically weigh less than nonsmokers and the majority who are able to quit do indeed gain weight when they stop. Back in 1982, for example, Judith Rodin and her colleague Jeffery Wack, writing in the *American Journal of Clinical Nutrition*, found that some smokers can actually consume several hundred more calories daily than nonsmokers. They speculated that nicotine resulted in less efficient storage of calories by changing the physiology of the gut (e.g. altering gastric motility or increasing the propulsive activity of the colon) or even by altering metabolism (e.g. increasing sympathetic activity and creating a thermogenic effect of increasing resting metabolic rate and burning more calories.)

Research on weight gain and smoking cessation, though, is often complicated. For one thing, tobacco contains thousands of other compounds and non-nicotine factors may contribute. Further, measures of smoking cessation and weight gain often come from self-report, rather than validated biochemically by either carbon monoxide in exhaled breath or by cotidine, a metabolite of nicotine and a biological marker of smoking, excreted in the urine.

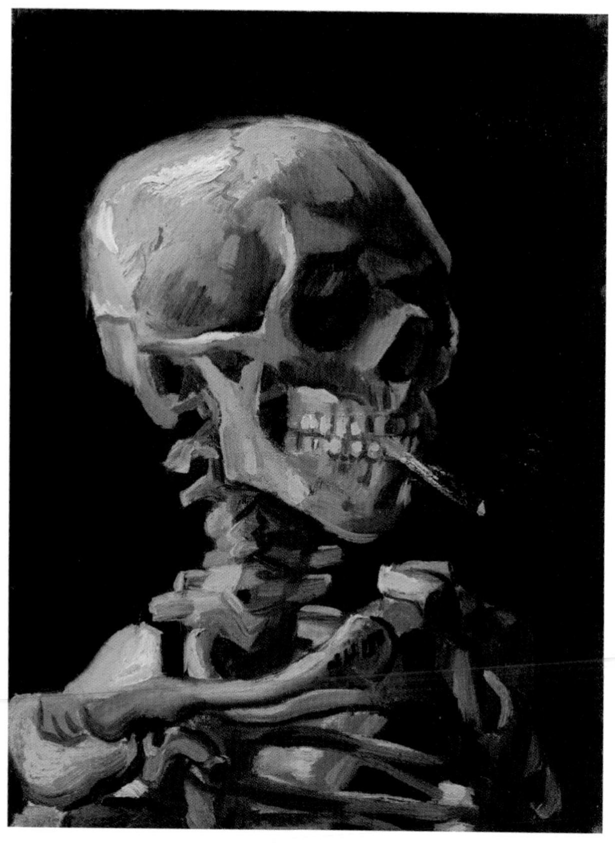

Vincent van Gogh, *Head of a Skeleton with a Burning Cigarette,* 1886. Van Gogh Museum, Amsterdam.
Wikipedia Commons/Public Domain.

T.S. Eliot wrote *Preludes,* from which my title comes.
Used with permission, istockphoto.com. Credit: traveler1116.

And data collected may be measuring what is called "point prevalence"—a "yes or no" response to smoking at a particular follow-up point—rather than measuring continued abstinence. Studies, as well, may be so heterogeneous in design and follow-up periods that a true meta-analysis becomes impossible.

Most researchers acknowledge there is great variability in the amount of weight gain after cessation, but younger ages, lower socioeconomic status, and heavier smoking are predictors of higher weight gain. (Filozof et al, *Obesity Reviews* 2004.) Weight gain seems also influenced by underlying genetic factors, as evidenced by twin studies where concordance for weight change is higher in monozygotic twins. The speculated mechanisms for weight gain post-cessation include increased energy intake, decreased resting metabolic rate, decreased physical activity, and increased lipoprotein lipase activity, but the underlying increase in weight (and specifically fat accumulation) is not well understood. Another theory is that nicotine may facilitate the reinforcing properties of food, and this effect may persist long after the individual stops taking it (Donny et al, 2011, *Physiology & Behavior.*) The effect of nicotine on food reinforcement may be related to an interaction between cholinergic and dopaminergic systems that have a genetic contribution as well (e.g. reduced availability of D2 dopamine receptors in some people). In other words, there may be individual differences at both the neurobiological and behavioral levels.

How much weight do people gain? Aubin et al, writing in the *British Medical Journal* (2012) performed a meta-analysis of 62 randomized controlled studies that addressed weight gain in those who quit smoking and

Used with permission/istockphoto.com. Credit: georgeclerk.

Ernst Ludwig Kirchner, *Couple in a Room,* date unknown.
Wikipedia Commons/Public Domain.

remained abstinent at one year. They found those who quit experience a mean weight gain of 4 to 5 kilograms (8.8 to 11 pounds) twelve months after quitting, with most weight gain occurring within the first three months of quitting—about 1 kilogram (2.2 pounds) a month— though the researchers found considerable variation: over the year, 16% actually lost weight; 37% gained less than 5 kilograms, 34% gained from 5 to 10 kilograms, but 13% gained more than 10 kilograms (22 pounds.) Interestingly, they noted that no study detailed how weight was measured (e.g. clothing worn; scales used.)

Many studies, often financed by pharmaceutical companies, have systematically reviewed interventions specifically targeting weight gain after quitting as well as those that were designed to help people quit. For example, varenicline (Chantrix), a partial agonist at nicotinic acetylcholine receptors that may block the reinforcing action of nicotine, has been used. A study just published in *JAMA* (Ebbert et al, 2015) notes the effectiveness of varenicline in some of those who are not quite ready to quit but want to reduce their usage—a "reduce to quit approach." Farley et al (2012, *Cochrane Data Base of Systematic Reviews*) found that interventions like bupriopion (Wellbutrin), nicotine replacement therapy (e.g. nicotine patch, gum, spray), and varenicline do reduce post cessation weight gain in the short term while subjects take the medication, but this is not necessarily maintained at 12 months after smoking cessation.

Yang et al (2013, *Addictive Behaviors*) did a systematic review of pharmacological interventions for smoking cessation to prevent weight gain. In general, about 80% of those who quit gain weight in the U.S. They noted that use of approved smoking cessation medications double the likelihood of quitting smoking, and combinations are particularly effective. Bupropion in combination with nicotine replacement therapy was effective in reducing weight gain post cessation; naltrexone, an opiate antagonist, (used for both drug and alcohol cravings) in combination with buproprion has been found to decrease weight gain compared to mono therapy with buproprion alone.

More recently, Stadler et al, in the *European Journal of Endocrinology* (2014) examined the metabolic changes, including insulin resistance and increase in the orexigenic hormone neuropeptide Y (NPY) in a group of long-term smokers after three months of smoking cessation. These researchers suggest that treating insulin resistance with a

Paul Cézanne, *Pyramid of Skulls,* circa 1901. Although weight gain may be an issue when someone stops smoking, no physician would suggest continuing to smoke.

Private collection, Wikimedia Commons, Public Domain.

Edvard Munch, *Self-Burning Cigarette,* 1895. National Gallery, Oslo, Norway.

Wikipedia Commons/Public Domain.

medication used to treat type 2 diabetes, metformin, could be beneficial in avoiding weight gain after quitting smoking. Other methods, including exercise programs, cognitive behavioral therapy, and very low calorie diets have all had varying degrees of effectiveness in preventing or reducing post cessation weight gain.

Bottom line: The relationship between weight and smoking is a complex one, and weight gain after smoking cessation is not completely understood. For the majority of those predisposed, it may be an inevitable occurrence, though there is considerable variation in how much weight gain occurs; there are some pharmacological combined treatments that may lessen the amount. Concerns about potential weight gain should never be an excuse to continue those "smoky days." The benefits of smoking cessation far outweigh any risks from subsequent weight gain. ■

CANCER RISK AND WEIGHT: OUR BODY AND "PATHOLOGIES OF SPACE"

What do we know about the relationship between excessive weight and cancer?

60—Posted March 21, 2015

Dr. George Papanicolaou, a Cornell Medical School physician years ago, devised the PAP test to detect cervical cancer.

Used with permission/istockphoto.com. Credit: KenWiedemann.

IN HER BOOK *Illness as Metaphor*, Susan Sontag writes, "Metaphorically, cancer is not so much a disease of time as a disease or pathology of space. Its principal metaphors refer to topography—cancer 'spreads' or 'proliferates' or is 'diffused'..." Ironically, body fatness (i.e., obesity or even overweight) can be seen as a "pathology of space" whereby fat 'spreads,' 'proliferates' or is 'diffused' throughout the body as well. Is there more than a metaphorical relationship between increased weight and cancer? There are many studies to suggest there is a strong association, and this potentially translates into a major public health concern.

With more than two-thirds of the adult population in the U.S. considered clinically overweight or obese, many researchers suggest it is imperative that the relationship between obesity and cancer be elucidated. For example, as prevalence rates for cigarette smoking continue to fall among some populations, obesity-related cancers "may become the largest attributable cause of cancer in women," according to Renehan et al, writing in a 2010 issue of the *International Journal of Cancer.* More recently, Booth et al, in the journal *Hormone Molecular Biology and Clinical Investigation* (2015) noted that it is now estimated that at least 20% (and "this may be an underestimation") of all cancers worldwide are caused by excess weight gain. Of course, though some researchers like Renehan and colleagues, use the word "cause" because of the time frame involved, consistency of findings, and plausibility of the association, actual causation is difficult to prove. The World Cancer Research Fund and the American Institute for Cancer Research note, instead, that body fatness is an "established and important *risk factor* for many cancers."

The mechanisms that underlie the association between greater body fatness and a higher risk of cancer are not completely understood. They seem to involve hormones such as insulin and insulin growth factors (IGF-1 and IGF-2) that lead to cancer promoting effects such as cell migration, invasion, and metastatic spread; the sex steroid hormones (e.g. estrogen, progesterone, testosterone); and even hormones produced by adipose tissue, a highly active endocrine organ itself that secretes many hormones including leptin that can have carcinogenic activity and adiponectin that can reduce carcinogenic activity. In general, chronic low-grade inflammation leads to increases in "pro-inflammatory" cytokines such as tumor necrosis factor-alpha and interleukin 6, that, in turn, stimulate the production of C-reactive protein (CRP), a systemic marker of inflammation. Essentially, the theory is that dysfunctional adipose (fat) tissue creates a microenvironment that is conducive for tumor development. The heterogeneity in the effects with different cancers and different patient subgroups, though, suggests that different mechanisms are involved.

Used with permission, istockphoto.com. Credit: LawrenceLong.

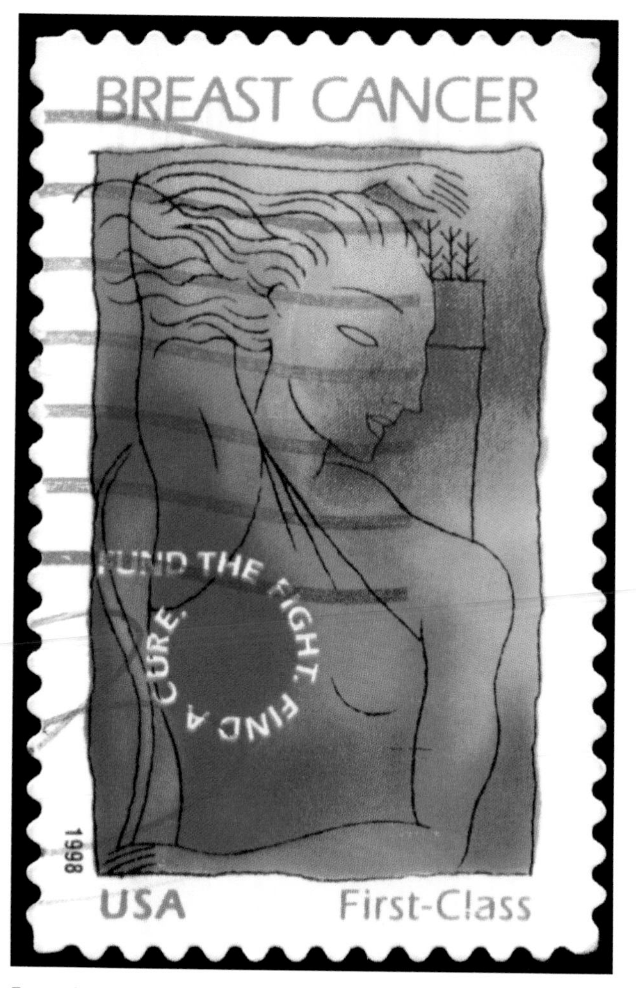

Excessive weight is a risk factor for many cancers, including postmenopausal breast cancer.

Used with permission, istockphoto.com. Credit: traveler1116.

Nimptsch and Pischon, though, in the journal *Hormone Molecular Biology and Clinical Investigation* (2015) explain that these "pathways are not exclusive, but rather interrelated with each other in a complex and not full elucidated manner." These researchers note that there is "convincing" epidemiological evidence that body fatness is associated with a higher risk of six types of cancer: colorectal, postmenopausal breast cancer, endometrial cancer, esophogeal adenonocarcinoma, renal cell carcinoma, and pancreatic cancer. And there is "growing evidence" that body fatness contributes to the development of ovarian cancer and advanced prostate cancer. For some cancers, like colorectal, abdominal (i.e., visceral) obesity, specifically, is an independent risk factor, though it is not yet certain whether it is an independent risk factor for other types of cancer. Bhaskaran et al (2014, *Lancet*) explored the relationship between body mass index (BMI) and the risk of cancer in a population-based cohort study of 5.24 million adults in the UK. These researchers found that BMI was associated with 22 different cancers. Each 5 kg/m^2 increase in BMI (over a normal weight of a BMI of less than 25 kg/m^2) was associated with cancers of the uterus, gallbladder, kidney, cervix, thyroid, and leukemia. "Assuming causality, 41% of uterine and 10% or more of gallbladder, kidney, liver, and colon cancers could be attributable to weight."

Most studies use BMI as their measurement of body fatness. James et al (*European Journal of Cancer*, 2015), though, cautions that BMI is only a "surrogate marker of body composition" and not a reliable measure of body fat because it does not take into account differences in the proportion of lean mass and fat. Furthermore, body composition is impossible to measure directly and accurately in a clinical setting, and it varies with sex, ethnicity, and age. Because fat in the abdominal area is more endocrinologically active, waist-to-hip ratio or waist circumference should be measured as well as BMI, but in many studies these measurements are not taken.

Lee et al (*Clinical Endocrinology*, 2014) note that "cumulative epidemiological evidence" would suggest that overweight or obese subjects are not just at increased risk of developing certain cancers: in those who have developed cancer, obese patients tend to have worse prognoses and are more apt to suffer recurrences. Diagnoses can often be missed or delayed in these patients, and there may be more surgical and radiotherapy complications. Back in 2003, in her classic study (published in *The New England Journal of Medicine)* of more than 900,000 U.S. adults, with over 57,000 deaths from cancer in those initially free from cancer (with 16 years of follow-up), Calle et al noted that those with BMIs of 40kg/m^2 or more had death rates from all cancers that were 52% higher in men and 62% higher in women than those of normal weight. Ungefroren et al (2015, *Hormone Molecular Biology and Clinical Investigation*) found that increased levels of insulin seen with obesity can interfere with the therapeutic effects of chemotherapy; furthermore, obese patients may not even receive the correct dosage of medication (i.e., may be under-dosed.)

Bottom line: There are still many unanswered questions regarding the connection between excessive weight and cancer. For example, we don't know the cumulative effects of excess body weight over several decades (including overweight and obesity beginning in childhood) as well as interactions with other risk factors. Nor do we understand all the mechanisms involved in sex differences and differences across ethnicities. And we don't know conclusively whether effective interventions to reduce BMI (e.g. such as with bariatric surgery) will have a protective effect from overall cancer risk. Nevertheless, body fatness is, like Sontag's description of cancer, a "pathology of space" and until proven otherwise, is a major risk factor for many forms of cancer.

Used with permission, istockphoto.com. Credit: ecliff6.

Note: See above for the stamp with an image of a Japanese Sumo wrestler. There is an actual pathological complex process called SUMOylation that seems to promote the development of certain cancers, though the mechanism is not completely understood. It is suggested that the protein SUMO itself may be a potential therapeutic target to treat cancer eventually. (Bettermann et al, *Cancer Letters*, 2012.) ■

16th century portrait (unknown painter), of English poet Edmund Spenser, author of *The Fairie Queene* and the phrase 'loathsome gluttony.'

Wikipedia Commons/Public Domain.

THE SELF "LOATHSOME GLUTTONY" OF BINGE-EATING DISORDER

When eating to live turns to out of control living to eat

61 — Posted April 21, 2015

THE GLUTTON, according to the *Oxford English Dictionary*, from the Latin word "to gulp down or swallow," is someone "who eats to excess or takes pleasure in immoderate eating." John Milton, in his *Paradise Regain'd* (1671) wrote of "sumptuous gluttonies, and glorious feasts." But it is Edmund Spenser's description of "loathsome Gluttony," in his *The Faerie Queene* (1590), or Shakespeare's "Gluttonlike she feeds, yet never filleth" from his poem *Venus and Adonis*, that best captures certain aspects of those who have binge-eating disorder (BED). For those patients, there is no pleasure in their immoderate eating and nothing poetic about their distress.

For years, overeating was not studied systematically. It was not until 1959 that one of the early researchers on obesity, psychiatrist Albert J. Stunkard, first described binge eating as a syndrome where "large amounts of food are consumed in an orgiastic manner at irregular intervals." It would take until 2013, though, with the publication of our *Diagnostic and Statistical Manual (DSM)-5*, that binge-eating disorder became an established psychiatric diagnosis.

Occhiogrosso (2008) in an essay in the edited volume *Food for Thought*, explored the longstanding and extraordinarily complex relationship that psychiatry has had with overeating. Occhiogrosso surveyed issues of the *American Journal of Psychiatry* since its inception in 1844 and found that the early psychiatrists—the so-called asylum superintendents—were more concerned with providing enough healthy food for their patients and dealing with those who refused to eat. In the mid-19th century, psychiatrists regarded "temperance in diet" as an important element for good health and even "went so far as to link habitual overindulgence in food and drink to the moral sin of gluttony and even the development of criminality." By the early 20th century, psychiatrists began to describe conditions (e.g. brain tumors; schizophrenia, atypical depression) that led to overeating. This was also the era of Freud's influence, and Occhiogrosso describes some analysts as portraying overeating in adults as reflective of "a flawed, inadequate personality structure" with "excessive orality." The first eating disorder to be included in psychiatric nomenclature was anorexia nervosa in the first edition of the *DSM* in the 1950s, and it would take another thirty years for bulimia nervosa, which included bingeing but also purging and preoccupation with weight and shape, to enter the official diagnostic compendium. Later editions of the *DSM* prior to our current edition included the wastebasket term "eating disorder not otherwise specified" for those who did not fit specific criteria.

How is binge-eating disorder diagnosed now? The diagnostic criteria are somewhat arbitrary (including how frequently a binge has to occur for the diagnosis to be made), but include: "eating within a discrete period of time (e.g within any two-hour period) an amount of food that is definitely larger than what most people would eat in a similar period of time under similar circumstances"; a "sense of lack of control over eating during the episode (e.g. eating very rapidly and when not hungry; eating well beyond

Georg Emanuel Opiz, *The Drunkard*, 1804. Many people who binge eat and drink become obese.
WikimediaCommons/Public Domain. Source: Beurret & Bailly.

fullness; eating alone because of feeling embarrassed)"; and "marked distress" (e.g. feeling disgusted, depressed) regarding the binge afterward. Binges typically consist of foods high in fat, salt, and sugar. People typically do not binge on vegetables. Those with binge-eating disorder can be of normal weight, overweight, or obese. Significantly, what distinguishes binge-eating disorder from bulimia nervosa is that those with bulimia engage in "compensatory" activities to rid themselves of the extra calories of their binges in order to avoid weight gain, such as purging (vomiting), excessive exercising, or use of laxatives. Those with bulimia also can have periods of restricting their food intake, including so-called "subjective binges" (i.e. what is to them excessive caloric intake but what would normally not be considered excessive.) In other words, according to Heaner and Walsh, writing in the journal *Appetite* (2013), they have a "dysregulation—in both directions—of the amount of food consumed." These researchers note that the disorders are "defined by disturbances in eating behavior for which there are currently no established etiologies" even though genetic, environmental, biological, and psychological mechanisms have been proposed.

Georg Emanuel Opiz, *The Glutton,* 1804. The glutton can eat with others; those with binge-eating disorder often overeat alone during their binges because they are embarrassed about the quantity of food they are consuming.

Wikimedia Commons/Public Domain. Source: Beurret & Bailly.

Amianto et al (2015, in the journal *BMC Psychiatry*) reviewed 71 studies of binge-eating disorder and concluded "longer and more structured follow-up studies" are needed. Many studies had high drop-out rates. These researchers found that the disorder often begins in the late teens or early 20s and is often co-morbid with other psychiatric disorders (e.g. mood disorders, substance abuse). They found a lifetime prevalence of 1.4 percent in the general population but a considerably higher rate in the obese population, with no marked gender differences, but also noted there is question on the "clinical stability" of the disorder.

Though binge-eating disorder is now part of our psychiatric nomenclature, not all psychiatrists agree it should be. Dr. Allen Frances, who had been the chairman for DSM IV, in his excellent book, *Saving Normal,* describes the "diagnostic inflation" that has occurred over the years such that we tend to medicalize conditions that he believes are part and parcel of the normal human condition. Frances describes how "gluttony has become mental illness" and notes that changes in public policy and not "phony psychiatric" labeling will be required to stem the tide of our obesity epidemic.

Though there seems enough evidence to support the diagnosis of BED, I concur with Frances that diagnoses can lead pharmaceutical companies to engage in major (and somewhat offensive) campaigns to market their products. This has recently occurred with the medication lisdexamphetamine (Vyvanse), initially released in 2007 for the treatment of attention deficit disorder with hyperactivity, but only recently approved in January 2015 for binge-eating disorder. In the past several months, several psychiatric publications have included a cover advertising attachment for Vyvanse that targets this newly diagnosed patient population. Treatment for BED, with the goal of limiting or eliminating binges and normalizing eating patterns, has often involved a multidisciplinary approach that has included psychological interventions such as cognitive behavioral therapy (CBT), and pharmacological interventions such as antidepressant medications (e.g. selective serotonin reuptake inhibitors (SSRIs) and even anti-epileptic medications such as topiramate, all with limited success in many cases.

Lisdexamphetamine is a Schedule II controlled substance, with considerable potential for abuse and the risk of both psychological and physical dependence. Other side effects include feeling jittery, insomnia, decreased appetite, dry mouth, increased heart rate, and constipation. It is not clear how long a person should remain on the medication, and it is likely that once discontinued, the binges will return. Clinical trials have been short-term (11 weeks) but have indicated that this medication (in doses of 50 mg or 70 mg) is statistically more effective than placebo in reducing frequency of binges and even resulting in some weight loss. For those interested in more details of the trials (that included a population that was mostly white, female and overweight or obese), see the 2015 report in *JAMA Psychiatry* by McElroy and her colleagues that recommended "further assessment of lisdexamfetamine as a treatment option."

Harry Herman Salomon, *Sir William Osler.*
Wikimedia Commons/Wellcome Trust Charitable Collection.

Bottom line: Overeating and specifically binge eating can cause serious psychological distress, lower quality of life, and weight gain. The great physician Sir William Osler (from *The Quotable Osler,* 2003) once wrote "The glutton digs his own grave with his teeth." I am reminded, though, of Osler's other words, "One of the first duties of the physician is to educate the masses not to take medicines." ■

Jacob van Loo, *Melancholy,* 17th century.

Wikimedia Commons/Public Domain.

THE MELANCHOLY OF ANATOMY: EXCESSIVE WEIGHT AND DEPRESSION

The complex relationship between weight and depressive disorders

62—Posted May 21, 2015

THE OXFORD SCHOLAR Robert Burton published his 2000-page treatise *The Anatomy of Melancholy* in the early 17th century. The book has been described as encyclopedic—an extraordinary combination of self-help book and medical textbook. Burton worked and reworked his tome, with multiple editions throughout his life, as a therapeutic means of dealing with his own melancholia. He defined "melancholia" as a mind anguished by "fear and sorrow." "Melancholy," from the Greek words for "black bile" refers to the ancient theory of Hippocrates whereby an imbalance of one of the four humors (black bile, yellow bile, phlegm, and blood) was believed to result in disease. For Burton, melancholia was a disease that affected mind, body, and soul and had many causes, including "bad diet," either in "substance" or "quantity."

Fast forward to our current psychiatric nomenclature, the 2013 edition DSM-5. Depression, of course, can be a transient symptom but the "depressive disorders" are defined by the "presence of a sad, irritable or empty mood, accompanied by somatic and cognitive changes that significantly affect" a person's functioning. There are several categories differentiated by "duration, timing, and presumed etiology." We now use the specified term "with melancholic features" to include particularly severe symptoms: "profound despondency and despair," marked agitation or psychomotor retardation, feeling worse in the morning, early morning awakening, excessive guilt, and loss of appetite with weight loss. Ironically, depression, though, can also be associated with weight gain. This is seen in so-called "atypical depressions," with the symptom cluster that includes "mood reactivity" (i.e. ability to be cheered up at least temporarily when presented with positive events), significant weight gain or increase in appetite, increased sleep (i.e. "hypersomnia"), "leaden paralysis," (e.g. heavy feeling in arms and legs) and particular sensitivity to rejection that affects someone socially and occupationally.

Does a depressive disorder lead to weight gain or does weight gain lead to a depressive disorder? Studies over the years have been confusing, inconsistent, and even contradictory. Psychiatrist Albert Stunkard, one of the pioneers in obesity research, and his colleagues (1998, *International Journal of Obesity*) noted that it should not be surprising that those who are weight-challenged would have psychological difficulties, including depression, because of the prejudice and overt discrimination to which these people are often subjected. But early researchers, including Stunkard, could not find "psychological characteristics" or a specific "distinctive personality" in those who were obese that could "consistently distinguish them" from those who were not. What these researchers found, though, is that those with excessive weight *who sought treatment* were more likely to suffer from depression and/ or anxiety. More recent research (Preiss et al, *Obesity Reviews,* 2013) has focused on risk factors associated with co-morbid obesity and depression and possible causal relationships and found more consistent associations between them in their systematic review of 46 studies. Preiss et al found, however, considerable differences in study methodologies, population characteristics, means of defining and even measuring depression, and even inconsistent reporting of results across the studies. Key factors associated with this relationship included the severity of obesity, particularly when a person's body mass index (BMI) is above 40 kg/m^2 (Class III obesity); socioeconomic status; body image, physical health, disordered eating (e.g. binge eating), and experience of stigma. For example, those in a higher socioeconomic class who are obese experience significant prejudice and discrimination that may lead to developing depression. Further, body image dissatisfaction may be an important risk factor and important target of treatment interventions.

Luppino et al (*Archives of General Psychiatry*, 2010) performed a systematic review and the first meta-analysis of longitudinal studies of 15 studies, including over 55,000 subjects. These researchers found "bidirectional associations" between obesity and depression: obese persons had a 55% increased risk of developing depression over time, whereas depressed persons had a 58% risk of becoming obese, with the association between obesity and depression stronger than overweight and depression, indicating a so-called "dose-response" association. Their longitudinal meta-analysis confirms this reciprocal relationship for both men and women, with follow-up as long as 28 years in one study. In other words, obesity increases the risk of depression and prior depression increasing the likelihood of obesity. (Lopresti et al, *Progress in Neuro-Psychopharmacology & Biological Psychiatry*, 2013.) Rather than thinking of the two conditions as co-morbid, researchers like Mansur and his colleagues (*Neuroscience and Biobehavioral Reviews*, 2015) think of a "bidirectional convergent relationship."Allison and his colleagues, though, caution (*American Journal*

Interior of Christ Church Cathedral, toward the organ, Oxford, England. Robert Burton was a student at Christ Church and was buried in its Cathedral.

Photo by David Iliff, licensed by Creative Commons SA 3.0, Wikipedia Commons.

Domenico Fetti, *Melancholy,* circa 1620. Louvre Museum, Paris.
Wikimedia Commons/Public Domain.

of Preventive Medicine, 2009), "Obesity and depression clearly co-exist, but the available data do not unequivocally demonstrate a causal relationship between the two."

What researchers have also found is that obesity can negatively impact treatment outcomes in mood disorders and those with depression often do less favorably with weight loss interventions: they tend to lose less weight and have more difficulty with long-term maintenance of any weight lost. The fact that depression causes an increase in weight may be due to neuroendocrine disturbances, (e.g. activation of the hypothalamus pituitary adrenal—HPA— axis and increased cortisol production), adoption of unhealthy lifestyle (e.g. lack of sufficient exercise), and use of antidepressant medications (particularly paroxetine, mirtazapine, and amitriptyline). Because weight gain may be a late consequence of depression, weight should be monitored in those with depressive disorders; furthermore, mood should be monitored with overweight or obese patients. (Luppino et al 2010)

Rossetti (*Frontiers in Psychology*, 2014) and her colleagues have suggested that the hormone leptin produced by adipose tissue "may represent a biological substrate underlying the pathogenesis of both obesity and depression." There is evidence that "impaired leptin signaling cascades" may be the biological mechanism that links obesity and depression, particularly when obesity is paired with compulsive overeating. Some researchers even describe a "metabolic mood disorder" (e.g. a predominantly depressive illness with an overrepresentation of atypical features, anxiety, and a chronic course.) (Mansur et al, 2015)

Bottom line: The relationship between obesity and depression is complex. Both disorders are heterogeneous with "overlapping pathologies" (Rossetti et al, 2014) and significant contributions from both genetic and environmental factors. Says Mansur et al (2015) "Considering the high impact of obesity and mood disorders in disability and morbidity, the co-occurrence of these conditions is incredibly relevant from a public health perspective." Clearly, there is need for further research to determine all the mechanisms involved especially because of the possible cumulative public health burden from both. ■

A POINT OF REFERENCE: WEIGHT AND THE CONCEPT OF SET POINT

Is the set point just another point of no return for weight-challenged people?

63—Posted June 22, 2015

Thomas Rowlandson, *A Little Tighter*, 18 May 1791.
Hand-coloured etching, Royal Collection Trust, London.

Wikipedia Commons, used with permission, Royal Collection Trust/Copyright: Her Majesty Queen Elizabeth II 2018.

In his less well-known poem of an imaginary childhood place, *Locksley Hall* (1842), Alfred, Lord Tennyson, the famous Poet Laureate of England, writes "Science moves, but slowly, slowly, creeping on from point to point." There are many points in science—including points of divergence and convergence, points of reflection and refraction, and point mutations among others, but the point most closely associated with weight is the set point. Is the set point a viable concept or merely one of Tennyson's "fairy tales of science"?" Most people have heard of this concept but few may really understand what researchers have in mind by it.

The concept of a set point for weight, i.e., an internal physiologically regulated system, though, was defined in series of papers by Dr. Richard E. Keesey and his colleagues in the 1970s and 1980s. Originally taken from an engineering model, it was seen as a homeostatic feedback control system (Mrosovsky and Powley, *Behavioral Biology*, 1977) and as analogous to a set point for body temperature or even blood pressure—though there is considerably more variability for weight among people than for blood pressure or body temperature. It grew out of the observation that remarkably our weight remains within a fairly constant range despite major fluctuations in our activity levels as well as in the varieties and quantities of food our bodies process both in the day-to-day short-term and long-term. For example, Jules Hirsch, (2003, *Dana Foundation lecture*), one of the early pioneers in obesity research at Rockefeller University, once observed that during our lifetime, our bodies process about 70 million calories or approximately 14 tons of food.

Speculation about a control mechanism that regulated the amount of fat in the body, though, began to occur in the 1950s. For example, GC Kennedy (*Proceedings of the Royal Society of London, Biological Sciences*, 1953), working with rats, suggested that fat itself might send a signal to the brain to regulate the amount of fat in our bodies. It would not be until the 1970s, though, that leptin, the hormone produced by adipose tissue, was isolated in the Rockefeller University labs of Jeffrey Friedman and "provided strong molecular evidence for such a feedback system." (Speakman et al, 2011, *Disease Models and Mechanisms)*

Other evidence for a set point grew out of human data that when people gain or lose weight (i.e. "the system is perturbed"), the body seems "to defend" the original weight. That is why after a weight loss, there is a tendency for many people to regain the lost weight. (Speakman et al, 2011) There is, though, an "asymmetry" in this process—namely that the body seems to defend against weight loss far more effectively than against weight gain, probably as an evolutionary advantage when food cycles were more variable. No one, though, has actually located the elusive set point or whether it is even one area, though there has been past simplistic speculation that it is in the hypothalamus.

Maclean and colleagues (2004, 2006), writing in the *American Journal Physiology: Regulatory, Integrated and Comparative Physiology*, studied obesity-prone rats and noted metabolic factors in rats are easier to study because there are not contaminating human motivational factors

Helen Mary Elizabeth Allingham, portrait of Alfred, Lord Tennyson, circa 1926. British Poet Laureate, and author of *Locksley Hall.*

Wikimedia Commons/Public Domain.

Thomas Rowlandson, *An Epicure,* dated 1788. Hand-coloured etching. Royal Collection Trust, London.

Wikipedia Commons, used with permission, Royal Collection Trust/Copyright: Her Majesty Queen Elizabeth II 2018.

(e.g. peer pressure to be thin; wishes for an ideal physique). These researchers found there was a "metabolic propensity" for their rats to regain weight after a period of caloric restriction and subsequent weight loss, both by an increased appetite and a decrease in resting metabolic rate. But MacLean et al noted that studies with humans can be inconsistent and found it "reassuring" that humans are able to counteract any metabolic tendency to gain weight by changing their behavior (e.g. consciously exercising, eating less, even taking medications for weight loss.) Levin (2004, writing in the same journal) summarized the controversy by noting that the regulation of fat accumulation in humans is very complicated and determined by "genetic, gender, perinatal, developmental, dietary, environmental, neural, and psychosocial factors."

William Bennett (*New England Journal of Medicine, Editorial,* 1995) noted that the set point seems to respond slowly "rather than rapidly to deviations from the internal ideal" and involves both eating and physical activity. Bennett clarified that although these two behaviors are thought to be "largely voluntary…there can be considerable ambiguity about the degree of volition" involved in either activity and "such behavior assumes a certain biologic inevitability."

The set point model has limitations. It does not explain why our set point is somewhat adjustable—i.e., why most people do gain some weight throughout their lives, particularly under certain environmental conditions, such as changes in marital status, age, social class, or even if they are "couch potatoes." In other words, though the set point theory is "rooted in physiology, genetics, and molecular biology," and "postulates an active feedback mechanism linking adipose tissue (stored energy) to intake and expenditure," it does not sufficiently explain the contribution of the so-called "obesogenic environment" and social issues that contribute to weight gain. (Speakman et al, 2011)

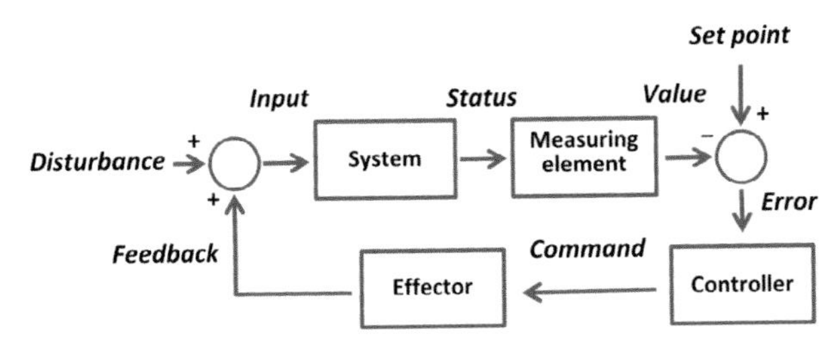

The set point concept came from an engineering model of feedback control.

Brews ohare, Licensed under Creative Commons CC 1.0 Universal.

There is another model, the *settling point model*, that proposes a passive (rather than active regulation) feedback system in which weight can "drift" (Farias et al, *Metabolic Syndrome and Related Disorders*, 2011.) This model does focus on environmental and social issues, but it does not, however, sufficiently focus on the more genetic and biological issues and hence the two models create an "artificial" divide between genetics and the environment. (Speakman et al, 2011) Another proposed model, the *general intake model*, emphasizes that there are "compensated factors" (e.g. primarily physiological) and "uncompensated factors" (primarily environmental) that impact weight regulation and can vary considerably from person to person, but it does not assume there is a set point. A fourth model is the *dual intervention point model*—a "more realistic version of the set point model" (Speakman et al, 2011) that can incorporate both genetic and environmental contributions in which there are upper and lower limits where "physiological regulation of weight and/or fat" becomes active.

Bottom line: In *Locksley Hall*, Tennyson wrote "Knowledge comes, but wisdom lingers." Whether we find the anatomical place or places for a set point remains to be seen. For many researchers, the concept of set point is too simplistic to explain the complexities and daunting science of weight control. For some people, though, a functioning set point becomes a kind of an anchoring point of reference for their weight; for those less fortunate, whose weight continues to climb and whose set point seems dysfunctional, it may be a point of no return.

Early theories suggested that the so-called *set point* for weight control was located in the hypothalamus. It is more likely that if it exists, it is not localized to one area of the brain.

Used with permission/istockphoto.com. Commons CC 1.0 Universal. Credit: acaimoron.

Note: I could not find many papers in very recent years on set point but the term continues to appear in the literature. For one paper, see Ravussin et al, *Molecular Metabolism* (2014), on rats given leptin exogenously and the relationship to set point. ■

THE QUICKSAND OF SELF-DECEPTION: THE NOCEBO EFFECT

Giving patients too much information about potential harm may be harmful itself

64 — Posted July 23, 2015

ONE OF THE FIRST TENETS OF MEDICINE "Do no harm," originates from Hippocrates's treatise *Epidemics* (section XI)—"as to diseases, make a habit of two things: to help or at least do no harm." It is also the title that neurosurgeon Henry Marsh chose for his extraordinary recently published memoir. There are times, though, when physicians can, perhaps inadvertently, cause more harm than good when they reveal all the potential side effects of a medication or therapeutic intervention, including those related to treatments for excessive weight and weight-related disorders.

There are many factors involved in the psychology of administering a medication or prescribing a treatment, including the quality of the physician-patient relationship itself, what doctors and patients believe about the particular treatment and how it works, what they each expect to happen, or specifically with a medication, the medication's physical attributes such as its color, shape, smell, method of administration, or even its name.

From the Latin "nocebo,"—"I will harm," the nocebo reaction (also called the nocebo "effect," "response," or "phenomenon"), was first described as the "contrary effect" to the placebo response by Walter P. Kennedy (1961, *Medical World.*) Kennedy described what he called the "unoriginal observation" that many of the subjects in his clinical trials exhibited non-specific adverse reactions. Kennedy noted that every physician "has met the nocebo reaction, even if he has not labelled it as such...." It has been described as placebo's "mirror image" (Cohen, *Bioethics*, 2014) or its "evil twin" (Schedlowski et al, *Pharmacological Reviews*, 2015.)

The nocebo reaction is a form of self-deception that involves the self-fulfilling prophecy. In other words, "People perceive what they expect to perceive." (Wells and Kaptchuk, *American Journal of Bioethics*, 2012) Many writers, as diverse as the Ancient Greek orator Demosthenes (3rd *Olynthiac*) and the American statesman and writer Benjamin Franklin (*Poor Richard's Almanack*), have recognized how easily we are able to deceive ourselves. Robert K. Merton (The *Antioch Review*, 1948), described the "self-fulfilling prophecy" as involving "fears that are translated into reality" and noted that its "specious reality.... perpetuates a reign of error." In fact, nocebo reactions can have far-reaching consequences: they can lead to "epidemic waves" of reports of adverse events and drugs being removed prematurely from market as well as a patient's psychological distress and even noncompliance with treatment. (Wells and Kaptchuk, 2012) Even reported complications observed when switching from a brand name medication to the generic can be related to nocebo responses. (Schedlowski et al, 2015)

The nocebo concept, though, suffers from a lack of a clear and precise standardized research definition. For Barsky et al (*JAMA*, 2002), nocebo responses describe negative treatment effects that are not directly attributable to a drug's pharmacokinetics. These nonspecific side effects are often "idiosyncratic and not dose-dependent." They are what Barsky et al call "misattributions" by the patient.

The scholars among you will know that my title derives from Shakespeare's play *Henry VI*. Portrait of Shakespeare, possibly by John Taylor, 1610, (known as the "Chandros Portrait," after a previous owner).

Peter Paul Rubens, *Hippocrates*, engraving, 1638. National Library of Medicine, Bethesda, Maryland.

Schedlowski et al (2015) define two variants: new or worsening of symptoms associated with a pharmacologically active drug, but also reduced efficacy due to negative expectations or prior experiences. Some researchers, as described by Symon et al (*Journal of Advanced Nursing*, 2015) include negative side effects that can also occur when a patient is given an actual placebo, i.e. a totally pharmacologically inert or harmless substance or therapy that is not expected to have either beneficial or harmful effects. Others define a nocebo effect generally when a treatment "expected to have beneficial effect has a detrimental outcome." (Dodd et al, *Journal of Clinical Psychiatry*, 2015) Jakovljevic (*European Neuropsychopharmcology*, 2014) notes that nocebo responses are "multifactorial, multidimensional, and etiologically complex," and researchers cannot even agree on whether these responses are either "the effect of" or "response to."

The prevalence of nocebo effects varies, depending on the intervention being prescribed or the condition being treated. Symon et al (2015) found a general prevalence internationally of 3 to 27% in a review of 48 empirical studies. More specifically, in a study of treatment for major depression, Dodd et al (2015) found that nocebo effects may explain almost 64% of treatment-emergent adverse effects and almost 5% of discontinuation in a large placebo-treated patient sample of over 2400 subjects.

Meynen et al *(American Journal of Bioethics*, 2012) maintain that the nocebo effect "shows us that mere information about potential harm is likely to be harmful itself." Says Meynen et al, "The physician's words not only describe reality, but they modify and create reality." These researchers, though, emphasize that, with the advent of the internet, the "treating physician is no longer the gatekeeper of medical information" with the result that patients can (and do) retrieve information about their treatments from many sources.

Little is known about the predictors of nocebo responses. (Schedlowski et al, 2015) Barsky et al (2002) report that those patients who expect to have distressing

Too much information can overwhelm a patient and create the *nocebo effect.* British publisher, Samuel William Fores, *Irish Stew: Not All Hot—Rather Too Much,* circa 1840, Lithograph. Metropolitan Museum of Art, NYC.

The Elisha Whittelsey Collection, The Elisha Whittelsey Fund, 2014, Public Domain.

side effects—i.e. negative expectation— are more apt to develop them. These responses occur more often in women (unlike placebo reactions that apparently occur equally in men and women) (Dodd et al, 2015). Psychological symptoms, such as depression and anxiety may make certain patients more prone to nocebo responses, as well as those with a type A personality (e.g. those competitive, with a sense of urgency and a tendency to hostility) or a pessimistic nature.

The mechanisms for nocebo reactions also remain poorly understood, but may involve conditioned learning and expectancy. For example, patients previously treated with chemotherapy can develop a conditioned nausea response just by seeing the nurse or entering the treatment room even before receiving another treatment. Allergic reactions have occurred in patients expecting to receive an allergen but receiving only saline. Further, nocebo reactions have been associated with dopamine, the endogenous opioid system, and cholecystokinin, and studies involving neuroimaging have shown that the "mere expectation of symptoms can activate brain structures." (Schedlowski et al, 2015)

Several sources, including Cohen, 2014, focus on medical ethical issues involved with the nocebo reaction, primarily the conflict between respecting a patient's autonomy and the principle of non-maleficence, i.e., "do no harm." (Wells and Kaptchuk, 2012) In other words, if disclosing detailed potential side effects or complications of a treatment can cause harm or adverse effects in a patient, should a physician give full disclosure, if that is even possible? On the other hand, failure to disclose possible effects can be

Joseph Siffred Duplessis, *Benjamin Franklin,* author of *Poor Richard's Almanac*. National Portrait Gallery, circa 1785, Washington, DC.

Wikipedia Commons/Public Domain.

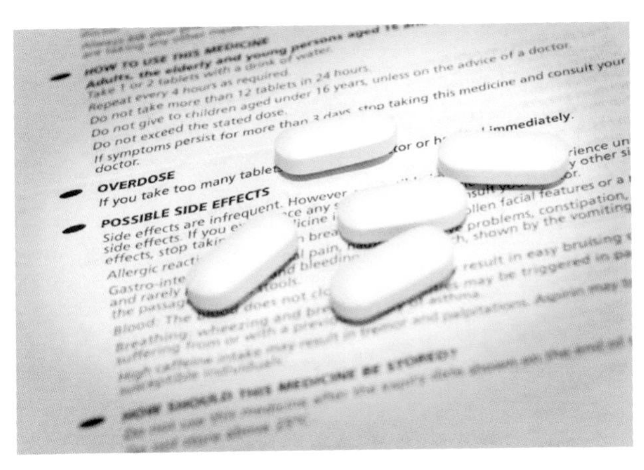

There is sometimes a fine line between giving a patient too much information about side effects and creating a paternal atmosphere where a physician, attempting to protect a patient, withholds information about side effects.

Used with permission/incamerastock/AlamyStockPhoto.

seen as paternalism. And hence there exists what these researchers call an inherent "ethical conundrum."

Bottom line: Factors contributing to the nocebo effect remain "elusive and are inherently difficult to profile and characterize." (Dodd et al, 2015) Patients can become mired in the quicksand of self-deception when burdened with too much clinical information that becomes self-fulfilling. Both physicians and patients should not assume that reported troublesome side effects necessarily stem from the pharmacological effects of a particular medication or intervention, especially when the symptoms are vague or when a patient is particularly anxious. Physicians must be sensitive to creating an optimal balance between respecting a patient's autonomy as evident in offering full disclosure of information and unnecessarily fostering an environment of paternalism, in the interest of protecting a patient from harm, by withholding it.

Note: Scholars of Shakespeare will know that my title derives from *Henry VI, part 3* (Act V, scene 4. 26): "What Clarence but a quicksand of deceit?" ■

SOME PHILOSOPHICAL MUSINGS ON FOOD

Only a philosopher can ask, 'What is the metaphysical coefficient of lemon?'

65—osted August 18, 2015

Louis-Jean Allais, portrait of French essayist Jean Anthelme Brillat-Savarin, who famously said, "Tell me what you eat, and I shall tell you who you are." 18th century, Palace of Versailles, France.

Wikipedia Commons/Public Domain.

"TELL ME WHAT YOU EAT, and I shall tell you who you are." So wrote the late 18th, early 19th century French essayist Jean Anthelme Brillat-Savarin in his classic book *The Physiology of Taste*. Of course, it is not so simple, as David M. Kaplan explains in the introduction to his book *The Philosophy of Food*, (2012), "Philosophers have a long but scattered history of analyzing food…Food is vexing. It is not even clear what it is." So predictably, says Kaplan, "There is no consensus among philosophers about the nature of food." He notes that even our most essential questions about food, such as what we should eat, whether food is safe, or what is considered good food are "difficult questions because they involve philosophic questions about metaphysics, epistemology, ethics, politics, and aesthetics." For example, Kaplan wonders what are the differences between natural and artificial foods, between food and an animal, between food and other things we take into our bodies such as water or medication. Or even how food can change its identity over time as it goes from raw to cooked to spoiled.

Kaplan describes food as nutrition (e.g. objectively required for the body); food as nature (e.g. the more natural the better); food as culture (e.g. with social and cultural meanings and significance, such as categories of good and bad, legal and illegal, ritualistic and symbolic foods); food as a social good (e.g. food distribution as a basic institution of society); food as spirituality (e.g. central to religious traditions); food as desire (e.g. object of hunger and cravings); and food as an aesthetic object (e.g. has taste and appeals to the senses.) To Kaplan, "Food is about life as well as luxury…It is a profoundly moral issue," especially when we consider the basics of not eating other humans and the responsibility to provide food for others, as well as the three food virtues: hospitality (e.g. being a good host); temperance (e.g. moderation in food and drink), and table manners (e.g. all cultures have rules that may involve health, enjoyment and community.)

Other than Hippocrates, whose body of work is replete with references to the importance of a healthy regimen involving a balance between food intake and proper exercise, Plato was one of the ancient Greek philosophers to address the importance of diet and its contribution to disease. Skiadas and Lascaratos (*European Journal of Clinical Nutrition*, 2001) review the many references to diet and even the dire health consequences of obesity throughout Plato's writings. For example, in *The Republic*, Plato writes, "…the first and chief of our needs is the provision of food for existence and life." In *Laws*, he writes, "For there ought to be no other secondary task to hinder the work of supplying the body with its proper exercise and nourishment" and he describes the obese as "an idle beast, fattened by sloth." In *Timaeus*, "…one ought to control all such diseases… by means of dieting rather than irritate a fractious evil by drugging." Skiadas and Lascaratos summarize Plato's contribution by noting that Plato's writings on diet reflected his general theory of moderation that had been a major concept dominating ancient Greek philosophy.

Michel Onfray has written a charming book, *Appetites for Thought: Philosophers and Food*, an *amuse-bouche*, if you will, or rather an *amuse d'esprit*—a book to stimulate our mind's palate. Originally published in the late 1980s, it has just recently (2015) been translated from the French. Onfray, who believes that one's food choice is really "an existential choice," imagines a "banquet of omnivores" where some of the world's greatest philosophers have come to dine.

For example, ancient Greek philosopher Diogenes, (born in the 400s B.C.), as typical of his group of Cynics, is "possessed of a resolute will to say no, to flush out the conformism of customary behavior," says Onfray. There are numerous reports of Diogenes' unconventional behavior, such as urinating, defecating, and even masturbating publicly. The first principle, though, of the Cynics (from the Greek word for "dog") is to eat only simple, pure raw foods. This is reflective of Diogenes' rejection of fire as a symbol of civilization—"limiting your needs to those of nature." One dies as one lives so it is not surprising to learn from ancient historian Plutarch that Diogenes risked his life in the process of eating a raw octopus.

Jules Bastien-Lepage, *Diogenes,* 1873. Diogenes was the quirky ancient Greek philosopher who famously ate a raw octopus. Musée Marmottan Monet, Paris.

Wikipedia Commons/Public Domain.

Painting (unknown artist) of Immanuel Kant, Prussian philosopher known for his *The Critique of Pure Reason* and *Metaphysics of Morals*. 18th century.
Wikipedia Commons/Public Domain.

Edvard Munch, Portrait of German philosopher Friedrich Nietzsche, 1906. Nietzsche reportedly had many digestive ailments and famously asked if there is a "philosophy of nutrition." Thiel Gallery, Stockholm, Sweden.
Wikipedia Commons/Public Domain.

Onfray describes 18th century philosopher Jean-Jacques Rousseau, famous for his treatise on education *Emile* as a "gastronomic self-denier" who developed a "spartan theory" of food whereby eating "is an imperative for survival, not for enjoyment." Rousseau apparently ate food that required the minimum of preparation: milk, bread, and water. Says Rousseau, in his autobiography *Confessions*, "I do not know know...any better fare than a country meal." In his novel *Julie; or The New Heloïse*, he writes, "In general I think one could often find some index of people's character in the choice of foods they prefer."

Eighteenth century Immanuel Kant, famous for his *Critique of Pure Reason*, distinguished between the "superior (and objective) senses" of touch, sight, and hearing from the "inferior (and subjective) senses" of smell and taste. In his *Metaphysics of Morals*, Kant writes, "Brutish excess in the use of food and drink is misuse of the means of nourishment...A man who is drunk is like a mere animal, not to be treated as a human being. When stuffed with food he is in a condition in which he is incapacitated for a time..." According to biographers, Kant suffered from irregular digestion and stomach problems throughout his life and admitted to being a hypochondriac. He apparently had nothing more than weak tea for breakfast and ate only one meal a day, at midday. In *The Conflict of the Faculties*, he writes, "...an impulse to have an evening meal after an adequate and satisfying one at midday can be considered a pathological feeling..."

Nineteenth century philosopher Friedrich Nietzsche, famous for his statement "God is dead" also suffered from digestive issues, among his many ailments. There has, in fact, been considerable speculation on the nature of his illness.

Hippocrates and his Four Humors. Hippocrates wrote extensively about the importance of certain foods for health in his volumes of *Regimen*. His four humors were blood, black bile, yellow bile, and phlegm.

Used with permission, Alamy stock photo.

Galen and Hippocrates, Inagni, Italy.
Wikimedia Commons, Licensed under Creative Commons 3.0. Photo by Nina Aldin Thune.

For a discussion of six possible hypotheses, see Tényi's 2012 paper in the journal *Psychiatria Hungarica*. Nietzsche wrote, in *Ecce Homo*, "I am much more interested in a question on which the 'salvation of humanity' depends far more than on any theologian's credo; the question of nutrition." In *The Gay Science*, he writes, "What is known of the moral effects of different foods? Is there a philosophy of nutrition? (The constant revival of noisy agitation for and against vegetarianism proves that there is no such philosophy.)" Onfray notes that Nietzsche tended to avoid restaurants because they "overfeed" their customers. "Know the size of one's

Reginald Gray, drawing for *The New York Times,* of Jean-Paul Sartre, French philosopher, 1965.
Wikipedia Commons/Public Domain

Attributed to Édouard Manet, an Impressionist lemon, 1880. Sartre famously asked, "What is the metaphysical coefficient of lemon?

Wikimedia Commons/Public Domain.

stomach," writes Nietzsche in *Ecce Homo.* According to Onfray, Nietzsche "never put into practice the dietetics of his theories" and again, in his *Ecce Homo*, Nietzsche writes, "I am one thing, what I write is another matter."

Twentieth century French philosopher Jean-Paul Sartre had a concept of the body that "was above all sick, mutilated, butchered, and unrecognizable," says Onfray, and Sartre had strong likes and dislikes among foods. He is, not surprisingly, the author of *Nausea*. Sartre's lifelong partner Simone de Beauvoir, quotes Sartre as saying, "All food is a symbol." Onfray notes that Sartre accepted only food that had been technically altered or prepared. Apparently, so unlike Diogenes, he disliked the natural and found "only manufactured, artificial products to his liking." De Beauvoir quotes him as saying, "Food must be the result of work performed by men. Bread is like that. I've always thought that bread was a relation with other men." In *Being and Nothingness*, he asks, "What is the metaphysical coefficient of lemon, of water, of oil?" Sartre thought it was for psychoanalysts to explore why someone "gladly eats tomatoes and refuses to eat beans, why he vomits if he is forced to swallow oysters, or raw eggs."

Over the centuries, philosophers never developed a consensus about food and eating but many have had strong opinions about both. For the 21st century, with widespread and pervasive obesity and overweight, perhaps a simple philosophy that we might agree upon is that we should eat to live, rather than live to eat. ■

Albert Edelfelt, *Louis Pasteur,* 1885. Pasteur said, "...everything is clear if this cause is known." Musée d'Orsay, Paris.
Wikipedia Commons/Public Domain.

TOWARD A 'KNOWLEDGE OF CAUSES...AND ALL THINGS POSSIBLE'

Traversing the rugged territory from observed associations to causation.

66—Posted September 17, 2015

"All things are hidden, obscure, and debatable if the cause of the phenomena be unknown, but everything is clear if this cause be known." So said Louis Pasteur in his *The Germ Theory of Disease and its Application to Medicine and Surgery.*

Though discussions of causality date back to the ancient philosophers, it was mid-19th Century German physician Jakob Henle and later his student Robert Koch, though, who developed postulates for assessing causation in acute infectious diseases such as tuberculosis, anthrax, and tetanus. In these cases, the offending agent was found to be present in every case, caused one specific disease, and could be isolated to cause re-infection with re-exposure. (Evans, *The Yale Journal of Biology and Medicine,* 1976) Over the years, though, physicians began to appreciate limitations to the Henle-Koch postulates, particularly when dealing with the complexities of viruses or with chronic syndromes such as obesity.

What is a cause? "In a pragmatic perspective," says epidemiologist Mervyn Susser (*American Journal of Epidemiology,* 1991), "a cause is something that makes a difference." More specifically, in their textbook on epidemiology, Rothman et al (2008, 3rd edition) define it as an event, condition, or characteristic that precedes the onset of a disease and is necessary for its occurrence. In general, though, typically epidemiologists have skirted the notion of what constitutes a cause, says Susser. Instead they focus on determinants, exposures, and risk factors "without facing the treacherous issues of the definition of a cause." For example, diseases can have *predisposing factors* (e.g. age, marital status, working environment); *enabling factors* that facilitate their development (e.g. climate, nutrition, availability of medical care); *precipitating factors* (e.g. exposure to a specific disease); and *reinforcing factors* (e.g. repeated exposure to an infectious agent.) (Porta, *A Dictionary of Epidemiology,* 5th edition, 2008)

Alternatively, epidemiologists categorize causes into "necessary" and/or "sufficient," "single" or "multiple," "direct" or "indirect." Furthermore, Mehta and Allison (*Frontiers in Nutrition,* 2014), in their discussion of the challenges specifically implicit in nutritional methodology and measurement, note, "Evidence of causation exists on a continuum."

Rothman et al speculate that causal criteria "have become popular, possibly because they seem to provide a road map through complicated territory." *Causation,* though, is to be differentiated from association or relationship (i.e., the probability of an occurrence of an event varies with the occurrence of another event), and *correlation* (i.e., the degree to which variables change together) Association, relationship, and correlation, though, are sometimes used interchangeably in epidemiology. (Porta, 2008) Importantly, correlation does not imply causation.

One researcher, considered by some to be the greatest medical statistician of the 20th Century, though trained neither as a physician nor as a statistician, (Doll, *Statistics in Medicine,* 1993), was willing to approach medical causation systematically. British-born Sir Austin Bradford Hill

Portrait of bacteriologist and physician Robert Koch, 1907. Koch received the Nobel Prize for Physiology in 1905. Photo from National Institutes of Health.

Wikipedia Commons/ Public Domain.

Sir Austin Bradford Hill, British epidemiologist and statistician. Hill presented nine "viewpoints" on causation. Portrait photo by Elliott & Fry, National Portrait Gallery, London, quarter-plate glass negative, 1950.

Used with permission/Copyright, National Portrait Gallery, London.

(1897–1991) began his epidemiological career with a focus on occupational medicine and observed convincing associations between environmental toxins in the work place and the subsequent development of disease. He investigated printers, bus drivers, and workers exposed to cotton, arsenic, or nickel. (Schilling, *Statistics in Medicine*, 1982) In the 1940s and 1950s, with his colleague Richard Doll, Hill was one of the first to report a strong link between exposure to smoking and lung cancer. Hill's work with Doll, though, brought sharp criticism from the famous statistician Ronald A. Fisher, known for his original agricultural experiments involving randomization, who took issue with this causal link between cigarette smoking and cancer. In what seems particularly ludicrous by current thinking, Fisher, for example, suggested that perhaps it was lung cancer that causes smoking (i.e., the disease causes mucosal irritation that is relieved by smoking) rather than the inverse. (Doll, *Perspectives in Biology and Medicine*, 2002)

Reportedly, it was in response to Fisher's criticism (Robbins, *Southwest Journal of Pulmonary & Critical Care*, 2012) that Hill wrote what became his most famous and now classic paper *The Environment and Disease: Association or Causation?* (*Proceedings of the Royal Society of Medicine*, 1965.) This year marks the fiftieth anniversary of the publication of that paper.

How did Hill approach the issue of causation? He wrote, "I have no wish, nor the skill, to embark upon a philosophical discussion of the meaning of 'causation.'" Instead, Hill presented nine "viewpoints" (never using the word "criteria") that explored "In what circumstances can we pass from this observed association to a verdict of causation?" Acknowledging that the cause of a disease might be either immediate and direct or remote and indirect, Hill delineated these considerations:

- **Strength:** "first upon my list;" size of the risk; example: death rate from cancer of the lung in cigarette smokers is nine to 10 times the rate in non-smokers, and the rate in heavy smokers is 20 to 30 times as great.
- **Consistency:** has the association been repeatedly observed in different people, places, circumstances, and time (especially when the results are reached both prospectively and retrospectively) and hence less likely to be due to chance?
- **Specificity:** limited to specific workers and to particular sites and types of disease, but diseases may have more than one cause
- **Temporality:** one of the most important viewpoints and particularly relevant when diseases take time to develop: does the disease follow the exposure? Example: does a particular diet lead to a disease or does the early stage of the disease lead to peculiar dietary habits?
- **Biological gradient:** is there a dose-response curve? Example: more cigarettes smoked, more likely cancer
- **Plausibility:** biologically should be consistent with (positively worded) current knowledge but often depends on the knowledge of the day.
- **Coherence:** should not seriously conflict (negatively worded) with generally known facts of the natural history and biology of the disease Example: Is the condition consistent with histopathological findings, etc.?
- **Experiment:** Can the conditions be varied experimentally, especially when evidence is obtained through the gold standard of randomized controlled trials?
- **Analogy:** For example, if one disease (e.g. rubella) or one drug (e.g. thalidomide) can cause birth defects, it is possible that another could do so as well.

John Vanderbank, after Unknown artist, Portrait of Francis Bacon, 1st Viscount St. Alban. 1731. Bacon wrote about "the knowledge of causes."

Hill appreciated that none of his nine "viewpoints" "brought indisputable evidence" for or against cause-and-effect, i.e., neither necessary nor sufficient for establishing causation. For Hill, the "fundamental question" was always, "Is there any other way of explaining the set of facts?"

Over the years, some have criticized Hill and others, such as Susser, have elaborated on his work, by subdividing "coherence" into theoretical, factual, biological, and statistical categories. Nevertheless, Hill's viewpoints are still used today when researchers consider causality. Recent examples include a paper by McCaddon and Miller (*Nutrition Reviews*, 2015) on the association between increased levels of homocysteine and cognitive decline and Frank Hu's comprehensive review on the connection between sugar-sweetened beverages and obesity-related diseases. (*Obesity Reviews*, 2013)

In her book *Illness as Metaphor* (1977), Susan Sontag writes, "The notion that a disease can be explained only by a variety of causes is precisely characteristic of thinking about diseases whose causation is not understood." With obesity, for example, we are dealing with neither one cause nor even one disorder. Says Hebert, Allison, and their colleagues, (*Mayo Clinic Proceedings*, 2013), "Obesity is not a single pathological condition but rather a sign of underlying primary pathological abnormalities...characterizing obesity as a distinct condition suggesting a single, straightforward, and unvarying etiology has biased all phases of obesity-related discourse." Along those lines, researchers and clinicians may need to differentiate between *the* cause and *a* cause. (Doll, 2002) For the most comprehensive, systematic approach to research on obesity, with an emphasis on the importance of conducting randomized controlled trials for gathering evidence, see the 2015 paper, in the *Critical Reviews in Food Science and Nutrition* by Casazza and her colleagues.

How much evidence, though, is enough? Years ago, researcher Douglas Weed (*International Journal of Occupational Medicine and Environmental Health*, 2004) raised the provocative question, "What is the least amount of evidence... about causation—needed to recommend a public health action?" Sometimes researchers, says Weed, employ what is called the "Precautionary Principle"—preventive measures are recommended even when causation is not established scientifically. This approach is disquieting, though, particularly for obesity and overweight, and makes researchers like Casazza et al (2015) understandably concerned that presumptions and myths are treated like scientific truths.

Unfortunately, epidemiologists have yet to agree on criteria for establishing causation (Rothman et al, 2008). It is indeed disheartening, then, to recall Francis Bacon (1627, *New Atlantis*) and appreciate the distance from his words, "The end of our foundation is the knowledge of causes... and the enlarging of the bounds of human Empire, to the effecting of all things possible." ■

THE "FALSE CREATION" OF BODY IMAGE DISTORTIONS

A pilgrimage toward understanding "of its own beauty is the mind diseased"

67 — Posted October 15, 2015

THE SENTIMENT "Beauty is in the eye of the beholder" originates from Plato's *Symposium*, an exposition on love written about 360 B.C., where Socrates speaks of "beholding beauty with the eye of the mind." For those with body image distortions, they see not beauty in themselves but imperfections and overt dissatisfaction with their body or parts thereof. Rather, what captures the mindset of these people is a verse from Lord Byron's early 19th century narrative poem, an autobiographical journey, *Childe Harold's Pilgrimage*, (*Canto IV, stanza CXXII*), "Of its own beauty is the mind diseased/And fevers into false creation." Those with body image disorders see themselves as though through a distorted prism, not unlike a Picasso painting.

German researchers Hewig et al (*Psychosomatic Medicine*, 2008) note that body image has two elements—the *mental image*, i.e., perceived size and shape of one's body and the *emotional aspect* of feelings and beliefs. Furthermore, there can be *perceptual distortions* (i.e., the inability to assess size and shape accurately) and *actual body dissatisfaction* (i.e., actual negative feelings about one's body or parts of it.) Didie et al (*Body Image*, 2010) note that the concept of body image is "multi-dimensional and encompasses perceptions, thoughts, feelings, and behaviors not only about physical appearance" but also about fitness and health.

Henry Pierce Bone, Portrait of Lord Byron, 1837. Byron, who had a congenital club foot, wrote of "false creation" in his narrative poem, *Childe Harold's Pilgrimage*. (From the original by William Edward West.) Enamel on copper.

Photographer: Christies. Wikipedia Commons/Public Domain.

Some researchers have inferred that our body image is "hard-wired" into the brain, particularly because not only those who have had a limb amputation but even those born without limbs can still experience a "phantom limb" syndrome, i.e., sensations including movement, temperature, pain, and touch in the missing limb(s). Furthermore, there is speculation that the right posterior parietal cortex "plays a critical role" in an integrated bilateral body image: lesions there can lead to neglect of the left side of the body. (For a more detailed discussion of neurological disorders of "self-embodiment," see Price, *Consciousness and Cognition*, 2006 and Giummarra et al, *Neuroscience and Biobehavioral Reviews*, 2008.) Price believes that a primitive sense of body image begins *in utero* with spontaneous movements of the fetus and corresponding sensory and proprioceptive (i.e., sense of position of the body) feedback.

Developmentally, though, one's conscious body image continues to evolve over the first ten years of life as sensory modalities such as vision and touch "contribute to both transient and permanent aspects of body image." (Price, 2006) Voelker et al (*Adolescent Health, Medicine, and Therapeutics*, 2015) emphasize that adolescence "represents a critical period for healthy body image development." Our body image can be influenced by comments from peers and family (e.g. teasing, bullying), as well as from content and images presented in a thin-and youth-obsessed culture. Media tend to promote highly unrealistic, ideal images, and lead adolescents (as well as people of all ages) to compare themselves unfavorably with these images. Dissatisfaction with one's body in adolescence has been shown to be the

Pablo Picasso, *Head of a Woman*, 1960. Picasso's painting reflects a distorted face, as may be seen with those with body image disorders.

Metropolitan Museum of Art, NYC. Gift of Mr. and Mrs. Leonard S. Field, 1990. Copyright 2018, Estate of Pablo Picasso, Artists Rights Society, NYC. Used with permission, Artists Rights Society and Art Resource. Digital image: Copyright The Metropolitan Museum of Art/ licensed by Art Resource, NY. Used with permission.

Raphael, Detail from *The School of Athens*, (Plato (on left) and Aristotle pictured) 1509. Plato reportedly was the first to express the sentiment of "beholding beauty with the eye of the mind." Apostolic Palace in the Vatican, Rome.

Wikipedia Commons/Public Domain.

"strongest predictor" of the development of eating disorders and disordered eating. (Voelker et al, 2015) Over the lifecycle, though, body image is hardly static and can be affected by pregnancy, menopause, actual disease (e.g. amputation of a limb or breast secondary to disease) or deformity (e.g. congenital anomalies). Of note is that Lord Byron, himself, incidentally, was born with the congenital deformity of a club foot (talipes) for which he was never treated successfully. (For those interested, see Strach's article in a 1986 issue of *Progress in Pediatric Surgery.*) Body image can also be affected by aging, when skin loses its elasticity, hair turns grey, adipose tissue accumulates, and people lose muscle tone. Kilpela et al (*Advances in Eating Disorders: Theory, Research, and Practice*, 2015) note that advertisements for products that supposedly prevent anti-aging communicate the importance of disguising our age and hence may induce conflictual feelings about our body image as we get older.

Body image preoccupations can range from minor concerns with actual "flaws" in our appearance to major fixations and distortions that substantially impair a person's functioning and reach delusional (i.e., fixed, false beliefs about one's body) proportions. Both men and women can be affected. A diagnosis of *body dysmorphic disorder* is made when bodily preoccupations are severe and incapacitating and involve clinically significant ruminative concerns with imagined or perceived slight defects that may not even be noticeable to others. Body dysmorphic disorder will be my subject in my subsequent blog.

What is the relationship of body image disorders to weight? Both excessive weight, as seen in overweight and obesity, and extremely low weight, particularly as seen in anorexia nervosa, can dramatically distort body image. For example, part of the very definition of anorexia nervosa includes a "disturbance in the way in which one's body weight or shape is experienced," (*DSM-5*, 2013) but any eating

British artist Thomas Rowlandson, *Amputation*. Many of those who have had an amputation experience "phantom limb pain." Publisher: William Hinton, February 17, 1786, hand-colored etching.

The Elisha Whittelsey Collection, The Elisha Whittelsey Fund, 1959. Metropolitan Museum of Art, NYC. Public Domain.

disorder, including bulimia and binge-eating disorder, can include significant preoccupations and overvaluation of our shape and weight and a failure to perceive our body accurately. (Grilo, *International Journal of Eating Disorders*, 2013).

Years ago, Stunkard and Mendelson (*American Journal of Psychiatry*, 1967) studied disturbances in body image in a randomly selected group of 70 obese people from medical and psychiatric clinics. These researchers found that not all obese people had body image disturbances and some viewed their obesity in a "thoroughly realistic manner." They also found that once a body image disorder developed, though, that person's weight became the central preoccupation, regardless of any other personal qualities, such as talent, wealth, or intelligence. Sarwer et al (*Journal of Consulting and Clinical Psychology*, 1998) found, in their study of 79 obese women who sought treatment, (with a control group of 43 nonobese, non- treatment-seeking), that the obese group scored "greater dissatisfaction with both global and specific aspects of their appearance" (e.g. stomach and abdominal area, thighs, lower body) that generated increased levels of depression and decreased self-esteem. Schwartz and Brownell (*Body Image*, 2004),

José de Ribera, *The Club-foot,* 1642, Louvre Museum. People may have real disabilities, such as a club foot, but still have body image distortions.

Wikimedia Commons/Public Domain.

though, found that one influence on body image is 'whether the person is gaining, losing, or maintaining weight." They also found that onset of obesity in childhood is typically (though not always) a risk factor for body dissatisfaction. They summarized their findings, "Body image and obesity are linked in ways that defy simple analysis," though they further note that body dissatisfaction rises with increased weight, and binge eating in the context of obesity is consistently associated with dissatisfaction with one's body.

In a recent study, Chao (*PLoS One*, 2015) performed a systematic review of seven studies and a meta-analysis of four studies to examine changes in body image in obese and overweight people who had been enrolled in programs to lose weight. Chao noted that body image "comprised aspects of subjective dissatisfaction, cognitive distortions, affective reactions, behavioral avoidance, and perceptual inaccuracy" and found that interventions for weight loss "may improve body image" among those who are overweight or obese. Chao further noted, though, that there was considerable variability among the studies, particularly regarding the type of intervention (no two studies used the same weight loss intervention program) and comparison groups (only one study used an actual control group), length of follow-up, type of body image assessment used, and the actual components of body image assessed.

Bottom line: Though there may be rudimentary beginnings of it *in utero*, our body image continues to evolve throughout the lifespan and is often highly influenced by the normal stages of our development, the culture and environment in which we live, as well as the diseases and disorders to which we are subjected. ■

ILLUSIONS AND "TROUBLED SENSES" OF BODY DYSMORPHIC DISORDER

The psychological distortions and "monstrous maladies" of "imagined ugliness"

68 — Posted November 14, 2015

"People say sometimes that beauty is only superficial... it is only shallow people who do not judge appearances." So wrote Oscar Wilde in his compelling (and some believe his most autobiographic work and only novel), *The Picture of Dorian Gray.* Dorian, who had "extraordinary personal beauty," makes a Faustian bargain to remain forever young as his painted portrait ages and becomes for him "the visible emblem of conscience." As his personality changes from an innocent Adonis, barely out of adolescence himself, to someone capable of extraordinary insensitivity and even murder, Dorian perceives changes in his portrait, and its facial expression bears evidence to him of his cruelty and ultimate depravity. "At first gazing at the portrait with a feeling of almost scientific interest," Dorian, though, struggles with wondering whether his portrait has actually changed or whether it is just a reflection of his imagination—"an illusion wrought on the troubled senses."

Napoleon Sarony, Portrait of Oscar Wilde, circa 1882.
Wilde wrote *The Picture of Dorian Gray.*

Image from the U.S. Library of Congress Wikipedia Commons/Public Domain.

Continues Wilde, 'Morning after morning he (Dorian) sat before the portrait wondering at its beauty...He would examine with minute care and sometimes with a monstrous and terrible delight, the hideous lines that seared the wrinkling forehead...wondering sometimes which were the more horrible, the signs of sin or the signs of age..."

Wilde's novel can be seen metaphorically as a backdrop for a discussion of body dysmorphic disorder, a diagnosis made when bodily preoccupations are severe and incapacitating and involve clinically "distressing preoccupations" with imagined or perceived slight defects that may not even be noticeable to others. For Dorian, his excessive preoccupations with his physical appearance were projected onto his portrait. This disorder, typically beginning in adolescence, is not uncommon, with a prevalence rate of 2.4%. It has been called the "distress of imagined ugliness" by Katharine A. Phillips, MD, who has written extensively on it. (*American Journal of Psychiatry*, 1991.)

Those affected often call their appearance "monstrous." (Of note is that I counted 15 times Wilde uses the word "monstrous" in his 229-page novel.) According to Phillips, about 80% of those suffering have suicidal ideation and almost 30% have actually attempted suicide. (Dorian Gray suicides at the end of the novel.) Those affected often have accompanying obsessive-compulsive symptoms whereby they may spend hours each day thinking about their perceived defects and checking them in the mirror, excessively grooming (e.g. combing, styling, plucking hair) and repeatedly attempting to camouflage the disturbing areas with clothing or makeup, though some may avoid mirrors altogether. (At one point, Dorian, disgusted by his preoccupations, flings a mirror onto the floor and crushes it "into silver splinters beneath his heel.") Phillips et al (*Depression and Anxiety*, 2010) emphasize that these behaviors are "time-consuming, typically difficult to resist or control,

Ivan Albright, *The Picture of Dorian Gray,* 1943–44.

The Art Institute of Chicago, Art Resource (NYC). Used with permission.

Pablo Picasso, *Girl Before a Mirror,* 1932. Those with body dysmorphic disorder often engage in frequent mirror checking and see "imagined ugliness" when they look at themselves.

Digital image copyright, The Museum of Modern Art, Licensed by SCALA/Art Resource, NY. Used with permission. Copyright 2018 Estate of Pablo Picasso/Artists Rights Society (ARS) NY.

and not pleasurable." When severe enough, some refuse to leave their homes, and many are embarrassed and quite secretive about their preoccupations. (Dorian would not let anyone see his portrait and kept it hidden behind a curtain in an unused, upstairs room that he kept locked at all times.)

Typically, those suffering have poor insight and some are even overtly delusional in their fixed beliefs of perceived imperfections. This disorder goes far beyond the common mild preoccupations many people have with their appearance. Most commonly, skin, hair, breasts, abdominal area, (particularly in women, in whom the disorder is more common) and muscle mass ("muscle dysmorphia" particularly in men) are the areas affected, but any area of the body can be a focus of attention. Freud's famous patient *The Wolf-Man,* for example, is described as so fixated on his nose that had been "ruined" by electrolysis (used to treat his obstructed nasal sebaceous glands) that he neglected his daily life, engaged in constant mirror checking, and "felt unable to go on living in… his irreparably mutilated state" though "nothing whatsoever was visible" to others. (Brunswick, *International Journal of Psychoanalysis,* 1928) Many patients actually seek cosmetic procedures, including surgery and dermatological procedures, often to no

Leonardo da Vinci, *A Grotesque Head,* between 1500–1505. People with body dysmorphic disorder often see themselves in highly distorted ways. Black chalk on paper. Christ Church, Oxford, England.

Wikimedia Commons/Public Domain.

avail and often leading to overt hostility toward the treating physician. Up to 15% of dermatology patients and up to 8% of those presenting for cosmetic surgery suffer from this disorder. (DSM-5)

Phillips (1991) reviewed the literature and found that this disorder "of a subjective feeling of ugliness" has a "rich tradition in European psychiatry"—first described in the Italian literature in 1891 (ironically, the same year as the publication of *Dorian Gray*), as "dysmorphophobia." It did not enter the U.S. psychiatric nomenclature until DSM III-R (1987), where it received its present name "body dysmorphic disorder." The classification of body dysmorphic disorder is controversial. It has features of an anxiety disorder as well as an obsessive-compulsive disorder. In psychiatry's DSM-5, (*Diagnostic and Statistical Manual*), it is found under obsessive-compulsive and related disorders, and in the newly released ICD-10

Photo of Freud's couch from his London office after he relocated from Vienna. Freud's famous patient "The Wolf-Man" had many symptoms suggestive of a body dysmorphic disorder (e.g. mirror checking and fixation on his "mutilated nose.")

Photographer: Robert Huffstutter, 2004. Licensed under Creative Commons Attribution 2.0 Generic.

(*International Classification of Diseases*), under somatoform, hypochondriacal disorders. There is also considerable psychiatric comorbidity, including anxiety (e.g. social phobia that predates the disorder), depression (the most common comorbid symptom), and obsessive-compulsive symptomatology in those with body dysmorphic disorder. (For a discussion of key considerations involved in its diagnostic classification for DSM-5, see the comprehensive review by Phillips et al, *Depression and Anxiety*, 2010.)

What is the relationship of weight to body dysmorphic disorder (BDD)? Kittler et al (*Eating Behaviors*, 2007) assessed weight concerns in 200 individuals with BDD. Of the participants, 29% had weight concerns and were more likely to be younger, female, and have more areas of body concern, greater body image disturbances, depression, and suicide attempts. In general, this subgroup of patients had poorer social functioning and was "a more severely ill and body-concerned group overall." Significantly, only 3.5% of this group reported that weight issues were their primary concern. Their conclusion was that the "diagnostic boundary" between body dysmorphic disorder and eating disorders is often not particularly well defined. Of note is that in their study, Sarwer et al (*Journal of Consulting and Clinical Psychology*, 1998) found that of the 79 obese, six had levels of distress, impairment, and preoccupation consistent with a diagnosis of BDD. Mufaddel et al (*The Primary Care*

Paulus Bor, Dutch artist, *The Disillusioned Medea,* circa 1640, Metropolitan Museum of Art, NYC.

Gift of Ben Heller, 1972, Public Domain.

Drapé, *Le Portrait de Dorian Gray,* the novel by Oscar Wilde. 2014.

Licensed under Creative Commons Attribution-Share Alike 3.0 Unported.

Companion for CNS Disorders, 2013) emphasize that while obesity and eating disorders may include disturbances in body image and even BDD, body image itself is distinct from it: those with disturbances in body image do not necessarily have the constellation of symptoms seen in BDD.

Hrabosky et al (*Body Image*, 2009) compared body image among 187 patients with eating disorders, body dysmorphic disorder, and controls in a multi-site study. These researchers found significantly higher levels of body image impairment in patients with anorexia nervosa, bulimia nervosa, and BDD as compared to matched controls. In general, though, those with BDD had fewer concerns with their weight and body shape than those with eating disorders and more concerns with their face and hair and spent more time seeking reassurance from others, mirror checking, and covering up their perceived imperfections. Further, these researchers found that those with BDD had greater overall disturbances with their body image and more impairment in their quality of life (i.e., "greater psychosocial dysfunction") than either controls or those with eating disorders.

Near the end of Wilde's novel, Dorian, "prisoned in thought," becomes more reclusive, avoids most of his friends, and seeks solace in opium dens. Before he ultimately commits suicide by knifing himself in the heart, he viciously stabs his own portrait. When discovered by his servants, Dorian is described as "withered, wrinkled, and loathsome." The "splendid portrait" of him, though, appears "in all the wonder of his exquisite youth and beauty." The reader is left to consider that it was, in fact, only his imagination, "grown grotesque by terror" and the "leprosies of sin," that had distorted Dorian's reality and contaminated what he saw in his portrait and that ultimately led to his demise.

Though there is still much to learn about BDD, today treatments include cognitive behavior therapy and medications such as the SSRIs (selective serotonin reuptake inhibitors.) There is some suggestion, though there are still few and inconsistent results, that those suffering from this disorder have abnormalities in visual processing and perceptual organization (e.g. over-attention to details), as evidenced by functional neuroimaging studies. (For a discussion, see Madsen et al, *Journal of Psychiatric Research,* 2013.) Body dysmorphic disorder may be a "brain-based disorder" and is not just a function of vanity, says Phillips (*Psychotherapy and Psychosomatics*, 2014) who emphasizes the need for appropriate treatment to prevent the considerable morbidity and mortality.

Note: For a general discussion of body image distortions, see blog 67, *The "False Creation" of Body Image Distortions.* ■

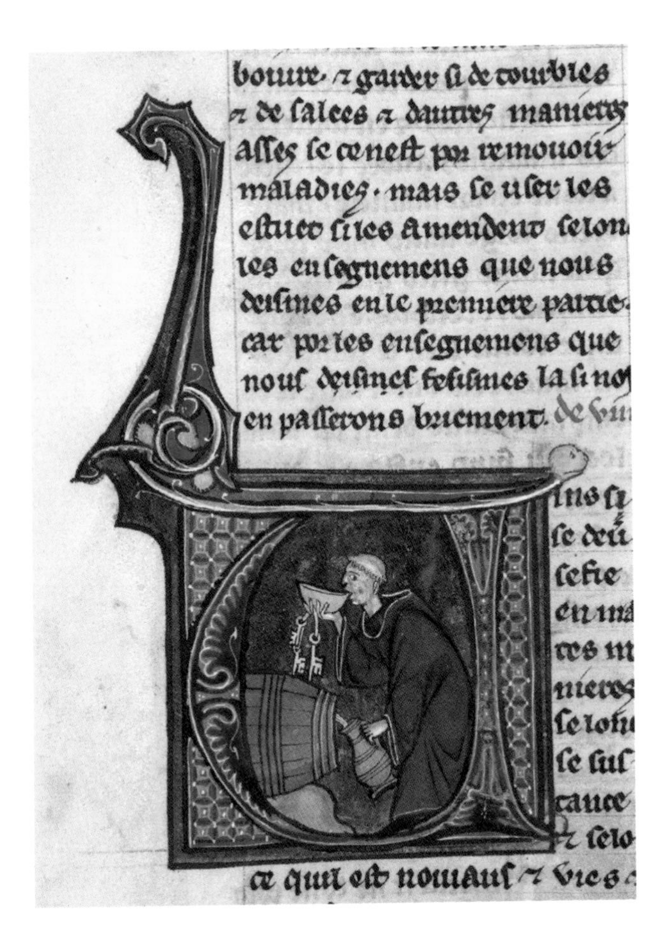

THE DOUBLE-EDGED SWORD OF ALCOHOL USE

The toxic and beneficial effects of drinking and its relationship to weight

69—Posted December 20, 2015

ALCOHOL USE, at least from archeological evidence of pottery shards that contain "fingerprint compounds—biomarkers" of traces of fermented beverages, from a Neolithic settlement in Henan province in China, dates back as early as 7000 B.C. (McGovern et al, *Proceedings of the National Academy of Sciences*, 2004)

Throughout history, the use of alcohol has had both positive and negative effects, often depending on the quantity and pattern of use. Alcohol has analgesic and disinfectant qualities and years ago was often considered much safer to drink than water, particularly when sanitation was poor.

The Bible is replete with references to alcoholic consumption. In the *New Testament*, for example, *John* (2:1–11) reports that Jesus's first miracle was the turning of water into wine when wine had run out at a wedding in Cana in the Galilee. There are, of course, allusions to alcohol's mind-altering and potentially destructive influences as well. For example, in the *Old Testament*, in *Genesis* (19: 35–37), after the annihilation of the cities of Sodom and Gomorrah, Lot's daughters, fearing that there were no men left with whom to procreate, get their father drunk enough with wine on two successive nights in order to have sex with him, become pregnant, and preserve the family line without his realizing that he has committed incest.

Monk drinking from a barrel, French, late 13th century. British Library, London.

Wikimedia Commons. License through Creative Commons CC01.0 Universal Public Domain.

In more recent years, the use of alcohol has been described as having overtly beneficial qualities but has been described as "analogous to the proverbial double-edged sword" because "perhaps no other health or lifestyle factor can cut so deeply in either direction—toxic or beneficial," say O'Keefe et al (*Mayo Clinic Proceedings*, 2014.) These researchers quote statistics from the World Health Organization that alcohol kills about 2.5 million people each year, with excessive consumption of alcohol being the third leading cause (after smoking and obesity) of premature death in the U.S.

O'Keefe et al report on studies that indicate that light to moderate alcohol use (one drink/day for women and one-two drinks/day for men) may lead to decreased hypertension, reductions in type II diabetes (and even may enhance insulin sensitivity, reduce inflammation, and elevate HDL cholesterol), and a decreased risk of heart failure. French et al (*Health Economics*, 2010), though, notes, "It is conceivable that moderate drinking is a marker for moderate living."

O'Keefe et al maintain that while red wine has been particularly implicated in beneficial effects because of the compounds of polyphenols, most evidence indicates it is the ethanol itself "rather than any other specific compound of a drink, that is the primary factor for both conferring health benefits and causing toxicity, depending on the pattern of consumption and dosing." Wine drinkers, though, may have a healthier lifestyle than those who drink primarily beer or hard liquor.

Giotto, (Detail from) *Marriage at Cana.* (From *Scenes from the Life of Christ)*, between 1304–1306. Cana was the site of the first miracle of Jesus in which he turned water into wine. Scrovegni Chapel, Padua, Italy.

Web Gallery of Art. Wikimedia Commons/Public Domain.

Scottish artist, Count Girolamo Nerli, portrait of Robert Louis Stevenson, 1892. Stevenson called wine "bottled poetry" in his travel memoir, *The Silverado Squatters.* Scottish National Gallery, Edinburgh.

Wikimedia Commons/Public Domain.

Heavier use (2.5 drinks/day for women and 4 drinks/day for men) has been associated with progressively increased risk of death in a dose-dependent relationship. O'Keefe et al, explain that at higher doses alcohol is clearly a cardiotoxin, and they caution that those who do not drink should not be encouraged to start because it is not possible to predict for whom regular use of alcohol will lead to that "slippery slope that many individuals cannot safely navigate."

What about the relationship, though, of alcohol to weight? In their recent update, Traversy and Chaput (*Current Obesity Reports*, 2015) reviewed both cross-sectional and longitudinal studies and found considerable variation among the findings, with some studies indicating an association between drinking and weight gain and others not finding any association. There are many reasons for the discrepancies, including differences in the pattern of drinking. They clarify that some studies do not necessarily control for physical activity of their participants and some do not even necessarily quantify how much alcohol is consumed (e.g. using a "yes/no" scale.) They summarize that while light-to-moderate drinking (in the *National Health and Nutrition Examination Study* –NHANES III—heavy drinkers have 4 or more per day and light-to-moderate have 1 or 2 drinks/day) is not associated with weight change or changes in waist circumference, heavy drinking is more consistently associated with weight gain. Traversy and Chaput also looked at experimental studies, often

Psiax, Greek vase painter of 520–500 B.C. Dionysus, god of wine, holding out a kantharos, on ancient Greek black-figure pottery. British Museum, London.

Photographer: Jastrow, 2006. Wikimedia Commons/Public Domain.

using wine or beer intake (e.g. adding alcohol to diets of those on a metabolic unit.) They found, again, that moderate alcohol use does not lead to weight gain, but they caution that studies are often short-term (e.g. 4 to 10 weeks) and may not be long enough for firm conclusions. Studies, as well, are often not consistent for comparison in their definitions of "light, moderate, and heavy" drinking and even more importantly, studies often rely on self-report (rather than actual objective measurements), not only of the amount of alcohol consumed but of height and weight and that can make a study fundamentally worthless due to the well-known distortion involved in self-reporting.

Sayon-Orea et al (*Nutrition Reviews*, 2011) performed a systematic review of 31 studies from 1984 to early 2010. They note that studies have had three different approaches: epidemiological (alcohol intake and body weight); psycho-physiological investigation (alcohol and appetite regulation); and metabolic studies (effects of alcohol on energy expenditure and substrate oxidation.) They found that a positive association between weight gain and drinking was more often associated with men, but caution "this difference might be an indirect consequence of habitual drink choice" as well as the fact that men tend to drink more than women in general. They also found that wine drinkers were less likely to have weight gain and speculated that compounds in red wine such as resveratrol may be protective of weight gain. They emphasize that inconsistences may arise due to a person's smoking history as well as changes in alcohol consumption over time. They noted it is currently unclear whether alcohol consumption is a risk factor for weight gain because studies have shown positive, negative, and no effect on weight, and the "precise effect of alcohol on weight remains to be determined." Some researchers, such as McCarty and Jéquier each writing articles in the *American Journal of Clinical Nutrition*, 1999, call it the "alcohol paradox" when alcohol does not lead to weight gain.

Why would alcohol theoretically lead to weight gain? First, alcohol is both caloric and nutritionally poor: it contains 7.1 calories per gram (for comparison, on average, fat has 9 calories, and carbohydrates and proteins each have 4 calories per gram.) Alcoholic drinks also vary considerably in their caloric count. Poppitt (*Nutrients*, 2015) notes that alcohol by volume can vary between 3 to 40% and is typically divided into three groups: fermented beers and ciders; fermented wine; and distilled spirits. In terms of calories, 12 ounces of a light beer contain 103 calories; regular beer, 153 calories; 80-proof of 1.5 ounces of gin, rum, vodka, or whiskey contains 97 calories, while 9 ounces of a Piña Colada contain 490 calories. (NIH statistics, retrieved 12/18/15: online on its website.)

Secondly, most people, when they do drink, do not necessarily substitute the liquid calories of alcohol for food and hence do not compensate by eating fewer calories. In fact, for many people, alcohol, especially when imbibed either prior to or during a meal, may stimulate a person's appetite and hence increase the number of calories eaten. Further, alcohol seems to have an effect on several of the hormones involved in satiety and control of hunger, such as inhibiting the effect of leptin and glucagon-like peptide-1 (GLP-1) or increasing CCK, a hormone that suppresses hunger. Both peripheral and central controls on satiety may be involved. Caton et al (*Current Obesity Reports*, 2015) also emphasize that alcohol has the psychological and physiological ability to promote an inability to resist temptation, i.e., an abandonment of restraint—and hence lead to overconsumption, and they have found that dieters and so-called "restrained eaters" are more susceptible to this effect. This has been referred to as the "what-the-hell effect" i.e.,"the "all-or-nothing" attitude that dieters sometimes have once they have broken their diet. (Herman and Polivy, 2004, in Baumeister's and Vohs' *Handbook of Self-Regulation.*)

So-called *Cobbe Portrait,* claimed to be William Shakespeare done while he was alive, unknown artist, circa 1610. In Shakespeare's *Macbeth,* the Porter says that drink "provokes the desire, but it takes away the performance." References to drinking in Shakespeare's 38 plays are so common that there is even a book by Buckner B. Trawick, *Shakespeare and Alcohol.*

Photographer: Getty Images. Wikimedia Commons/Public Domain.

Alcohol also inhibits fat oxidation. Alcohol cannot be stored in the body so it is preferentially metabolized (Sayon-Orea et al, *Nutrition Reviews*, 2011). What this means is that alcohol is oxidized immediately and the oxidation of fat and carbohydrates is delayed so that while alcohol is being metabolized, fat may tend to accumulate.

Is there any truth to the widespread belief that beer drinking promotes the abdominal accumulation of fat that is typically called the "beer belly?" Schütze et al (*European Journal Clinical Nutrition*, 2009) conducted a prospective study over an over 8-year period to assess waist circumference change in over 7800 men and over 12,700 women from the Potsdam area of Germany. Beer consumption ranged from less than 250 mL/day (about 8.5 ounces, "very light") to over 1000 mL/day (about 33 ounces, "heavy" consumption.) These researchers concluded that there is no support for the common belief in the development of a beer belly and the "observed" beer bellies may be related to the "natural variation in fat patterning and not from the fact of drinking beer." Bendsen et al (*Nutrition Reviews*, 2013) conducted a systematic review (from 1950 through 2010, including 35 observational studies and 12 experimental studies) to assess whether beer leads either to specific accumulation of fat in the abdominal area or even general obesity. These researchers acknowledged conflicting, inconsistent results among studies, often due to confounding factors, such as differences in lifestyle eating, smoking, and exercising habits. For example, they reported on one Danish study in which those who purchase beer were more likely to buy butter, sausages, ready-cooked meals, and soft drinks, whereas wine drinkers were more apt to purchase more vegetables and products with lower fat content.

The actual mechanism through which beer or other alcoholic drinks promotes abdominal fat accumulation is not known. Bendsen et al summarize by noting there is a "strong theoretical basis for the fattening effect of beer consumption" but due to experimental studies of low quality and conflicting observational studies, "the present data provide inadequate scientific evidence" to assess whether beer at moderate levels (less than 500 mL/day) is associated with either abdominal or general obesity though the researchers cannot rule out that larger quantities of beer might lead to increased fat accumulation (and even particularly abdominal obesity.)

Bottom line: Though it makes theoretical sense that excessive calories from alcohol will lead to increased weight, studies to date have yielded conflicting evidence, i.e., the so-called "alcohol paradox." Nevertheless, during this holiday time, moderation in both food and drink obviously makes weight gain less likely. ■

IN NUTRITION, WHERE DOES SCIENCE STOP AND FANTASY BEGIN?

Paleolithic diets: Hardly a "Garden of Earthly Delights"

70—Posted January 14, 2016

WHILE THERE ARE THE FAMOUS "cave paintings" found in France and Spain, some of the most characteristic art from Paleolithic times depicts obese women. In fact, a photo of the sculpture "Venus of Willendorf" (approximately 29,500 years old, now in the Natural History Museum of Vienna) appears in many books on the history of obesity and indicates "obesity has been known to humans…for more than 20,000 years." (Bray, *An Atlas of Obesity and Weight Control*, 2003) These statuettes date from about 40,000 to 10,000 years ago and are called the "Venus" women, possibly because to some archeologists, they may represent fertility symbols, though no one knows their significance. This is an anachronistic designation since Venus was a Roman goddess, and this Paleolithic era is many centuries before the classical times of the ancient Greeks and Romans. Many more than 17 of these figurines, including the Venus of Vestonice and the Venus of Laussel, have been found throughout Europe and the Ukraine and are carved from bone, ivory, wood, soft stone, or clay. They are all massively obese, whether presenting with abdominal obesity or with enormous pendulous breasts and protruding buttocks. These Stone Age figurines are some of the most iconic images of the Paleolithic period. It becomes particularly ironic that the so-called "Paleolithic diet," then, is touted by some as a particularly healthy one that can prevent obesity and its metabolic abnormalities.

Original *Venus from Hohlefels,* around 35,000 to 40,000 years old, found in cave near Schelklingen, Germany. Many of these so-called Venus figurines were found in Europe and even as far as Siberia. No one knows their significance, though they may be fertility symbols. Photographer: Thilo Parg, 2015.
Wikipedia Commons. Licensed under Creative Commons Attribution Share-Alike 3.0.

The Paleolithic epoch, often considered synonymous with the Stone Age, spans a vast time period, from roughly 2.6 million years ago until about 10,000 years, with the advent of agriculture and the beginning of the Neolithic period. The first reference to using our Paleolithic ancestors—the hunter-gatherers—as a model for a contemporary diet came from the 1975 book *The Stone Age Diet*, by Dr. Walter Voegtlin, a gastroenterologist. (The complete book can be downloaded online.) One of Voegtlin's most important messages is "There is no such thing as a part-time dieter." He dedicates his book to those "who can still resist the specious authority of food merchants, their lavish advertisements and spectacular television commercials, and retain sufficient intellectual independence to think for themselves." His primary focus was on the importance of a carnivorous diet (including eating all the fat around the meat), with a very low percentage of carbohydrates (and no raw vegetables.) Voegtlin believed humans were "strictly carnivorous" until 10,000 years ago, something we now know to be patently false.

The idea of the Paleo diet, though, first appeared in a peer-reviewed journal in 1985 when Eaton and Konner published their article "Paleolithic Nutrition" in *The New England Journal of Medicine*. The authors did acknowledge that these early humans demonstrated the considerable

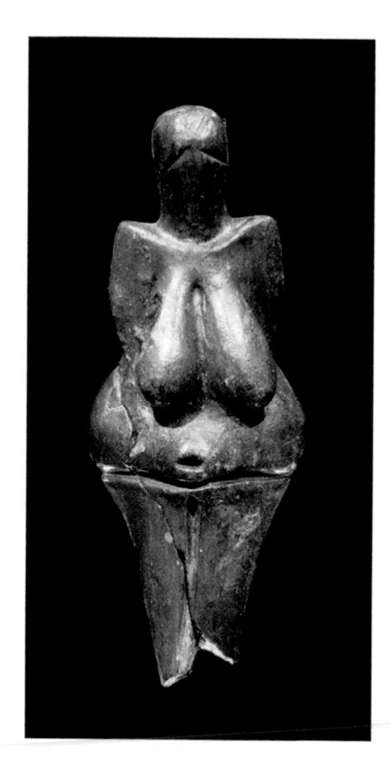

▲ Paleolithic cave painting of animals from Lascaux Cave, south-western France, around 17,000 years ago.

Alamy stock photo. Used with permission.

◀ Venus of Dolní Vêstonice, about 26,000 years old, found in the Moravian Basin, now Czech Republic.

Photographer, Petr Novák, Wikipedia, 2007. Licensed under Creative Commons Attribution-Share Alike 2.5 Generic, Wikipedia Commons.

Venus of Lespugue, about 23,000 years old. Musée de L'Homme, Paris. Photographer: José-Manuel Benito, 2006.

Wikimedia Commons/Public Domain.

"versatility of the omnivore" in that the foods available to them varied widely depending on the specific time within the Paleolithic period, geography, location, and seasonal changes. These ancestors ate meat, including wild game, bison, horses, mammoths, and deer, but the authors acknowledged that it is more difficult to assess their consumption of vegetables since plant material is not as well preserved over the centuries. Even so, Eaton and Boyd believed that this diet could be a "reference standard" for modern nutrition, and they speculated that our modern Western diet contributes to "diseases of civilization," such as diabetes and atherosclerosis.

Interest in a Paleolithic diet has grown considerably over more recent years, as evidenced by articles about the value of a predominantly protein and fat-based diet (in various proportions) in popular magazines such as *Science* (2008), *Scientific American* (2013), *The Atlantic* (2014) and *The New Yorker* (2014.) On a recent search on barnesandnoble.com, I found over 500 books that now feature some version of what authors consider the Paleo diet. Konner and Eaton revisted "Paleolithic nutrition" 25 years later (*Nutrition in Clinical Practice*, 2010) The authors maintained "We did not then and do not now propose that Americans adopt a particular diet and lifestyle on the basis of anthropological evidence alone" and instead they called for more research. They acknowledged that our "flexibility in adaptation may have been central to human evolution" but emphasized that they believed the potential dangers of eating meat today were related to the high proportions of both total and saturated fat in commercially available meats.

The problem, though, is that this notion of an "idyllic past"—some imagined food Eden or "Garden of Earthly Delights," if you will, (to use the title of the famous Hieronymus Bosch painting) is a "fabricated myth," explain Nesse and Williams in their 1994 book *Why We Get Sick*. Only to a limited degree can we reconstruct the diet of our Paleolithic ancestors. First, we are dealing with an enormous time span of hundreds of thousands of years that involved humans living in vastly different geographies and climates. There was never "one" Paleolithic way of eating.

The word *paleofantasy* captures the sentiment of Nesse and Williams exactly. The word was first used by professor of anthropology Dr. Leslie Aiello, president of the Wenner-Gren Foundation for Anthropological Research, in one of her University College of London lectures in the 1990s (personal communication.) For Dr. Aiello, it means quite simply "People's fantasies about what the past was like," and it made her think "about where science stops and fantasy begins."

Dr. Marlene Zuk, a professor of ecology, evolution, and behavior, has borrowed the concept directly from Dr. Aiello in her book *Paleofantasy: What Evolution Really Tells Us about Sex, Diet, and How We Live* (2013). Zuk explains,

"Neither we nor any other species have ever been a seamless match with the environment. Instead, our adaptation is more like a broken zipper, with some teeth that align and others that gape apart." She notes that the concept of *paleofantasy* implies that we humans, "were at some point perfectly adapted to our environments." It reflects an "erroneous idea" and "misunderstanding" about how evolution works. Zuk notes that *Homo sapiens* (i.e. "anatomically modern humans") appeared about 100,000 years ago, but there was never a single lifestyle "any more than there is a single Modern lifestyle."

How do researchers attempt to reconstruct the diet of our Paleolithic forbearers? They seek out clues from remains from archeological sites that have human bones and teeth. For example, Zuk notes that researchers have "studied the plaque clinging to the teeth" ("which has survived in our pre-flossing days") of these early humans and found evidence of seeds, date palms, and other plants, as well as gelatinized starch grains indicating these people may have harvested grain and cooked their food, "all of which calls into question the various forms of the so-called Paleo diet." Furthermore, they do studies of carbon isotopes that can distinguish different paths of photosynthesis, particularly in tooth enamel and to a certain extent, in bone collagen although collagen is not preserved beyond 200,000 years. (Kuipers et al, *Nutrition Research* Reviews, 2012) Researchers also seek out evidence from the few hunter-gatherer populations still in existence, such as the !Kung or the Maasai of Africa or primitive tribes in Australia, but even these populations are vastly different, depending again on their locale and climate. Kuipers et al (2012) also make the point that hunting "leaves the most prominent signature in the archeological record." As a result, the importance of big game hunting as a food source may be "exaggerated" in these fictionalized accounts of Paleolithic life (Nesse and Williams, 2003) and in all likelihood protein

Hieronymus Bosch, *The Garden of Earthly Delights,* between 1480 and 1505. The left panel is of the Garden of Eden. Paleolithic diets have recently been idealized in ways that reflect a misunderstanding of evolution and what has been called "paleofantasy," a word first coined by professor of anthropology Dr. Leslie Aiello. Dr. Marlene Zuk, professor of ecology, evolution, and behavior, took the word for the title of her book.

Prado Museum, Madrid. Wikimedia Commons/Public Domain.

The famous Paleolithic figurine Venus of Willendorf, around 29,000 years old. Naturhistorisches Museum, Vienna.

Bridgeman Images, used with permission.

sources could also have come from gathered nuts, tubers, and small animals. (Kuipers et al, 2012)

Zuk asks, "Even if we wanted to have a more paleo-diet, could we?" The answer is no. Virtually all of the foods available to us today are vastly different from those eaten in Stone Age times. Many of those foods, for example, are now extinct, such as the woolly mammoths. And a white-tailed deer has 2.2 grams of fat while extra lean ground beef has 18.5 grams of fat. Nesse and Williams explain, "We would find most of the game strong-tasting and extremely tough" and quite "tedious" to prepare the carcass for consumption. Even ripe wild fruits would be sour or bitter to our tastes.

The issue is not so much whether the Paleolithic diet is efficacious, though there are small short-term studies such as the recent ones by either Masharani et al (*European Journal* of *Clinical Nutrition*, 2015) or by Bligh et al (*British Journal of Nutrition*, 2015) on metabolic health benefits of a Paleolithic-style diet. Zuk herself emphasizes that her focus, though, was not on comparing Paleo diets with others but rather whether a contemporary version really could possibly replicate what our Stone Age ancestors actually ate.

In a comprehensive review of various diets, Katz and Meller (*Annual Review of Public Health*, 2014) note there have never been any rigorous, long-term comparative studies that use methodology that is completely free from bias and confounding. In the absence of such, say Katz and Meller, it is "not so much evidence of absent benefits but a relative lack of evidence." A pattern of eating that avoids highly processed foods and an excessive intake of sugar, while increasing our intake of vegetables, fruits, nuts and seeds, and lean meats, certainly makes sense but calling this some kind of precise replication of Stone Age diet does not. Rather, we should focus more on "guiding principles." Say Katz and Meller, "Claims for the established superiority of any one specific diet over others are exaggerated," and they add, "We need less debate about what diet is good for health." Instead, we should be more attuned to public health issues of how to get people "in the direction of optimal eating." ■

THE CARE AND FEEDING OF MYTHS: BREASTFEEDING AND WEIGHT

Separating out science from fiction in breastfeeding research on obesity

71 — Posted February 15, 2016

THE ORIGIN OF OUR MILKY WAY began with a reluctant breastfeeding. Or so the Greco-Roman myth goes, as described in an ancient Byzantine text *Geoponica*: the god Jupiter had an illicit affair with a mortal and wished his offspring Hercules to obtain immortal powers. To bring this about, Jupiter gave the infant to nurse at the breast of his sleeping goddess wife Juno, but when Juno awakened, she became furious and violently pulled her breast away, only to have her milk spill out across the sky and form the Milky Way. This myth is itself immortalized in two paintings, the 1575 painting by Tintoretto that hangs in the National Gallery in London, and the Rubens painting (1636-38) hanging in the Prado Museum in Madrid. Of course, for a more scientific version of the origin of the Milky Way, see Lisa Randall's 2015 book *Dark Matter and the Dinosaurs.*

Some of the myths surrounding breastfeeding, though, may be just as fanciful as our Milky Way saga. In her new book *Lactivism*, Courtney Jung writes of the "moral righteousness" that has enveloped breastfeeding, "Breastfeeding is no longer just a way to feed a baby; it is a moral marker that distinguishes us from them—good parents from bad." It has become a "talisman to ward off evil." It is in this present climate that breastfeeding and the commodity of breast milk specifically have been endowed with "super," almost immortal powers.

Breastfeeding, of course, has always been around—it is one of the defining characteristics of being a mammal, i.e., nursing young at the breast. There are even several Biblical breastfeeding references, such as in *Exodus* when Pharaoh's daughter employs a wet nurse to feed the infant Moses whom she has rescued. Over the years, interest in breastfeeding has waxed and waned among women. According to the most recent CDC statistics, seventy-nine percent of women in the U.S. start breastfeeding, and forty-nine percent are still breastfeeding at 6 months. Many factors influence a woman's decision to breastfeed, as well as to her decision to stop, including the health of either mother or infant; family support and expectations; and even issues of convenience related to family or job. (Woo and Martin, *Current Obesity Reports*, 2015) Sometimes, though, breastfeeding is contraindicated, (e.g. when a mother takes medication or has an infection such as HIV that would dangerously transfer to the infant or an infant is born with deformities that would obviate breastfeeding.) But when a mother does not produce enough milk, she is labeled pejoratively, says Jung, a "lactation failure."

What about harms in breastfeeding? Lehmann et al (*Critical Reviews in Toxicology*, 2014) reviewed potential health risks to breastfed infants from environmental chemicals (e.g. heavy metals, phthalates, bisphenols, perchlorates, etc.), including synergistic mixtures of these that can accumulate in breast tissue of a mother throughout her life. Neither the "critical window" of exposure nor the dose-response information for these pollutants is known, and animal studies are limited because these chemicals can have different half-lives in animals and differences in the duration of exposure. Of course, Lehmann et al acknowledge that bottle-fed infants can also be exposed through the environment, and given these caveats, they still conclude that the natural process of breastfeeding is best.

Jacopo Tintoretto, *The Origin of the Milky Way,* circa 1575, National Gallery, London.
Wikimedia Commons/Public Domain.

Madonna Litta, attributed to Leonardo da Vinci, circa 1490s. Our Milky Way began, so the myth goes, with a reluctant breastfeeding. Hermitage Museum, St. Petersburg.
Wikimedia Commons/Public Domain.

What are some of the advantages, particularly of exclusive breastfeeding (i.e. no other food) for the first six months as is recommended by the American Academy of Pediatrics? (Kramer and Kakuma, *Cochrane Database of Systemic Reviews*, 2012) (But note, even this number is somewhat arbitrary as I did not find any studies in which mothers stopped breastfeeding at 4 and ½ or 5 months.) Human breast milk has been called "the optimal first food." (Woo and Martin, 2015.) Breastfeeding establishes a strong intimate bond between mother and infant (Jelliffe and Jelliffe, *New England Journal of Medicine*, 1977) and conveys immunity to the infant (e.g. preventing less morbidity from infectious disease.) (Kramer and Kakuma, 2012) Exclusive breastfeeding, as well, can also act as a contraceptive in delaying the onset of ovulation ("prolonged lactation amenorrhea," Jelliffe and Jelliffe, 1977; Kramer and Kakuma, 2012) and hence enable increased spacing between children. Furthermore, since it burns 500 more calories a day, it can be beneficial in helping women lose their pregnancy weight gain.(Stuebe, *Seminars in Perinatology*, 2015) Breastfeeding, as well, is also considerably safer than using formula when water sanitation is poor in many countries.

The science, though, is considerably weaker, at least at this point, in regard to linking breastfeeding with long-term

Peter Paul Rubens, *The Birth of the Milky Way,* between 1636–37, Prado Museum, Madrid.
Wikimedia Commons/Public Domain.

control of overweight or obesity during childhood or into adulthood. The first reports of a link between weight in adolescence and breastfeeding occurred in the early 1980s (see the discussion by Bartok and Ventura, *International Journal of Pediatric Obesity*, 2009). Many studies have followed, but research comparisons across these studies are often problematic when there are different populations, for example, with different inclusion and exclusion criteria, and even different definitions of "exclusive breastfeeding" and "nonexclusive breastfeeding." (Marseglia et al, *Women and Birth*, 2015) Furthermore, crucially, Cope and Allison (*Obesity Reviews*, 2008) have found that some breastfeeding researchers use the "inappropriate" language of cause and effect and fail to differentiate possible associations from actual causation. In their comprehensive and critical review of the 2007 World Health Organization report on breastfeeding, Cope and Allison also found evidence that researchers failed to take into full consideration of so-called *publication bias*, i.e., "when the probability of a study being published depends on the outcome of the study" and most often, "involves statistically significant studies having a higher likelihood of being published than do studies with results that are not statistically significant." When these events happen, a compromise to research is inevitable.

Cope and Allison (2008) acknowledge that conducting research on breastfeeding is difficult. For example, since there are indeed recognized benefits to breastfeeding, it is unrealistic and unethical to conduct randomized controlled studies that would entail assigning mothers to a non-breastfeeding group. Further, there are potential

Poster promoting breast feeding, *Nurse the baby,* between 1936-1938. WPA poster. Throughout the years, breastfeeding has come in and out of fashion.

U.S. Library of Congress's Prints and Photographs Division. Author: Erik Hans Krause. Wikimedia Commons/Public Domain.

confounding variables (e.g. mother's own weight, education, and race) that are challenging, if not almost impossible, to separate out entirely: those who breastfeed are more likely in recent years to come from a higher socioeconomic class, have more social support, and to be more health-conscious (e.g. exercise and diet more regularly and be less likely to smoke.) (Bartok and Ventura, 2009; Stuebe, *Seminars in Perinatology*, 2015.) For those statistically-minded, see the 2015 paper by Allison and colleagues (Pavela et al, *European Journal of Clinical Investigation*) on so-called "packet randomized experiments—PREs"—an intermediate design between randomized controlled trials and observational studies that can be used to strengthen

Stamp of Belarus with breastfeeding mother, 2005.

Author: Post of Belarus, Mihail Sevruk. Wikimedia Commons/Public Domain.

potential causal relationships in research design, including research on breastfeeding. From their review, though, Cope and Allison (2008) conclude that the benefits of breastfeeding "outweigh the harms for most people " but they found it "unwarranted at this time" to conclude that breastfeeding "causally reduces the risk of overweight or obesity." More recently, Casazza et al (*New England Journal of Medicine*, 2013) underscore that it is one of the many myths about obesity that breastfeeding is protective against obesity. In fact, the one major randomized study (called "cluster randomization" since randomization was done by site) from Belarus, (the PROBIT study—Promotion of Breastfeeding Intervention Trial) involving over 17,000 mothers (31 different hospital sites and their clinics) followed over 11.5 years found there was **not** a clear and consistent relationship between breastfeeding and weight in infants and children over time, but cautioned that the Belarus population results might not generalize to other populations. (Martin et al, *JAMA*, 2013)

Why could breastfeeding, though, theoretically lead to reduced weight in an offspring? One theory is that since intake is harder to monitor in breastfed babies, a mother has to rely on cues such as an infant's turning its head away rather than an external cue such as an empty baby bottle. (Woo and Martin, 2015) As a result, infants may learn to self-regulate their amount of food intake. Further, breast milk contains many hormones from the mother, such as insulin, insulin-like growth factor, leptin, adiponectin, etc that may be influencing "neuroendocrine circuits" involved in appetite control in the infant (Marseglia et al, *Women and Birth*, 2015) and "of critical importance in the metabolic development of the infant." (Çatli et al, *Journal of Clinical Research in Pediatric Endocrinology*, 2014) Breastfed infants, as well, have different levels of certain bacteria (e.g. Bifidobacteria) in their gastrointestinal tracts that ultimately may have effects on long-term weight.

Over the years, there have been considerable, though unsuccessful, attempts by formula manufacturers to "humanize" their product. (Jelliffe and Jelliffe, 1977). Infant formulas, particularly in this country, are (and have been for years) highly regulated. Such was not the case in China when the toxic levels of the chemical melamine tainted

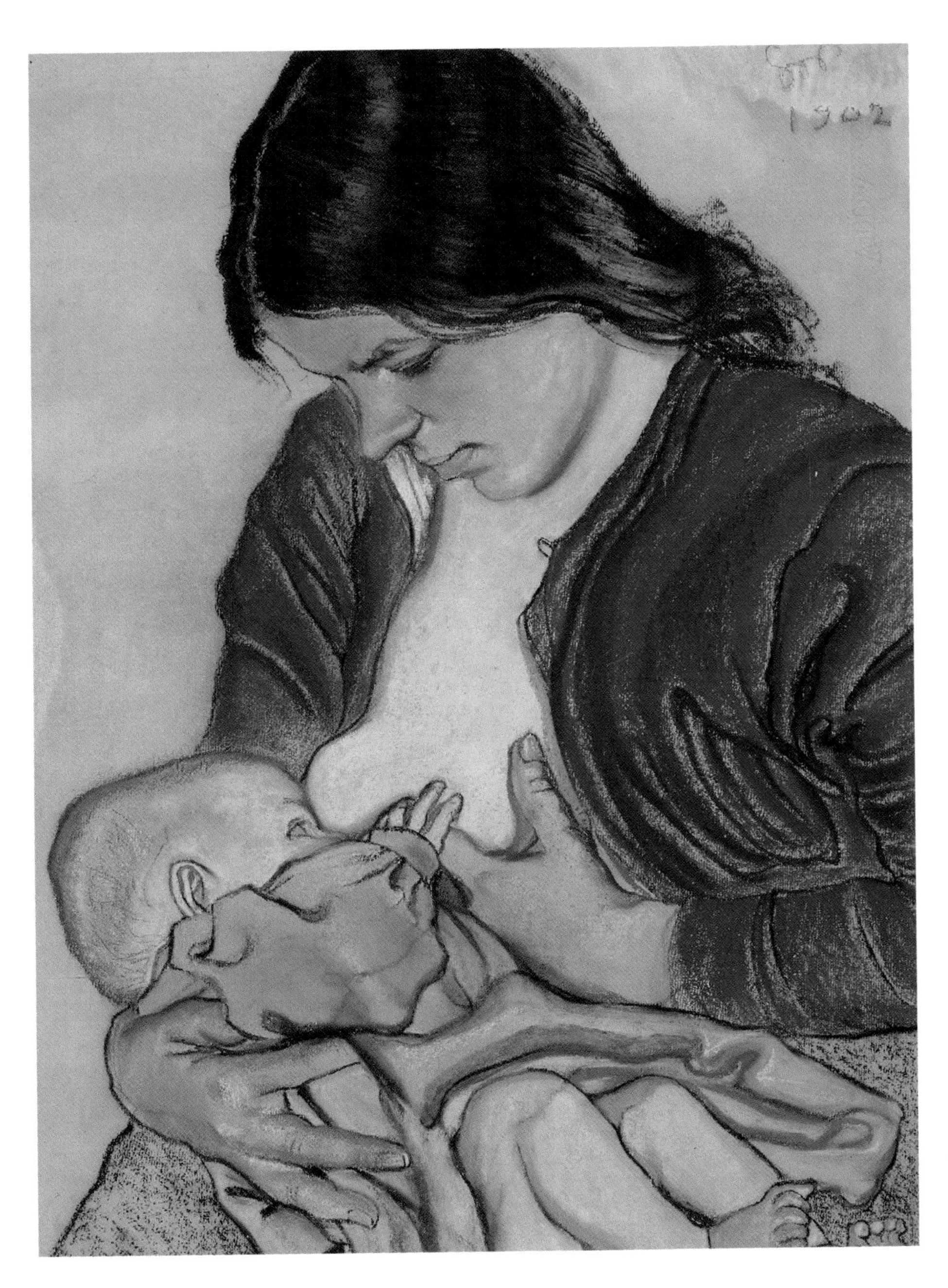

Stanislaw Wyspiański, *Motherhood*, 1902, pastel on paper, National Museum in Warsaw.
Wikipedia Commons/Public Domain.

formula there. For a history of infant formulas, see the most recent article by Wargo (*Journal of the Association of Official Analytical Chemists International*, 2016) as well as Jung's discussion of the Nestlé Company's tactics to encourage women to use formula. Despite the fact that most formula today derives from cow's milk, "formula's macro-and micro-nutrient composition still contains key differences from human breast milk." (Bartok and Ventura, 2009) Further, mother's milk varies with her diet so that breastfed infants may be exposed to and develop a taste for novel flavors whereas the taste of a formula does not vary. (Woo and Martin, 2015)

Bottom line: Breastfeeding is clearly a natural and beneficial process. Woo and Martin (2015), though, summarize its connection to obesity, "It is not sufficient to continue to ask whether obesity is associated with breastfeeding. Future research should refine the question to 'under what circumstances does feeding human milk affect growth patterns, or body composition, or metabolic processes in the short, medium, and longer term.'" ■

ADOLPHE QUETELET AND THE EVOLUTION OF BODY MASS INDEX (BMI)

A 19th century 'Renaissance man' devised a ratio we use today

72—Posted March 18, 2016

PROCRUSTES (literally the "one who stretches") was a robber who had an inn beside a road that led away from ancient Athens. He boasted that his bed could fit anyone who came to stay the night but instead of making the bed fit the person, he made the person fit the bed. So for those travelers who were too tall, he amputated their legs and for those too short, he stretched them to fit his one-size-fits all bed. In both scenarios, so the ancient Greek myth goes, the unlucky traveler was killed. But Procrustes got his due—Theseus, of Minotaur and labyrinth fame, killed him in the same way he had killed his guests, i.e., by making him fit his own bed, and according to one version, decapitated him. The myth is referenced by the Greek historian Plutarch in *Parallel Lives* and by the Roman poet Ovid in *Metamorphoses*, as well as on Greek red figure pottery. Nassim Taleb used this myth as inspiration for his book —*The Bed of Procrustes*, a book of aphorisms that relate to situations of changing the wrong variable.

Detail from an Ancient Greek red figure amphora of Theseus killing Procrustes, by Alkimachos, 470–460 B.C. Procrustes was famous in mythology for making his visitors fit his bed by either stretching them or cutting their limbs. Bavarian State Collection of Antiquities, Munich.

Wikipedia Commons/Public Domain.

Procrustes, though, with his focus on a one-size-fits all mentality, may have been the first in history to mandate standardization. In his new book, *The End of Average*, Todd Rose writes how society has used standards and norms as a means of understanding individuals. From our regulation of size proportions of military uniforms and airplane cockpits, cut-offs for test scores in education and college admissions, and the selection of applicants for employment, Rose notes that we have created an emphasis on conformity and the rise of "averagarians." Instead, we should focus on the "science of the individual" that involves appreciating that our behavior is often context-dependent and acknowledging that people don't all have to follow the same path for success.

Where, though, did this concept of average originate? Rose discusses numerous sources, but for our purpose here, Adolphe Quetelet deserves much of the responsibility and for Rose, some of the blame.

Quetelet (1796–1874), though, was responsible for far more than a concept of average. Belgian born, he has been described as a 'Renaissance man' (Rössner, *Obesity Reviews*, 2007), with equal interests in the arts and sciences and reportedly fluent in six languages. (Eknoyan, *Nephrology Dialysis Transplantion*, 2008) Early on, he dabbled in painting and poetry (Landau and Lazarsfeld, *International Encyclopedia of the Social Sciences*, 2008) but received his doctorate in mathematics at age 23. (Faerstein and Winkelstein, *Epidemiology*, 2012) He was a prodigious letter-writer and influenced the thinking of people as diverse as Karl Marx, Emile Durkheim, Francis Galton, Goethe, and Florence Nightingale. (Jahoda, *Springerplus*, 2015; Landau and Lazarsfeld, 2008) Until he had a stroke in his later years, he was extraordinarily productive. Interested in astronomy, he established the Brussels Observatory and was its director for fifty years, but his major interest was statistics. (Porter, *British Society for the History of Science*, 1985) He established the first international conference on

Charles Fraikin, 19th century sculptor, *Adolphe Quetelet.* (1796–1874). Brussels. Quetelet was considered a "Renaissance Man," and one of the founders of statistics as a scientific discipline. The body mass index (BMI) is referred to as the Quetelet Index.

Creative Commons Share Alike 3.0 Unported. Author: Klever.(GNU Free Documentation License. Wikimedia Commons/Public Domain

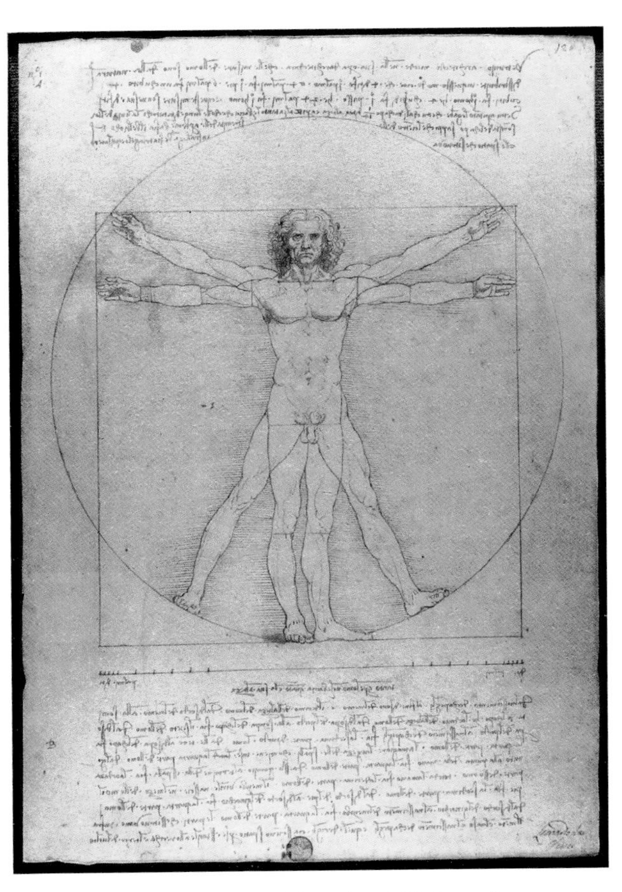

Leonardo da Vinci, *Vitruvian Man,* circa 1492. Drawing pen, ink, and wash on paper. Academia of Venice.

Photographer: Luc Viatour. Wikimedia Commons/Public Domain.

statistics, and some consider him one of the founders of statistics as a scientific discipline. He was most fascinated with regularity in statistical patterns (Desrosières, *The Politics of Large Numbers*, 1998) and collected data on rates of crime, (with an interest in what he called "moral anatomy"), marriage, mental illness, and mortality, including suicides. (Porter, 1985) He believed that conclusions come from data of large numbers—populations—rather than from a study of individual peculiarities. For Quetelet, perfection in science was related to how much it could rely on calculation. Many of these original ideas are found in his classic *A Treatise on Man and the Development of his Faculties*, initially published in French in 1842 and not translated into English until recent years by R. Knox of Cambridge University Press.

Perhaps as a result of his interest in painting, Quetelet became absorbed in measurements of the human body. (Eknoyan, 2008) At the time, he was most known for his concept of the *l'homme moyen*—the "average man." For Quetelet, this average man was hardly the "average" (read "mediocre") that is our present connotation. *L'homme moyen* was an ideal. Says Quetelet, "If the average man were completely determined, we might consider him as the type of perfection; and everything differing from his proportion or condition, would constitute deformity or disease…or monstrosity." He gathered information on the height and weight of different populations. Most notably, though he had no particular interest in the study of obesity, (Eknoyan, 2008) Quetelet was the first to devise the equation that relates weight to height, i.e., w/h^2 (with weight in kilograms and height in meters squared),(Caponi, *História,Ciências,Saúde-Manguinhos,* 2013) now known as our own standard for indicating obesity, the body mass index (BMI) and called quite appropriately, by those in the field, *Quetelet's Index.* (de Waard, *Journal of Chronic Diseases*, 1978; Garrow and Webster, *International Journal of Obesity*, 1985)

Throughout the years, researchers have grappled with standardizing the measurement of overweight and obesity as well as with comprehending obesity's medical implications. It was at the beginning of the 20th century that scales

Pieter Bruegel the Elder, *The Fight between Carnival and Lent* (detail), 1559. Kunsthistorisches Museum, Vienna. An artistic rendering by Bruegel of a fight between the fat and the lean.
Wikimedia Commons/Public Domain.

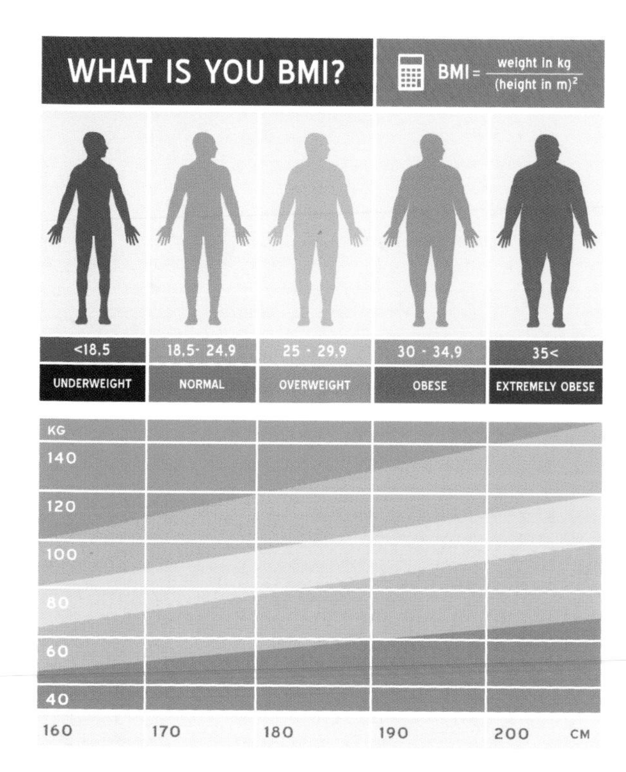

Despite its limitations, BMI has become a standard for indicating our body's level of overweight or obesity. The original ratio, made popular by researcher Ancel Keys, was devised by 19th century Adolphe Quetelet.
Used with permission, istockphoto.com. Credit: elenabs.

became available for home use and insurance companies began to associate excessive weight with decreased life expectancy. (Harrison, *Annals of Internal Medicine*, 1985; Pai and Paloucek, *Annals of Pharmacotherapy*, 2000) These early tabulations, though, were hardly random samples: they were data compiled on customers who had purchased life insurance policies during a particular time period. Furthermore, there was absolutely no attempt at standardization. Some of those in the sample reported their own height and weight, often notoriously inaccurate. Those who were actually measured wore their own clothing and shoes that could distort both measurements. In the early 1940s, one of the companies, the Metropolitan Life Insurance Company, had developed tables of "desirable weight" that did not include a person's age and introduced an initially arbitrary and subjective measure of body "frame"—small, medium and large. (Pai and Paloucek, 2000) The Metropolitan Life Insurance Company revised its tables over the years, and some may remember these were very popular benchmarks, particularly in the late 1950s and 1960s, that were used by physicians to assess "ideal weight" in their patients. During these years, Quetelet's Index was apparently lost to history.

The term "index of bodily mass," also referred to as the "ponderal index," first appeared in the 1940s book *The Varieties of Human Physique* by William H. Sheldon, famous for his division of body types into ectomorph, endomorph, and mesomorph. Sheldon used a different ratio, of height in meters/weight in kilograms[3] that he described as "long been used in attempts at bodily classification…(but) by no means an infallible index." The first reference to the term "body mass index" (even using the initials BMI) appeared in a 1959 paper (Di Mascio, *Psychological Reports*) on the somatotypes of dogs, but the ratio used was also not the one devised by Quetelet, but rather the ratio of weight in kilograms to height in meters cubed (w/h^3). References to the different indices (including mentioning *Quetelet's Index* and a simple w/h ratio) continued to appear in the scientific literature during the 1960s. Quite presciently, Billewicz et al (*British Journal of Preventive and Social Medicine*,

Figure of Dido as she kills herself. (Book IV, Vergil's *Aeneid.)* Quetelet collected statistical data on crime, marriages, mental illness, and even suicide. Tin-glazed earthenware, 1522. Painter of the *Judgment of Paris.*

Metropolitan Museum of Art, NYC. . Public Domain. Gift of V. Everit Macy, in memory of his wife Edith Carpenter Macy.

1962) wrote in the early 1960s that no formula that related weight to height could actually measure fat.

It was not until 1972, though, when researcher Ancel Keys and colleagues popularized the use of Quetelet's original index, claiming it was superior to other indices after they compared the index with measurements of fat by skin calipers and underwater weighing (body density) in an analysis of over 7400 healthy men in five countries. (Keys et al, *Journal of Chronic Diseases*) In this paper, Keys and his colleagues proposed that Quetelet's ratio, w/h^2 be termed *body mass index*. In that paper, Keys and colleagues refer to Quetelet but ironically, despite an extensive bibliography, do not directly reference any of Quetelet's many papers. They also note Quetelet never actually advocated his ratio as any kind of general measure of body 'build' or fat. Belgium, though, issued a stamp honoring Quetelet in 1974.

Since Keys and his colleagues' classic paper, body mass index (BMI) has become the standard indicator for obesity, though cut-off values have gotten more stringent over the years and have led to more people being labeled obese. At

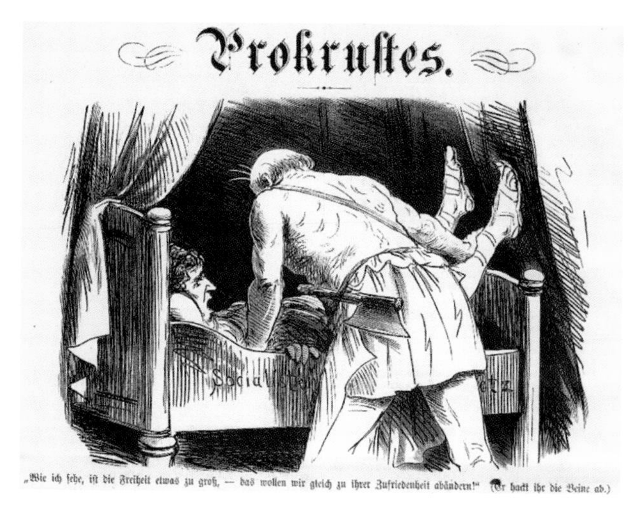

▲ English caricature from a 19th century German satirical magazine, *Berliner Wespen*. Title: *Procrustes*. 30 August 1878.

No author given. Wikimedia Commons/Public Domain.

◀ *A Renaissance man gestures coolly at all the knowledge that lies available to him.* Etching by G.M. Mitelli, circa 1700.

Wellcome Images, Wellcome Trust. Licensed under Creative Commons Attribution 4.0 International. Wikimedia Commons.

present, those with a BMI of 30 kg/m^2 or higher are considered obese, and those over 25 kg/m^2 to 29.9 kg/m^2 are considered overweight. But as noted, BMI is just an estimate of the amount of adipose tissue we have; it does not differentiate fat from muscle and can be particularly inaccurate in certain populations such as athletes or those who are very tall or very short. One reason for its popularity is that it is convenient to use: a physician, who often now has a BMI chart in the office, requires no more than a balance scale for weight and a tape measure for height. There is even a means of converting our ratio in pounds and inches to the metric system by multiplying by 703. More recently, researchers have suggested using waist-to-height ratios as an indicator of health risk. (Ashwell and Gibson, *British Medical Journal*, 2016)

There are, of course, more accurate means of assessing body composition, such as underwater weighing (densitometry), MRIs, CT scans, or DXA (dual-energy X-ray absorptiometery, used for bone density evaluation), but these require a laboratory setting or special equipment and cannot be used in all populations (e.g. pregnant women) if radiation is involved. (Karasu and Karasu, *The Gravity of Weight*, 2010)

Despite all the progress we have made in science since Quetelet's 19th century index, we are still far from being able to measure our body's fat conveniently and accurately in a physician's office. Body Mass Index is one approximation we have at present but sometimes it may seem like the modern day Procrustean equivalent of attempting to force people into simple paradigms. ■

MARIJUANA AND WEIGHT: A PLANT WITH VIRTUES TO BE DISCOVERED?

The "world's most used illegal drug" has paradoxical effects on body weight

73 — Posted April 25, 2016

ENGLISH NOVELIST and poet Rudyard Kipling (1865–1936), in his 1923 lecture *Surgeons and the Soul*, given to the Royal College of Surgeons, said, "I am, by calling, a dealer in words; and words are, of course, the most powerful drug used by mankind." For Kipling, who won the Nobel Prize for Literature, words may have been most powerful, but there are many millions who tout the power of the drug marijuana. It has been estimated, for example, that 178 million people age 15 to 64 worldwide used marijuana at least once in 2012 (Whiting et al, *JAMA*, 2015). With the growing movement to legalize marijuana nationally, there is every reason to believe the rates are increasing.

The scientific name for marijuana is *Cannabis sativa*, but not all *Cannabis sativa* is marijuana: hemp is also in this family and has the same five-pointed leaf but does not have its psychoactive properties. The "defining difference" between the two is in their cultivation: marijuana tends to grow in hotter climates while hemp is cultivated in the north and in cooler climates. (Chasteen, *Getting High: Marijuana Through the Ages*, 2016, p. 13) Throughout history, hemp has been grown for cloth, jewelry, and even canvases for artwork.

The major psychoactive ingredient in marijuana is delta-9 tetrahydrocannabinol or THC, the most researched of one of 100 cannabinoids in the cannabis plant. There are preparations from the herb (marijuana), cannabis resin (hashish), or cannabis extract (hashish oil.) (Reuter and Martin, 2016, *Clinical Pharmacokinetics)*. Marijuana can be smoked, drunk in tea, eaten in food such as the proverbial brownies, or inhaled in an e-cigarette vaping system.

Cannabis sativa (marijuana plants).

Used with permission, istockphoto.com, mrhighsky.

Cannabis is a fat-soluble molecule that crosses both the placenta and the blood-brain barrier and accumulates in adipose tissue where its half-life can from last days to more than a week. (Haelle, quoting Schizer, *Clinical Psychiatric News*, 2015)

Historian Chasteen (2016) notes historically that marijuana has been used globally, particularly for spiritual rather than recreational use—"more as a hallucinogen than a euphoriant." (p. 5) He reports it came into the U.S. only around 1900, probably from Mexico. Chasteen writes that marijuana initially was seen as a "minor poison by comparison to alcohol," (p. 19) whose prohibition came into effect in 1919, and it did not come into U.S. consciousness until the 1930s when it was considered the "devil's weed."(p.4) There are many slang terms for marijuana, but the *Oxford English Dictionary (OED)* notes the word "weed" was first mentioned for marijuana back in the late 1920s.

Harry Anslinger, then head of the Federal Bureau of Narcotics, touted the dangers of marijuana use and made marijuana illegal (and included hemp!) with the *Marihuana Tax Act of 1937*. Anslinger's negative assessment was depicted in the 1930s propaganda (and later 1960s cult classic) film *Reefer Madness*. It was not until the countercultural movement of the late 1960s and early 1970s that marijuana use in the U.S. reached its heyday and became the "world's most used illegal drug." (Chasteen, p. 42, 2016)

Though the word "munchie" first appeared in English in the early 1900s to signify chocolate, according to the *OED*, it later came to represent any snack food and by the late 1960s, specifically the "powerful feelings of hunger" after taking marijuana. Anecdotally, those taking marijuana described strong cravings and feelings of hunger as far back as ancient Hindu texts (Kirkham, *International Review of Psychiatry*, 2009). The symptoms of the "munchies" led researchers to become curious about the effects of marijuana on weight, including how it would affect those who

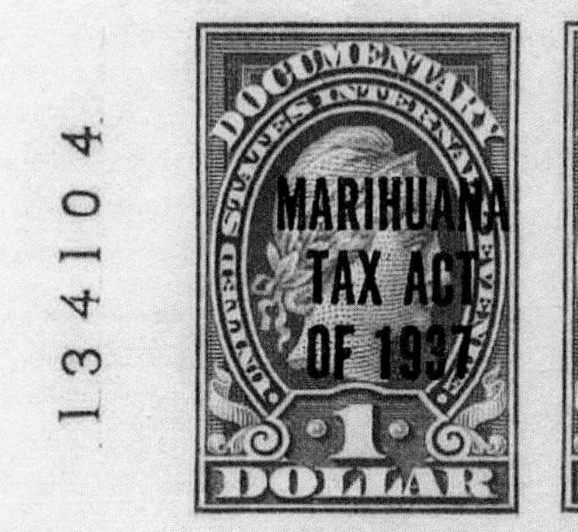

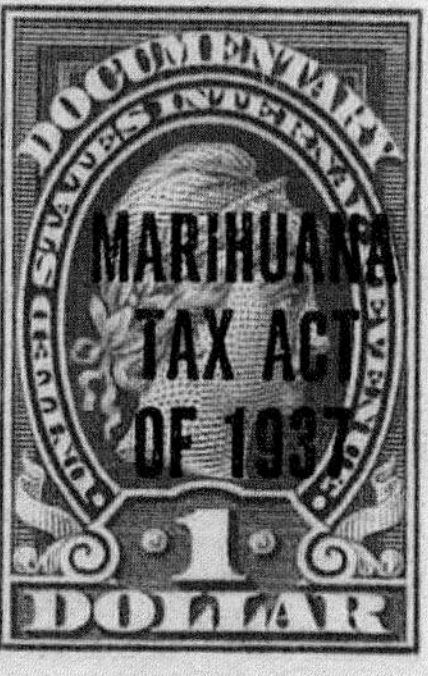

Anslinger was responsible for the *Marihuana Tax Act of 1937* that made both marijuana and even hemp illegal in the U.S.

Source: U.S. Government, Dept. of Internal Revenue, Smithsonian National Poster Museum, U.S. Bureau of Engraving and Printing: Imaging by Gwillhickers. Wikipedia Commons/Public Domain.

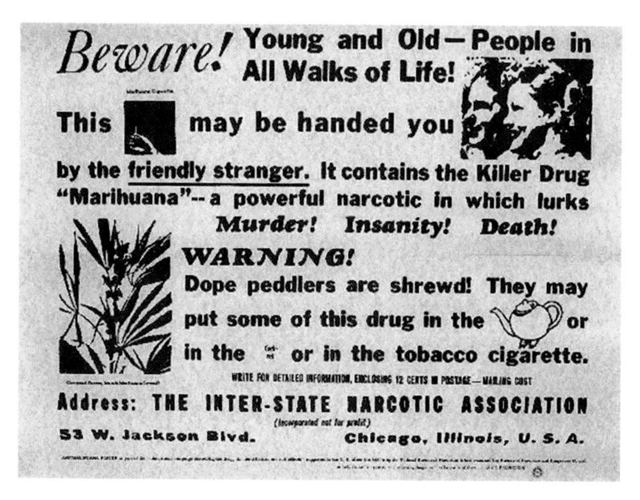

Advertisements that warned about the dangers of marijuana surfaced in the 1930s, particularly by Harry Anslinger, then head of the Federal Bureau of Narcotics. Source: Federal Bureau of Narcotics, 1935.

Wikipedia Commons/Public Domain.Wikipedia Commons/Public Domain.

suffer from cachexia, i.e., the reduced appetite and wasting due to cancer or HIV AIDS. The use of cannabis for medical conditions such as pain and nausea, though, dates back thousands of years B.C. in China and India. (Farrimond et al, *Phytotherapy Research*, 2011).

In the late 1980s, researchers found an endogenous cannabinoid signaling system within animals and humans that included cannabinoid receptors CB1 and CB2. Romero-Zerbo and Bermudez-Silva (*Drug Testing and Analysis*, 2014) report that CB1 receptors are found predominantly at nerve terminals and responsible for the psychotropic effects of THC, whereas CB2 are found mostly in immune cells. The CB1 receptors are the ones responsible for the hyperphagic ("munchies") effects of the cannabinoids. In fact, the CB1 antagonist medication rimonabant was developed years ago as an anti-obesity medication but was ultimately withdrawn from the European markets even before approval in the U.S. because of unwanted psychiatric side effects including suicidal ideation and depression.

It is believed that the endocannabinoid system may be involved in the maintenance of our body's homeostasis (food intake and energy balance, including lipid metabolism and glucose control) as well as in modulating our reward system in the brain. Kirkham (2009) reports that the endocannabinoids may be involved in food seeking and eating initiation, as well as in food "liking"—the hedonic aspect of eating and that CB1 gene variants might eventually be used to assess endocannabinoid functioning and the predisposition of some to overeating, overweight, and obesity. Medications like rimanobant make food less pleasurable. There is some suggestion that those with a weight problem may have a "dysregulated endocannabinoid system and targeting this system may ultimately be an effective anti-obesity medication." (Romero-Zerbo and Bermúdez-Silva (2014) Farrimond et al, 2011 report that since marijuana increases both "appetitive behaviors" that decrease time between meals and "consummatory behaviors" that regulate the size of a meal, these different behaviors can be eventually "manipulated differently."

While medications that block the CB1 receptor lead to weight loss, there is some suggestion that cannabis itself can lead to decreased weight. Le Foll et al (*Medical Hypotheses*, 2013), for example, speculate that *chronic* use of THC or some combination may lead *paradoxically* to weight loss "under carefully controlled circumstances" but cautioned against the use of smoked cannabis for obesity control because of health consequences associated with inhaling the smoke.

The literature, though, is confusing. In reviewing the connection between marijuana and body weight, Sansone and Sansone (*Innovations in Clinical Neuroscience*, 2014) found that a single marijuana cigarette tends to have no effect on food intake, whereas higher doses of two or three marijuana cigarettes increase caloric intake between meals. Dronabinol, a synthetic compound with the same effect as marijuana, is used medically to increase food intake but over time patients have developed tolerance to its effects. In general, marijuana can have negative effects in those compromised by illness such as cancer or AIDS: fatigue, somnolence, and even hallucinations. Recent studies are complicated because of the considerable individual variability among subjects as well as clinical trials for evidence-based research on work with cancer patients, for example, have investigated only oral formulations and not different modes of administration. (Reuter and Martin (*Clinical Pharmacokinetics*, 2016)

Sansone and Sansone speculate that contradictory findings regarding the relationship of weight to marijuana may

Brahma with attendants and musicians, unidentified artist, late 16th century, Korea. Hanging scroll on hemp. Hemp has been used throughout the years as a fabric for clothing and artworks.

Gift of Mrs. Edward S. Harkness, 1921. Metropolitan Museum of Art, NYC. Public Domain.

Daniel Chester French, Bust of Ralph Waldo Emerson, who asked, "What is a weed? A plant whose virtues have yet to be determined." 1879, Metropolitan Museum of Art, NYC.

Gift of the artist in 1907. Public Domain.

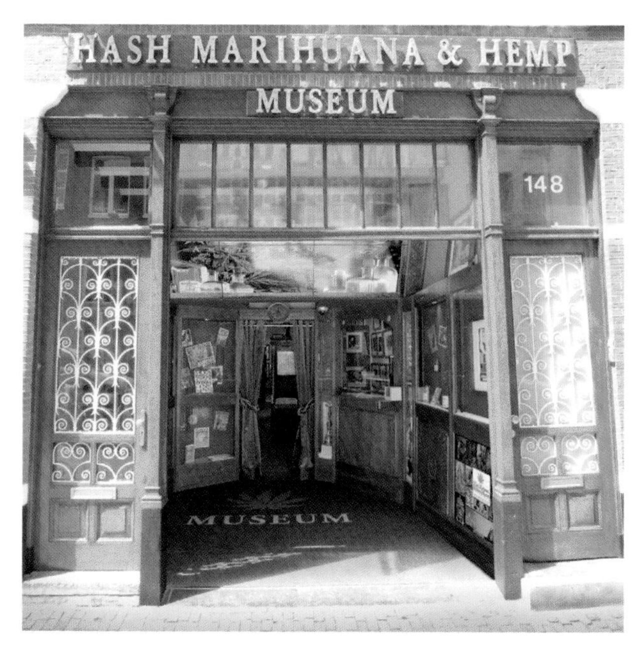

Hash Marihuana & Hemp Museum in Amsterdam where marijuana has been legal for years.

Didier le Ger, 2013. Licensed under Creative Commons Attribution-Share Allike 3.0 Unported. Wikimedia Commons.

be related to potentially confounding factors such as varying dose, frequency, other components of marijuana, various exercise levels, other medications prescribed or other drugs used, as well as differences between short and long term use of marijuana. Further, both marijuana and food compete for the same reward centers in the brain. They further speculate that marijuana may have a "broad spectrum regulatory effect"—increasing weight in those who are underweight but not in those with normal or increased weight.

Bersani et al (*Drug Testing and Analysis,* 2016) report on a trend on the internet of an increased use of cannabis for weight loss—reporting "appetite reduction and eating pattern modifications." These authors speculate that this paradoxical effect may be due to the development of tolerance or to the presence of other compounds and caution against the health risks associated with this behavior.

Cluny et al (*PLOS One*, 2015), in experiments on mice, found that chronic exposure to THC led to reduced food intake in their mice and prevented high fat diet-induced increase in weight and fat accumulation. These researchers speculate that THC has direct effects on the microbiome of the gut (and not likely due to changes in gastric motility) and may be responsible for the decreased weight: those mice exposed to *chronic* THC had an increased ratio of Firmicutes to Bacteroidetes, as well as an increase in a mucin degrading bacteria involved in the regulation of mucus in the gastrointestinal tract.

In their comprehensive 2016 review in *JAMA Psychiatry* on the effects of marijuana on human behavior, Dr. Nora Volkow, Director of the National Institute on Drug Abuse (NIDA), and her colleagues write that the national trend to legalize marijuana "could trigger a broad range of unintended consequences." They note that there have been decades of "ill-informed and porous legal and illegal drug regulations that have exacted a devastating public health toll from our society."

Volkow et al suggest that the recent efforts to legalize and even normalize cannabis use are "being driven largely by a combination of grassroots activism, pharmacological ingenuity, and private profiteering, with a worrisome disregard for scientific evidence..." Further, over the years, the concentration of THC has increased dramatically from that of the 1960s, and cannabis has become potentially more dangerous. These researchers appreciate there are clearly gaps in our knowledge and support the need to clarify scientifically the aspects of cannabis exposure (e.g. age at initiation, quantity used, frequency of use, duration of use and potency of cannabis used) that may lead to increased risk.

Reports of troublesome effects, especially in certain vulnerable populations such as adolescents or those prone to psychosis, though, have been reported over the years. For example, psychiatrists Kolansky and Moore, in a series of articles in *JAMA* in the 70s (1971, 1972, 1975) presciently warned of a deleterious and toxic amotivational (though reversible over time once exposure ceased) syndrome they had found in adolescent and young adult patients who were heavy, chronic users of marijuana referred to them for consultation. Of note, Kolansky and Moore reported that their patients tended to be physically thin.

Both the benefits and risks of cannabis use remain controversial all these years later. Ralph Waldo Emerson (1803–1882), American essayist, lecturer, and abolitionist, asked in his 19th century lecture *Fortune of the Republic,* "What is a weed? A plant whose virtues have not yet been discovered..." A weed, though, according to the *OED,* is a herbaceous plant not valued for its use or beauty, growing wild and rank..." In that context, "weed," then, becomes a paradoxical slang term for marijuana. Evidence-based scientific research will ultimately tell whether marijuana will have other virtues, including for weight control.

For those interested in research on the uses of marijuana as a medical therapy to treat disease or reduce symptoms (e.g. nausea and vomiting due to chemotherapy for cancer; chronic pain or spasticity due to multiple sclerosis), see the comprehensive review by Whiting et al (*JAMA* 2015) or research done by Reuter and Martin (*Clinical Pharmacokinetics*, 2016.) There is also a new book *Marijuana and Mental Health*, edited by Michael T. Compton, MD, MPH. (2016, American Psychiatric Association Publishing, Inc.) ■

John Collier, portrait (1891) of Rudyard Kipling, Nobel Prize Winner for Literature, who famously said he was a "dealer in words."

Wikimedia Commons/Public Domain.

BODY WEIGHT IN THE TIME OF CLIMATE CONTROL

Creatures of comfort: Entering the monotony of the TNZ—the thermoneutral zone

74—Posted June 3, 2016

Honeywell's iconic model thermostat, *The Round,* Smithsonian Institute.

THE HEAT in the unnamed Caribbean seaport in *Love in the Time of Cholera*, Gabriel García Márquez' steamy novel set in the last years of the 19th century and early years of the 20th, is "devastating," "sweltering," "infernal," "deadly," "suffocating," "dusty" and "blazing." Likewise, Harper Lee, in her powerful classic *To Kill a Mockingbird*, set in Alabama in 1932, describes "men's stiff collars wilted by nine in the morning...(and) Ladies bathed before noon... by nightfall were like soft teacakes with frosting from sweating and sweet talcum." These were the years before air conditioning became commonplace. In each novel, the external heat is an insidious and dramatic backdrop for the subsequent unfolding events.

History is replete with examples of human attempts to control our ambient environment. In the 18th century, even American diplomat, scientist and inventor Benjamin Franklin became interested in the mechanisms for cooling our bodies. In a letter written in 1758 (June 17th, in Franklin, *Experiments and Observations on Electricity*) to Scottish-born physician John Lining—who was himself interested in metabolism and famous for being the first to report on yellow fever in the colonies and to keep *A Diary of Weather* (Mendelsohn, 1960—*Chicago Journal, History of Science),* Franklin speculated on the complex mechanisms (e.g. continual sweating and evaporation of the sweat) that the human body develops to cool itself. After one of his experiments involving ether and "spirits," Franklin wrote to Lining, "...one may see the possibility of freezing a man to death on a warm summer's day."

Though there were earlier precursors, the modern thermostat, an automatic device for regulating temperature around a set point, was patented as a "heat governor" by a Scottish scientist Andrew Ure in the 1830s. (*Oxford English Dictionary*) The "air conditioning revolution," itself developed slowly over about the past 70 years, though there were earlier unsuccessful attempts. (Arsenault, *The Journal of Southern History,* 1984.) Engineer Willis H. Carrier is considered the inventor of the world's first air-conditioning system (controlling both temperature and humidity) at the turn of the 20th century, and according to Arsenault, Carrier's "ingenuity and vision" guided the industry for the next fifty years.

Initially air conditioning was used primarily in industries to standardize the working environment. Later, movie theaters, railroad cars, and banks were air-conditioned. Air conditioning in hospitals did not begin to become more common until the 1940s and 1950s, and air-conditioned automobiles, after World War II. Most courthouses and federal buildings in the South did not have air conditioning until the 1950s. (Remember the scene of the almost palpable sweltering heat in the un-air-conditioned courtroom of the film version of *To Kill a Mockingbird.*) Arsenault notes that by 1955 one out of every twenty-two American homes had air conditioning, and one out of ten in the South, and once air conditioning "invaded the home and automobile, there was no turning back." He credits air conditioning with changing the southern way of life, "influencing everything from architecture to sleeping habits."

In the South, with some of the highest rates of obesity in the U.S., central air conditioning increased in from 37 to 70% of homes just between 1978 and 1997 (Keith et al, *International Journal of Obesity*, 2006); by 1997, 93% of Southern households had some form of air conditioning. (Healy, *Building Research & Information*, 2008) U.S. census data for the presence of air conditioning in completed new single family homes in the U.S. found 49% in 1973 and 91% by 2014.

Healy (2008) writes of the "thermal monotony" or "homogeneity" of air-conditioned environments that we create "via scientifically delineated norms of thermal comfort" throughout the world. Johnson et al (*Obesity Reviews*, 2011) note that we now have "increased expectations of thermal comfort." For example, not only have bedroom temperatures increased over the decades but also hallways and other rooms are more likely than in the past to be kept at homogeneous temperatures. Further, temperatures in the workplace have increased.

Does all this homogeneity of our ambient environment have an effect on us? Being homeothermic, we humans are able to maintain a fairly stable body temperature (around 37 degrees Centigrade, or 98.6 Fahrenheit), despite the air temperature around us. This ability, though, comes at a metabolic cost—it requires energy. When humans are

Benjamin West, *Benjamin Franklin Drawing Electricity from the Sky,* circa 1816. Franklin was not only an American diplomat, but also was renown for his scientific experiments, including taking an interest in how the body cools itself.

Mr. and Mrs. Wharton Sinkler, 1958. Philadelphia Museum of Art, Wikipedia Commons/Public Domain.

exposed to cold, they conserve heat by peripheral vasoconstriction; they also may begin shivering that entails an increase in energy expenditure—generating heat through involuntary contraction of skeletal muscles. There is also non-shivering or so-called "adaptive thermogenesis" that defends body temperature by generation of heat in tissues, and is also initiated by eating. Diet-induced thermogenesis is the "energy cost" involved in the digestion and absorption of food. Further, metabolically active brown fat, once thought to be present only in infants (and is responsible for their not shivering), is now thought to play a much greater role in thermogenesis than once believed. It appears in adult humans with mild cold exposure. For more on brown fat, see blog 21, *Special Delivery: What Can Brown (Fat) Do for You?*

There is the suggestion that there are seasonal differences in the prevalence of active brown fat. Furthermore, Gerhart-Hines and Lazar (*Endocrine Reviews*, 2015) suggest that brown fat is under circadian rhythm control, and they hypothesize that the circadian clock was evolutionarily programmed to "turn off" brown adipose tissue during sleep when most animals are in a shielded environment. Now with a 24/7 lifestyle of highly caloric diets eaten later, there may be less calorie burning at night. Johnson et al (2011) maintain that as we seek "thermal comfort" and have less exposure to cold, we will reduce the frequency and duration of occasions when "cold-induced expenditure" (e.g. brown fat) is utilized.

In an attempt to explain the burgeoning rate of overweight and obesity globally in recent years, researchers

Willis Carrier in 1915, from the Carrier Corporation. Cornell engineer Carrier is credited with designing the first air conditioning system.

Wikipedia Commons/Public Domain.

Michelangelo, Ceiling of the Sistine Chapel, from 1508–1512, Vatican. The Carrier Company was commissioned to help preserve this masterpiece by designing a groundbreaking ventilation, heating, and air conditioning system.

Wikipedia Commons/Public Domain. Creative Commons Attribution Share Alike 3.0, unported. Photographer: Antoine Taveneaux, 2014)

Paul Gauguin, *The Siesta,* 1890s. Even the traditional siesta is less common in some places because of air-conditioned, climate-controlled environments. Metropolitan Museum of Art, NYC.

The Walter H. and Leonore Annenberg Collection, Gift of Walter H. and Leonore Annenberg, 1993, Bequest of Walter H. Annenberg, 2002, Public Domain.

have begun to explore "the roads less traveled."(Keith et al, 2006, after poet Robert Frost) Since our genetics have not changed during this brief period in evolutionary history, investigators have focused primarily on the environmental effects of excessive caloric intake and decreased physical activity due to our sedentary lifestyles as possible contributions—the so-called "Big Two" to explain the rising obesity rates. (Keith et al, 2006) Increased energy intake and decreased energy expenditure, no doubt, have had a major impact, but there are many other "putative contributions" (Keith et al, 2006) that are "biologically plausible" explanations (Johnson et al, 2011.) Note that the operative word is "contribution" rather than "cause." No one, of course, really knows what factors are responsible for recent increases in "globesity." Allison and his colleagues, though, have written two exceptionally comprehensive reviews that suggest that exposure to environmental toxins (e.g. endocrine disruptors), medications that lead to weight gain, (e.g. antipsychotics and antidepressants) reduced smoking in the population, advanced maternal age, decreased sleep time, exposure to viruses, and for our purposes here, more time spent in what is called the *thermoneutral zone (TNZ)* (Keith et al, 2006; McAllister et al, *Critical Reviews in Food Science and Nutrition*, 2009) may be involved.

Pieter Bruegel the Elder, *The Hunters in the Snow*, 1565. As we spend more time in climate-controlled environments, we may be exposed much less to naturally cold weather. Some researchers believe this may be one factor contributing to the increased prevalence of obesity.

Kunsthistorisches Museum, Vienna. Wikipedia Commons/Public Domain.

For humans, the thermoneutral zone is a range of ambient temperatures in which we do not need to utilize energy to maintain a constant body temperature, i.e., "thermal homeostasis." (Mavrogianni et al, *Indoor and Built Environment*, 2013) Daly (*Obesity*, 2014) maintains this zone is 25 to 30 degrees C (77–86 degrees F when *unclothed* or 20.3 to 23 C (68.54–73.4 degrees F for *clothed* humans, while Mavrogianni et al (2013) maintain that it is from 25 to 27 degrees C (77–80.6 degrees F) for *unclothed* humans. When we are not in this zone, either above or below it, our bodies require modifications in energy intake and expenditure. (McAllister et al, 2009) These investigators speculate that since we are spending considerably more time in environments of climate control (e.g. air conditioning and central heating) over the course of the past forty years that this may be one "modest, yet significant contributor to the recent increase in the prevalence of obesity." (McAllister et al, 2009) They have reviewed evidence from both animal and human studies. For example, rats had more adipose tissue when raised in a thermoneutral environment than those rats raised at cooler temperatures whose bodies had to utilize energy to maintain their body temperature. Weight is often reduced when animals, including chicken, pigs, and cattle, are subjected to warmer temperatures, both because these animals eat less at high temperatures and there is an "increased metabolic cost" to maintain a constant body temperature under these conditions. Likewise, when humans are subjected to cooler ambient temperatures, their metabolic rate is lower and more subject to weight gain,

Maurice Brazil Prendergast, *Central Park*, circa 1914–1915. Metropolitan Museum of Art, NYC.

George A. Hearn Fund, 1950, Public Domain.

and when they are subjected to higher ambient temperatures, they tend to eat less, have a higher metabolic rate, and lose weight. These effects, though, can be mitigated in the presence of highly palatable foods.

Back in 1981, Dauncey (*British Journal of Nutrition*) exposed nine women to two different temperatures (a month apart) in a controlled environment and concluded that environmental temperature may "play a more important role than was previously recognized in energy balance."

Over the years, large epidemiological studies have been conducted worldwide, including in England (Daly, *Obesity* 2014); Korea (Yang et al, *PLOS One,* 2015); Spain (Valdés et al, Obesity, 2014); and Nepal and Tibet (Sherpa et al, *International Journal of Environmental Research and Public Health,* 2010). Though different researchers have found significant associations between obesity and ambient temperature, they acknowledge they cannot determine a cause-effect relationship since most studies are cross-sectional and do not account for whether subjects had moved to or from other areas. Further, often the studies rely on potentially unreliable self-report of activity levels and diet. In one study, subjects reported lower indoor temperatures than assessed by thermostat verification. Discrepancies among studies may be reflective of different ranges in mean temperatures, different altitudes (e.g. low oxygen levels at higher altitudes), different ethnic groups that are not representative of diverse populations, or even unspecified confounding variables, as well as methodological differences in data collection. Most of studies done have been in residential rather than work environments (Moellering and Smith, *Current Obesity Reports,* 2012) Furthermore, these authors note there are many factors that influence temperature perception such as air speed, humidity, body size, age, body fat, clothing, diet, and activity, and we all have behaviors to deal with uncomfortable temperatures, such as by donning protective clothing in winter. No single temperature is perfect for everyone.

Thermal control, though, is a "socio-cultural construct." (Mavrogianni et al, *Indoor and Built Environment,* 2013) These authors note that due to climate change, it is likely that outdoor ambient temperatures will increase and reduce heating demand in the winter and increase cooling demand in the summer. They are apprehensive that as people become used to certain comfort, they will become less adaptable to changes in temperature in the future and prefer a "narrower" range of comfort.

Bottom line: How and to what extent almost continuous exposure to climate-controlled ambient temperatures affects us, particularly our body weight, remains open to question. Ambient temperature can clearly affect both energy intake and energy expenditure but may affect people differently. Moellering and Smith (2012) believe that we need more studies that reflect the "daily cycles" of life as compared to the artificial environments of clinical laboratories. These researchers suggest that just as we need exposure to variable diets for health, so too perhaps we need exposure to a varied, natural range of ambient temperatures. ■

ADVISE AND CONSENT

Ethical dilemmas in clinical investigation

75 — Posted July 4, 2016

A DEVASTATING PLAGUE overwhelms the fictional island of St. Hubert in Sinclair Lewis' Pulitzer Prize-winning novel *Arrowsmith.* Physician-researcher Martin Arrowsmith, who has been instrumental in isolating a potential treatment called "phage" at his institute, loosely based on the turn-of-the century institute of Rockefeller University in New York City, is sent to the island to administer his possible cure. Before Arrowsmith leaves, his mentor says, "If I could trust you to use the phage with only half your patients and the others as controls, under normal hygienic conditions but without the phage, then you could make an absolute determination of its value…" Martin swore that "he would not yield to compassion" and would uphold the experimental conditions. When, though, Arrowsmith saw those afflicted "shrieking in delirium" with "sunken bloody eyes," and particularly after his beloved wife who has accompanied him to the island, dies of the disease, he says, "Damn experimentation!" and gave the treatment to everyone who asked.

Mentor Dr. Max Gottlieb was insisting that Arrowsmith conduct a clinical trial—essentially an experiment involving human subjects to assess the efficacy of at least one specific treatment intervention that is conducted to advance general scientific knowledge and not necessarily for the benefit of an individual patient. Subjects in clinical research can sometimes have a "therapeutic misconception," namely that their individualized needs will determine their treatment allocation and have an "unreasonable appraisal" that they will necessarily receive "direct therapeutic benefit" from participating in a research study (i.e., "misconception regarding the process or goals of the research.") (Swekoski and Barnbaum, *Ethics & Human Research,* 2013) A clinical trial is based on the concept of *equipoise* in which either the researchers themselves "genuinely do not know what is the best way to treat their patients" (Doll, 1982, *Statistics in Medicine)* or a state of "genuine uncertainty" within the medical community about the therapeutic merits of a particular treatment. (Freedman, *NEJM,* 1987)

Throughout the centuries, there have been reports of several "clinical trials," reviewed in detail by both Bhatt (*Perspectives in Clinical Research,* 2010) and Vandenbroucke, (*Journal Chronic Diseases,* 1987). Possibly the very first one involved comparing diets when King Nebuchadnezzar of Babylon in the Old Testament's *Book of Daniel,* by request, allowed Daniel and some of his men to take only vegetables and water and the others, only meat and wine. (Those who ate only the vegetables apparently fared better.) There was also the famous 1747 study by Scottish ship surgeon James Lind, in another diet comparison, who discovered the anti-scurvy benefits that sailors gained from a diet that included lemons and oranges. (Bhatt, 2010)

In Lewis's novel, Martin Arrowsmith is confronted with the difference between the experimental conditions within the "security of the laboratory" and what one actually has to contend with in the trenches of human misery.

The clinical decision, though, namely to whom to administer the unproven treatment, was not as difficult in the 1946 British Medical Research Council's (MRC) groundbreaking study of streptomycin as a treatment for pulmonary tuberculosis because the medication was in short supply. This landmark clinical trial was conducted under the auspices of Sir Austin Bradford Hill (1897-1991), then director of the MRC's Statistical Research Unit. (I have written previously about Sir Austin, known, as well,

Founders Hall, Rockefeller University (formerly referred to as "the Institute"), founded in 1901, and the model for Sinclair Lewis's *Arrowsmith.*

Photo by KaurJmeb, Wikipedia Commons, Licensed under Creative Commons Attribution 1.0 Generic.

Sinclair Lewis, author of the Pulitzer-Prize winning novel *Arrowsmith* that explores conflicting goals between the researcher-scientist and the clinician. Of note, Lewis refused to accept the Prize.

Photographer: Arnold Genthe, 1914, Wikipedia Commons/Public Domain.

Sir George Chalmers, painting of 18th century Scottish physician James Lind, who recognized that lemons and oranges alleviated the symptoms of scurvy in sailors.

Wikipedia Commons/Public Domain.

for his "viewpoints" on causation and his linking of smoking with lung cancer.) This trial is often considered the first strictly controlled and most importantly, *randomized* trial that "ushered in the new era of medicine." (Hill, *Controlled Clinical Trials*, 1990). Initial randomization is crucial as it prevents *selection bias*—"The aim, then is to allocate those admitted to the trial in such a way that the two groups—treatment and control are initially equivalent." (Hill and Hill, *Principles of Medical Statistics,* p. 219, 1991) Hill, incidentally, had himself been bedridden years earlier for almost two years with tuberculosis that he had contracted in Greece during World War I and his own illness had precluded his career aspirations of becoming a physician. (Hill, *British Medical Journal,* 1985)

Streptomycin had been discovered in the U.S. in 1944 and was not then readily available in a Britain that was impoverished by the War. (D'Arcy, *British Medical Journal,* 1999) Said Hill (1990), "...the shortage of streptomycin was the *dominating feature*" and made it ethically possible to consider a clinical trial in which a potentially beneficial (but not clearly proven) treatment was withheld from half the "desperately ill" patient population. (Hill, *British Medical Journal,* 1963). (The control population received the standard treatment of that era—bed rest.) Significantly, though, the trial was not double-blinded (i.e., both physicians and patients knew which treatment was administered, though the two independent radiologists who read the chest films were "blind" to which group each patient belonged.) Nor was it placebo-controlled (D'Arcy, 1999.) The rationale for this protocol ("no need to throw common sense out the window,"said Hill, *British Medical Journal,* 1963) was that streptomycin administration then required four intramuscular injections a day for four months, and the researchers did not want to subject their control patients to four injections of saline water daily for the duration of the experiment. (Doll, *Statistics in Medicine,* 1982)

Though Hill was always concerned about the ethics of clinical trials (e.g. the use of placebos; withholding of treatment), he did believe that the question of consent should be, "*When* is it necessary to ask the patient's consent to his inclusion in a controlled trial?" (Hill, *British Medical Journal,* 1963) He believed that giving patients too much information, especially about the uncertainty of a treatment, potentially undermined their trust in their physicians. In effect, he was emphasizing that how physicians frame information can have a detrimental effect on the treatment—what we now call the *nocebo effect*—giving patients too much information about potential harm may be harmful itself. (For more on the nocebo effect, see blog 64, *The Quicksand of Self-deception....*)

William Blake, *Nebuchadnezzar,* 1795. Nebuchadnezzar, of the Bible's *Book of Daniel,* is cited as one of the first to allow a trial of two different diets. He later went mad and here is depicted by Blake as half-man/half beast.

Minneapolis Institute of Art. Wikimedia Commons/Public Domain.

Hill noted that ethical standards were different in the 1940s: researchers did not obtain the patient's or anyone's permission. Nor did Hill and his researchers even tell patients they were part of a trial. In fact, Hill believed it was "wrong to shift the entire consent-giving responsibility onto the shoulders of patients who cannot really be informed." (*Controlled Clinical Trials,* 1990) Hill, incidentally, was criticized for his view on consent in a letter by "a member of the uninitiated public" (Hodgson, *British Medical Journal,* 1963) and in an editorial in the *British Medical Journal* that same year.

Bhatt notes that the framework for protection of human subjects had its origins in the Hippocratic Oath—i.e., *Do no harm.* (Bhatt, 2010), but it was not until after the egregious medical experiments, conducted in the name of science by the Nazis, became known that the *Nurenberg Code* of 1947 highlighted the importance of "voluntariness" in giving consent. (Bhatt, 2010) Primo Levi, in his moving *If This Is A Man,* about his experiences as a survivor of Auschwitz, wrote, "We are slaves, deprived of every right, exposed to every insult, condemned to almost certain death, but we still possess one power, and we must defend it with all our strength, for it is the last—the power to refuse our consent." (p. 37) And it was in 1964 that the *Helsinki Declaration* by the World Medical Association established "general principles and specific guidelines on use of human subjects in medical research." (Bhatt, 2010; World Medical Association *Declaration of Helsinki* in *JAMA,* 2013)

Poster promoting early treatment of syphilis, 1941. Poor, disenfranchised Black subjects in Tuskegee, Alabama never gave consent and were left untreated for syphilis for years in an egregious violation of human rights.

Work Projects Administration Poster Collection (Library of Congress). Wikimedia Commons/ Public Domain.

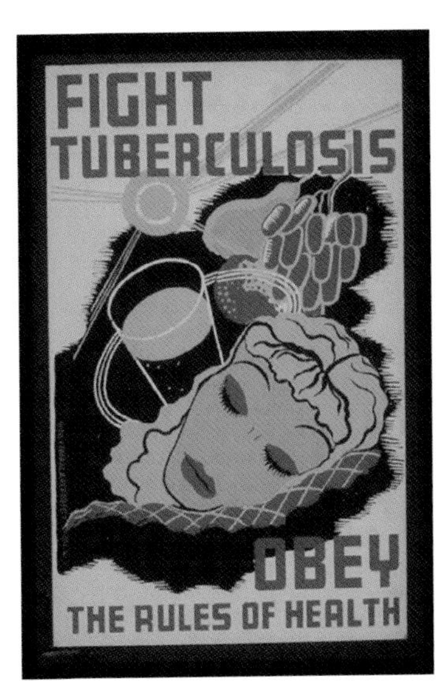

Poster promoting health, 1941. Tuberculosis was the scourge of Britain after World War II. The major treatment before streptomycin became widely available was bed rest. Sir Austin Bradford Hill conducted the first randomized controlled trial of streptomycin.

Work Projects Administration Poster Collection (Library of Congress), Wikipedia Commons/ Public Domain.

Beecher, a professor of anesthesiology at Harvard, reviewed "a variety" of 22 unethical or questionably ethical practices in medicine and noted the "unfortunate separation between the interests of science and the interests of the patient." Though he believed that an experiment is either ethical or not initially, Beecher also acknowledged, "Consent in any fully formed sense may not be obtainable. Nevertheless…it remains a goal toward which one must strive…There is no choice in the matter." (Beecher, *NEJM*, 1966)

Significantly in the last edition (12th edition) before his death of his textbook *Principles of Medical Statistics,* written with his son, Hill not only had a chapter on ethics but included an *Appendix* on ethics and human experimentation that included the principles of the *Helsinki Declaration.* (1991 Hill and Hill)

Throughout the years, despite international attempts at regulating experiments on human subjects, there have continued to be alarming abuses, (some even initiated by U.S. government agencies), particularly on the disenfranchised, such as Blacks (e.g. untreated syphilis studies in Tuskegee, Alabama that did not stop until 1972); mentally impaired (e.g. hepatitis study at Willowbrook State School for the Retarded in NY, also until 1972) and prisoners (e.g. testicular irradiation of state prisoners in Oregon conducted until 1974).

Bottom line: Since exposure of these abuses, the U.S. government has mandated the establishment of Institutional Review Boards (IRBs) that regulate any federally funded research. This "checks and balance" system, as evident in the concept of "advise and consent" is far from perfect but considerably better than any previous time in history. For a discussion of IRBs, see the 2015 book *The Ethics Police?* by Robert Klitzman and for more details on a history of decades-long abuses and more recent regulations, as well as concrete suggestions to improve protection of human research volunteers, see the two-part article by Dr. Marcia Angell in November and December 2015 issues of *The New York Review of Books.* ■

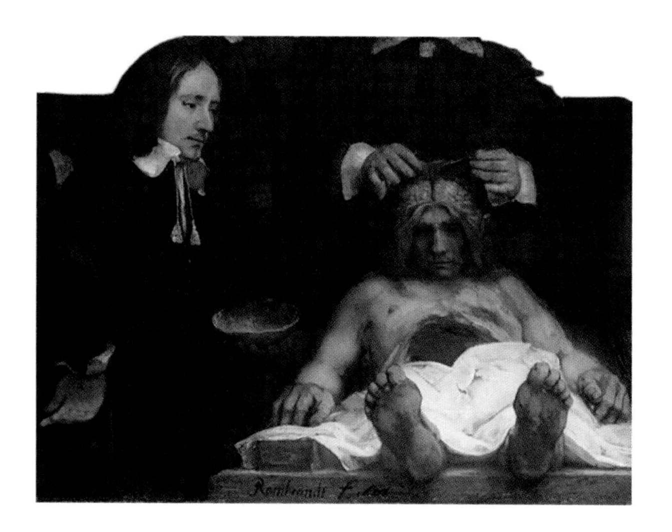

THE "ELECTROCHEMICAL PERSUASION" OF NEUROMODULATION

Was the treatment of obesity ever "brainless?"

76—Posted August 13, 2016

In the totalitarian future of the novel *We*, written in the 1920s by Russian novelist Yevgeny Zamyatin, the Benefactor rules One State; there is no privacy, and people, known only by numbers, live in glass houses. The main character D-503, a mathematician, commits the sin of falling in love. He is suffering from the "illness of imagination," whose cure is "exclusively a matter of surgery" i.e., "a special brain operation invented to excise imagination forever."

Likewise, in George Orwell's *1984*, Winston Smith's torturer says, "We control matter because we control the mind." He adds, "Power is in tearing human minds to pieces and putting them together again…we burn all evil and all illusion out of him…we make the brain perfect…" As a result of his torture, Winston believes "as though a piece had been taken out of his brain." Ironically, in the language of Newspeak, "there was indeed no word for Science."

These two dystopian novels tap into the feelings most people have that there is something particularly disturbing and even creepy about attempts at changing behavior through manipulation of the brain. But this is not just science fiction: it was not so many years ago that physicians attempted to modify a patient's behavior with the psychosurgical procedure of frontal lobotomies, and many remember the vivid scene in the film *One Flew Over the Cuckoo's Nest* when the Jack Nicholson character received ECT for his recalcitrant behavior.

Rembrandt, *The Anatomy Lesson* of Dr. Jan Deijman, 1656, Amsterdam Museum. Much of the original was destroyed by fire.

Wikipedia Commons/Public Domain.

Is there a place for a kind of behavior control, as it were, through either direct or indirect stimulation of the brain in the treatment of obesity and eating disorders? Some researchers think so. After all, these disorders are difficult to treat and our current treatments, including psychological ones, have "only limited efficacy." (Schmidt and Campbell, *European Eating Disorders Review,* 2013) Researchers are now turning to the "emerging neurotechnologies" of both invasive and non-invasive brain stimulation. In other words, treatment is definitely not "brainless." (Schmidt and Campbell, 2013; McClelland et al, 2013, (*European Eating Disorders Review)*

Back in 2005, Spiegel et al (*Nature Neuroscience*) wrote an article "Obesity on the Brain." These researchers proposed to focus on "neurobiology, behavior, and obesity" in developing their strategic plan for obesity research in their attempt to understand why some people overeat and remain sedentary. They emphasized that imaging studies, such as MRI and positron emission tomography "thus becomes the cornerstone of efforts to understand the biology of human eating behavior."

Along those lines, Alonso-Alonso and Pascual-Leone wrote (*JAMA*, 2007), "The Right Brain Hypothesis for Obesity." These authors suggest that the right prefrontal cortex is "preferentially" involved in decision-making and question whether deficits in this area of the brain "may contribute to the inability of some obese patients to commit to weight loss interventions long term…" as well as to their failure to consider the adverse consequences of their "obesogenic" behaviors. This article sparked a lively debate in which Bachman et al (*JAMA, 2007*) wrote, "Before adding the stigma of 'brain damaged' to the high physical and social burden obese persons already bear, we would like to see more compelling data to support the conclusion." Pascual-Leone et al defended their position: "Finding a pattern of

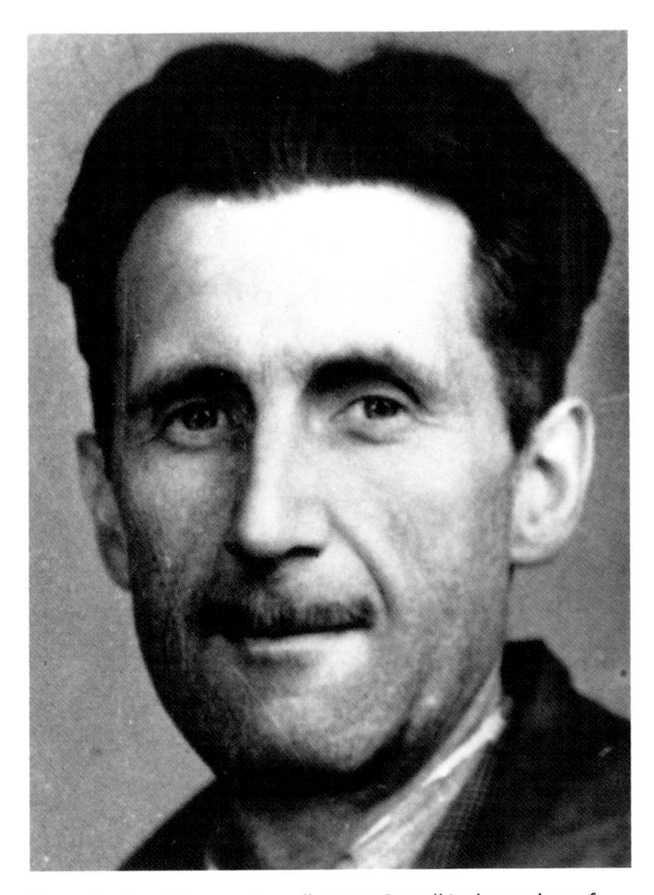

Press photo of George Orwell, 1933. Orwell is the author of the extraordinary novel *1984*.

Unknown photographer. Wikipedia Commons/Public Domain.

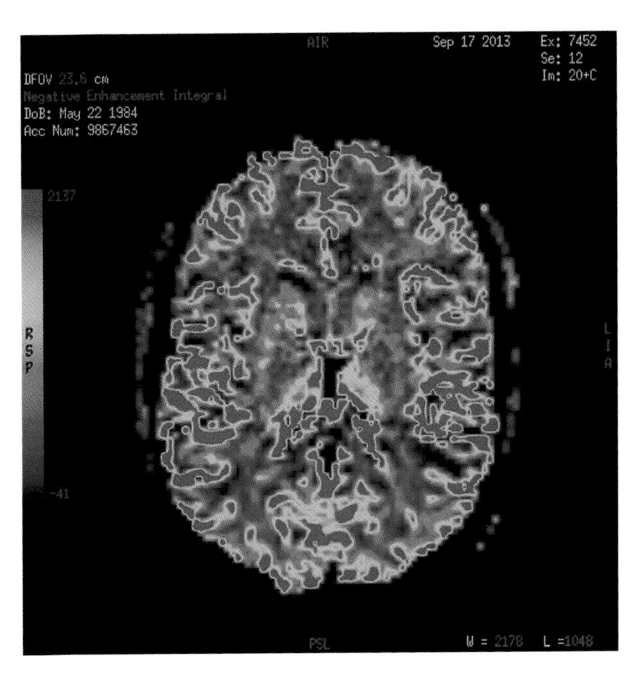

Perfusion MRI of a normal brain. The "color-drenched" (Satel and Lilienfeld) brains seen on MRI scans.

Used with permission/istockphoto.com. Credit : wenht.

brain activity associated with a given behavior or disorder should not be taken as revealing 'brain damage,'" but rather could lead to brain "targeted therapies." (*JAMA*, 2007)

The situation, though, is complex, and "theories of aberrant neural responses" for both obesity and eating disorders appear to conflict with each other: both increased and decreased responsivity to the reward circuit has been found. (Burger and Berner, *Physiology & Behavior*, 2014) Further, Pawel K. Olszewski and Allen S. Levine (*Pharmacology, Biochemistry and Behavior*, 2016) emphasize the important complexities of food intake within a social context. For humans, meals have a social component, and social interaction itself can have an impact on appetite and how much we eat. These researchers are studying the "prosocial" neurohormone oxytocin, known to decrease food intake. Furthermore, though there are two major aspects of control of food intake—*homeostatic* (eating for survival, i.e., hypothalamic, reflexive) and *hedonic* (i.e., eating for pleasure, i.e., cognitively reward-based, reflective), the "degree to which homeostasis and hedonic systems interact to influence intake has yet to be fully understood." (Burger and Berner, 2014)

Satel and Lilienfeld, in their excellent 2013 book *Brainwashed*, raise the question of whether "neurodeterminism" is "poised to become the next grand narrative of human behavior." They write that for many, "brain-based explanations appear to be granted a kind of inherent superiority over all other ways of accounting for human behavior"—what these authors call "neurocentrism," and they call the pictures from MRIs "color-drenched brains."

McCabe and Castel (*Cognition*, 2008), in their study, "Seeing is Believing…" demonstrate Satel's and Lilienfeld's argument. They explored how images of the brain, because they appeal to people's need for "reductionistic explanations of cognitive phenomena" have a "particularly persuasive influence" and artificially lend a "great deal of scientific credibility:" those studies that incorporated brain images (rather than bar graphs or no images) became more scientifically credible to their subjects even when there were obvious errors in the data.

Likewise, Weinberger and Radulescu (*Am. J Psychiatry*, 2016) assert that findings that may seem "neurobiologically meaningful" may, in fact, represent, "artifacts or

A young Atlantic salmon. Bennett et al (2010) demonstrated that even an fMRI of a dead salmon's brain lit up with a false positive response. Researchers must be cautious not to read too much into fMRI scans.

Photo: U.S. Fish and Wildlife Service. Credit: William W. Hartley. Wikimedia Commons/ Public Domain.

epiphenomena of dubious value." These authors emphasize, for example, that MRIs are "not a direct measure of brain structure—an MRI is a "physical-chemical measure" and as such, many things (e.g. body weight, medications, alcohol use, nicotine use, exercise, hydration, pain, etc.) can affect variations in MRI signals and anatomical measurements. Furthermore, even head motion can be a "corrupting influence." The main readout of an MRI is the blood-oxygen-level-dependent (BOLD) response that is only a "proxy used to indicate activity." (Burger and Berner, 2014)

An article that highlights the potential for false positives in functional neuroimaging reads like a classic *Monty Python* skit. Bennett and his colleagues (*Journal of Serendipitous and Unexpected Results*, 2010) *invited* a mature *dead* Atlantic salmon (*Monty Python* might have used a dead parrot) to participate in an fMRI scan. There was no mention of whether consent was obtained but a "mirror directly above the head coil *allowed* the salmon to observe the experimental stimuli" and foam padding prevented excessive movement "but proved to be largely unnecessary as subject motion was exceptionally low." The salmon "was *asked* to determine which emotion the individual in the photo must have been experiencing." What the researchers found is that certain areas of the dead salmon's brain actually lit up on fMRI—clearly a false positive. (We do not know, of course, whether a repeat experiment with a *dead* wild-caught Alaskan salmon would yield similar results.) Nevertheless, the authors *confidently* recommend continued fine-tuning of fMRI technology.

Given these caveats, what is neuromodulation? It is a technique, intended to change behavior, by targeting areas of the brain, (sometimes specific areas such as the dorsolateral prefrontal cortex, amygdala, or nucleus accumbens), either non-invasively or invasively, by various means. Parpura et al (*Journal of Neurochemistry*, 2013) explain that the "brain operates through complex interactions in the flow of information and signal processing within neural networks" and the "wiring" of these networks can sometimes "go rogue in various pathological states." Neuromodulation, they note, "attempts to correct such faulty nets." They describe its potential effects as "electrochemical persuasion."

Neuromodulation procedures can be used either to stimulate or inhibit neural activity via different parameters such as frequency, duration, intensity, number of sessions and stimulation sites. (McClelland et al, *European Eating Disorders Review*, 2013) Val-Laillet et al (*NeuroImage: Clinical*, 2015) reviewed the two most common non-invasive

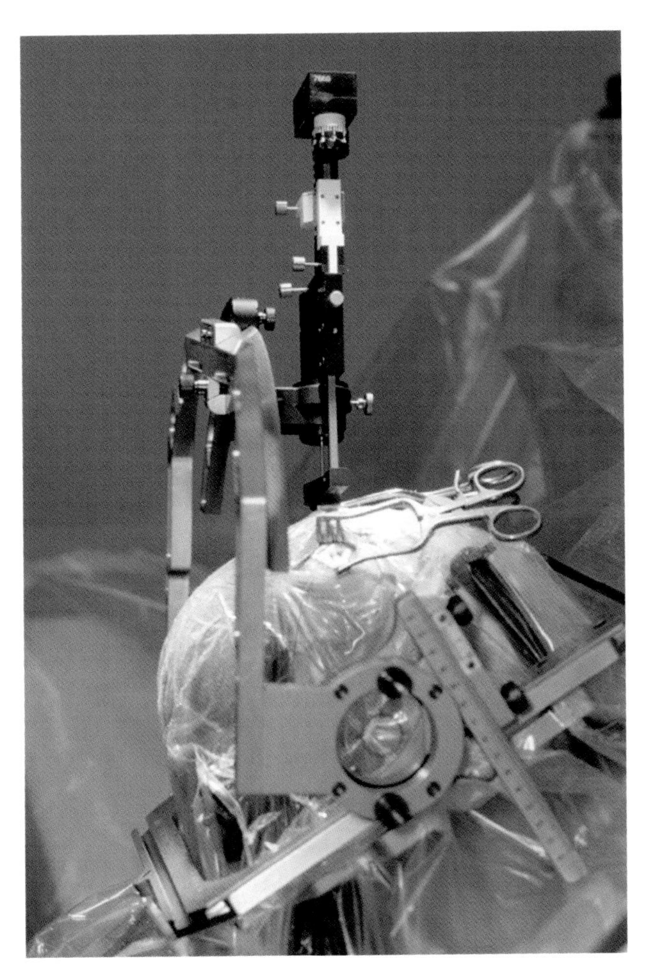

Insertion of electrodes for deep brain surgery, in this case, for treatment of Parkinson's Disease, is hardly an innocuous procedure.

Licensed under GNU Free License, Creative Commons Attribution Share Alike, 3.0, Unported. Wikimedia Commons/Public Domain.

Orwell's "Big Brother," poster from his iconic novel *1984*.

Free Art License, Copyright Frederic Guimont. Wikimedia Commons/Public Domain.

neuromodulation techniques—-transcranial magnetic stimulation (TMS) and transcranial direct current stimulation (tDCS), as well as the two more invasive procedures—deep brain stimulation (DBS) and vagal nerve stimulation (VNS). Most studies involving either tDCS or TMS evaluate effects on food craving, subjective appetite, and food intake, including more recently, binge-eating. (Burgess et al, *International Journal of Eating Disorders*, 2016.) In this study, the researchers suggest that those people with the "cognitive mindset" to restrict caloric consumption may be particularly primed to respond to tDCS.

Val-Laillet et al (2015) note that the specifics of neuromodulation, including the most efficacious protocols "remain to be defined." Furthermore, there is variability among the reactions of patients, including that sham procedures can evoke placebo responses. The techniques are not necessarily "ready for prime time." For example, Gluck et al (*Obesity*, 2015) studied transcranial direct current stimulation (tDCS) administered to the dorsolateral prefrontal cortex of their subjects to assess caloric intake and potential for weight loss. They note that some current "does penetrate the brain" when they applied low amplitude direct currents with electrodes placed on the scalp: with cathode stimulation, nerves are less excitable and less likely to fire while anode stimulation increases their excitability and makes them more likely to fire. Unfortunately, in the first arm of their research, the anode and cathode leads had been inadvertently reversed, and the study had to be halted.

In general, though, Val-Laillet et al (*NeuroImage: Clinical*, 2015) note side effects with the non-invasive types are usually mild and can include neck pain, headache, tingling, itching, but the "most worrisome" is the induction of a seizure. They emphasize, though, "the current state of the science is far from being conclusive" in "the possibility of manipulating the human brain."

McClelland et al (2013) reviewed 60 human and animal studies involving the four available (two non-invasive and two invasive) neuromodulation techniques: research indicates that neuromodulation has the potential for altering food intake, body weight, and disordered eating (e.g. as found in those with anorexia, bulimia or those who have strong food cravings.)

Studies from lesions have implicated three potential brain targets associated with increased food intake for

Ink Stand with Madman Distilling his Brains, circa 1600, Italian, probably Urbino, Maiolica. Present-day manipulations of the brain may eventually seem as primitive as that depicted by this "madman." Metropolitan Museum of Art, NYC.

Rogers Fund, 1904. Public Domain.

consideration for the surgically invasive technique of deep brain stimulation (DBS): the lateral hypothalamus (appetite regulation), ventromedial hypothalamus (often considered the "satiety center" of the brain), and the nucleus accumbens (value of food, regardless of appetite.) (Kumar et al, *Annals of Neuroscience*, 2015) With DBS brain surgery, complications can be significant and include stroke.

The other invasive procedure is the newly FDA-approved vBloc, a reversible vagal nerve blockade that is far less invasive than bariatric surgery. The *ReCharge Trial* is ongoing (though many of those involved have conflicts of interest in that they have an affiliation with the manufacturer of this device.) To date, this therapy resulted in significantly greater weight loss than sham surgery, though those in the sham control group lost more weight than expected (Ikramuddin et al, *JAMA* 2014; Morton et al, *Obesity Surgery, 2016*) It is described as "a device that controls hunger and feelings of fullness by targeting the nerve pathway between the brain and the stomach." Side effects include nausea, pain at the implant site, heartburn, vomiting and surgical complications.

Both Val-Laillet et al (2015) Ho et al (*Cureus, 2015)* raise the bioethical issues involved in neuromodulation, as a "behavior-altering treatment." For example, Ho et al question whether manipulating the rewarding aspect of food intake may have unforeseen consequences on a person's ability to experience other pleasures. Clearly that was the intent in the "Big Brother" society of *1984*.

Bottom line: Neuromodulation remains poorly understood, whether it involves "depth electrodes (i.e. DBS), arrays of microelectrodes penetrating the brain cortex, or cortical surface electrodes placed outside or inside the dura." (Parpura et al 2013) These researchers emphasize that these techniques as practiced currently are "all insultingly inelegant for interacting with such a marvelous structure as the nervous system." ■

IS EXTREME CHILDHOOD OBESITY 'NUTRITIONAL NEGLECT'?

How much are parents responsible for their children's excessive weight?

77 — Posted September 14, 2016

"...There was once more great dearth throughout the land." So wrote the brothers Grimm in their classic tale *Hansel and Gretel.* We all know the story: A wretched stepmother "scolds and reproaches" a father until he relents and allows this woman to send his two young children deep into the forest. Wandering for days, with essentially nothing to eat, Hansel and Gretel discover a house made of sugar and cakes (and in some versions, gingerbread.) The house's owner, though, is a wicked witch who initially feeds them "milk and pancakes, with sugar, apples, and nuts," but alas, her goal is to fatten Hansel in order to cook him and eat him for her "feast day." The story, with its ultimately happy ending, is beautifully depicted in the opera by 19th century composer Engelbert Humperdinck.

Aside from the obvious cannibalistic element in the story, we are confronted with a dreadful stepmother who is willing to starve and abandon children (child neglect) and a wicked witch who wants to fatten her child victims for own her ulterior motives (child abuse).

Hansel and Gretel, by the Brothers Grimm features an example of child neglect and child abuse. Credit: Arthur Rackham, illustrator, 1909 edition of *The Fairy Tales of the Brothers Grimm.* *Wikimedia Commons/Public Domain.*

Is it child neglect or abuse to overfeed a child to the point where he or she becomes extremely obese? Some would think so. Former restaurant critic and now political columnist for *The New York Times* Frank Bruni, in his 2009 charming memoir *Born Round*, describes an experience his mother remembers when he was 18 months old: "My mother had cooked and served me one big burger, which would have been enough for most carnivores still in diapers." Not only had toddler Frank wanted a second burger, but according to his mother, he started banging on his high chair tray for a third one. His mother was able to resist, despite a temper tantrum of "histrionic" proportions. Says Bruni, "A third burger isn't good mothering. A third burger is child abuse." (p. 10)

The topic of childhood obesity has become an impassioned one in recent years. Even the popular media outlets have entered the dialogue: *Time* featured an article "Should Parents of Obese Kids Lose Custody? (Faure, October 16, 2009) and Jon Stewart of *The Daily Show,* quipped, "If you are like me, every year we make a New Year's Resolution to make *our children* lose weight." (January 3, 2011)

How do we define overweight and obesity in children and adolescents? There is some variation among researchers but most use the growth charts of Centers for Disease Control and Prevention (CDC): Since body mass index (BMI) is age- and sex-specific for this population, we don't use the adult BMI table. Instead, overweight is defined as a BMI (i.e., weight in kilograms divided by height in meters squared) between the 85th and 95th percentile for children and adolescents of the same sex and age; similarly, obesity is defined as a BMI at or above the 95th percentile, and extreme obesity is defined as a BMI at or above the 120% percentile (Ogden et al, *JAMA*, 2016).

Color engraving, 1815. Theresia Fischer, a girl weighing 151 pounds.

Images of Wellcome Trust, London. Wikimedia Commons. Licensed under Creative Commons Attribution 4.0.

Master Wybrants, an infant weighing 39 pounds, on his mother's knee, 1806, Colored etching by C. Williams.

Images of Wellcome Trust, London. Wikimedia Commons. Licensed under Creative Commons Attribution 4.0.

Given these norms, how prevalent is obesity in children and adolescents aged two to 19 years in the U.S? Ogden and her colleagues (2016) have been measuring over 40,700 children and adolescents since the late 1980s through the nationally representative National Health and Nutrition Examination Surveys. Their most recent data (years 2011-2014) found an obesity prevalence of 17% and extreme obesity of 5.8% among children and adolescents. Obesity rates, though, can differ significantly by ethnic and social class categories and even by state. For details, see the Ogden et al 2016 report or the CDC website.

Despite the fact that rates may be leveling off in some subgroups, there is no reason for complacency. In their editorial accompanying the Ogden et al report, Zylke and Bauchner (*JAMA* 2016), describe the situation as "neither good nor surprising." These statistics are indeed alarming because obesity in childhood and adolescence is associated with increased medical morbidity, including abnormal glucose levels (and even overt type 2 diabetes), abnormal lipid levels, hypertension, sleep apnea, orthopedic problems, as well as psychological morbidity such as depression and stigmatization that often result from bullying. (Jones et al, *Trauma, Violence & Abuse*, 2014) Furthermore, we know that obese children tend to "track" into adulthood, with continuing potential for increased morbidity and early mortality.

While there are potentially damaging physical effects of childhood obesity, does the situation warrant some intervention by authorities? Murtagh and Ludwig (*JAMA* 2011) write that while "poor parenting is analogous to secondhand smoke in the home, there is a "well-established constitutional right of parents to raise their children as they choose."

"The issue of governmental interference with parental rights for decision-making, such as the need to combat childhood obesity…has become a national debate that remains unresolved," writes lawyer Denise Cohen (*Cardozo Public Law, Policy, and Ethics Journal*, 2012) Ms. Cohen maintains that though parents "have a unique role in reducing childhood obesity" they have not yet been "sufficiently engaged in this effort." As a result, government may need to protect "the best interests of the child" by interfering with the private parental-child relationship. The situation is complex in that currently there are no federal or state laws that deal specifically with assessing and treating childhood obesity. Furthermore, many factors may contribute to the development of childhood obesity, including both genetic and environmental factors, and not all these factors are related to parental behaviors or "within parental control." (Garrahan and Eichner, *Yale Journal of Health Policy Law and Ethics*, 2012) Yanovski et al (*JAMA*

Juan Carreño de Miranda, *Eugenia Martínez Vallejo,* circa 1680. Carreño was the court painter. He painted this six-year-old girl, called *La Monstrua,* both with clothes as shown here and without clothes. It was speculated that her obesity was due to a hormonal disorder. Prado Museum, Madrid.

Exploitation of Children, 1883.

Cedias. Credit : G. Julien. Wikimedia Commons/ Public Domain.

2011) emphasize extreme childhood obesity is not *de facto* evidence of deficient or neglectful parenting and suggest that these children may be more likely to have "unrecognized single-gene defects" that may be contributing to their obesity. Harper (*Pediatric Clinics of North America,* 2014) highlights that obesity is rarely the result of only one factor but rather a combination of medical, genetic, and socioeconomic influences and requires "multidimensional assessment and management."

Programs have addressed school physical fitness and school lunch, but these programs "fail to place any requirement on parents who are undoubtedly…the most influential people in their children's lives." (Cohen, 2012)

Parents can influence their children's eating habits by their own food preferences and behavior (i.e. modeling) (Anzman et al, *International Journal of Obesity,* 2010) Further, parenting styles can have a significant impact on encouraging children to eat healthy foods: In their systematic review, Shloim et al (*Frontiers in Psychology,* 2015) found "indulgent and uninvolved parenting and feeding styles were associated with a higher child BMI; parents who were *authoritative,* i.e., "who combined expectations about adherence to a healthy diet and set limits on certain foods,"—parents who are responsive but not psychologically pressuring—(Sleddens et al, *International Journal of Pediatric Obesity,* 2011), fostered healthier children and lower BMIs. Too much control and restriction, though, can backfire: *authoritarian* parents who pressure children to eat nutritious foods may promote overweight and even discourage healthy eating. (Scaglioni et al, *British Journal of Nutrition,* 2008) Likewise, Bergmeier et al (*American Journal of Clinical Nutrition,* 2015) found in their systematic review that children had a higher BMI when parents disparaged certain foods and made negative comments about them.

For a discussion of strategies to avoid (e.g. arguing with children; using food as a reward in an *if-then* scenario; forcing children to eat; cooking meals exclusively for children; and eating separately from children), see Russell et al, *Appetite* (2015) and Gibson et al, *Obesity Reviews* (2012.) Berge et al (*JAMA Pediatrics,* 2013) found that disordered eating was more likely to result when parents emphasized *weight* rather than healthy eating. Interestingly, Wolfson et al (*The Millbank Quarterly,* 2015) found that both men and women did attribute blame and responsibility to parents for childhood obesity though they also felt there should be "broader policy action" such as school-based programs.

Cohen reviewed legal cases in which the specific rights of parents have been addressed; these fall under the liberty and privacy aspects of the Fourteenth Amendment. She notes that courts "have generally accorded great deference" to parents. This right to privacy can be challenged, though, when there is concern that a child has been neglected or abused, and definitions vary by state but fall "within the range provided by the *Child Abuse Prevention and Treatment Act (CAPTA)*" of 1974. There is a long history of children being removed when they have been severely malnourished; more recently some extremely obese children have been removed from the home for neglect when "childhood obesity becomes life threatening" and there is a "compelling interest in preventing harm." (Cohen, 2012) Further, many parents of obese children are obese themselves and "treating the obese child of an obese parent as neglected has the effect of criminalizing obesity among parents." (Cohen, 2012) In a large study, Trier et al (*PLOS One,* 2016) found that the prevalence of obesity or overweight among the parents of children and adolescents presenting to a childhood treatment program can be as high as 80%.

Lawyers Garrahan and Eichner (2012) focus on the "increasing need" for state courts to intervene in cases of childhood obesity when parents "negligently fail to address the medical needs of their morbidly obese children." Jones et al (*Trauma, Violence, & Abuse,* 2014) note that a state can order removal from a home but must "prove imminent danger to the physical health or safety of a child, determine whether it is contrary to the welfare of the child to remain at home and make reasonable efforts to prevent the removal." Only recently has overfeeding been seen as being as "harmful as underfeeding." They further note that based on recent court cases, "…the designation of childhood obesity as a form of child abuse or neglect is quickly becoming a legal reality in the U.S." In other words, "morbid obesity is essentially nutritional neglect." Getting child protective services involved early on should not be seen as punitive but rather about educating and helping a family. Cohen (2012) makes the point, though, that it is not always clear that removing a child to a foster home is necessarily a better option. Siegel and Inge (*JAMA*, 2011) found there are actually few data that support using foster care for obese children, and note that foster care may even be an "obesogenic environment" itself.

The obesogenic environment, though, is potentially everywhere. For example, should local governments be able to prohibit restaurants from providing free toys with children's meals that do not meet established nutritional guidelines? When San Francisco and Santa Clara County passed their *Healthy Food Ordinance* in 2010, they were accused of fostering a paternalistic "nanny state," (with concerns of a "paternalistic slippery slope") not unlike the response to former New York City Mayor Michael Bloomberg's proposal to limit the size of sodas. (Etow, *American University*

Fernando Botero, *The Family*, 1966, Botero Museum, Bogota, Colombia. Obesity often runs in families, as depicted by Botero, and obese parents may be more likely to have obese children, with both genetic and environmental contributions to the development of their children's obesity.

Renato Granieri. Used with permission/ Alamy.com.

Bartholomeus van der Helst, *Gerard Andriesz Bicker, Lord of Engelenburg,* circa 1642, Rijksmuseum, Amsterdam. Children who are allowed to become obese often become obese adolescents and eventually obese adults. Bicker is portrayed by the artist at age 20.

Rijksmuseum, Amsterdam, Public Domain.

Law Review, 2012) (For more on the NYC soda-size ban, see blog 23, *Supersizing and the Tyranny of the Soda Police?)* *The Healthy Food Ordinance* even garnered the attention of *The Daily Show with Jon Stewart* (January 3, 2011): correspondent Aasif Mandvi recommended that a "Crappy Meal" be equipped with the Periodic Table of Elements, CPR instructions, and an action figure of the then Head of Health and Human Services.

Bottom line: Clinicians should be "mindful" of the potential role of abuse or neglect in contributing to childhood obesity (Viner et al, *British Medical Journal,* 2010), but just because a child fails to lose weight alone does not constitute potential negligence or abuse. After all, weight management programs for childhood obesity do not necessarily lead to weight loss. On the other hand, there are red flags when parents fail to keep appointments, refuse to work with professionals, or actively sabotage weight management treatment plans. Further, it is essential for there to be a full multi-disciplinary evaluation, encompassing all aspects of a child's medical, physical, and emotional environment.

For an excellent book on children and eating, see also Bee Wilson's *First Bite: How We Learn to Eat* (2015). ■

INTERESTS CONFLICTED: A 'WICKED PROBLEM' IN MEDICAL RESEARCH

Along a 'continuum of moral jeopardy' for scientific researchers

78—Posted October 12, 2016

A PHYSICIAN SUSPECTS his town's water supply is contaminated and sends a sample of the water to the University for a "precise chemical analysis." When the results confirm his suspicions that the water is indeed poisoned and a menace to the townspeople's health, the physician believes he has a moral duty to share the information with the town leaders. The town's livelihood, though, depends on its water reserve: its baths have become its "little gold mine" as "crowds of invalids" flock to the town. Repairing this water source is estimated to take two years. To the physician's astonishment, the Mayor, the physician's own brother, essentially forbids him from releasing the report because the information will result in the financial ruin of the town. The physician is hailed, not as the hero he had expected to be, but as a pariah, "*a real enemy of the people*." The story, of the same name, is, of course, by 19th century Norwegian playwright Henrik Ibsen (1828–1906), and it depicts a typical conflict of interest scenario, i.e., between "public duty and private interest." (Newton et al, *Biomedical Central Public Health*, 2016)

Eilif Peterssen, Portrait of Henrik Ibsen, the Norwegian playwright who wrote, *An Enemy of the People*, 1895, private collection, Oslo.

Wikimedia Commons/Public Domain.

The *Oxford English Dictionary* defines conflict of interest as "a situation in which an individual may profit personally from decisions made in his or her official capacity." Financial gain is often the focus in a discussion of conflict of interest, particularly in scientific research, because it is more objectively quantifiable but there are other types of gain. Researchers, for example, may champion a position or conduct investigations that will enable them, their colleagues, or even their institution to achieve professional recognition, advancement, and fame. (Cope and Allison, *Acta Paediatrica*, 2010)

A conflict of interest is a *wicked problem*. This term was popularized and defined in the 1970s, in the context of government planning, by Berkeley professors of design and city planning, Horst W.J. Rittel and Melvin M. Webber. (Rittel and Webber, *Policy Sciences*, 1973) Wicked (i.e., "malignant" or "tricky") problems have certain distinguishing properties. They are "essentially unique." Wicked problems have no stopping rule: one stops with the sense that "That's the best I can do" or "That's good enough," and they are neither true nor false, but more likely "better or worse." There is neither any immediate nor any ultimate test of a solution to a wicked problem. As a result, there is no opportunity to learn by trial and error. Wicked problems are "incorrigible," and they defy "efforts to delineate their boundaries and to identify their causes." (Rittel and Webber, 1973)

There are no official boundaries on what could be a reason for a conflict of interest," writes Boston University Professor of Epidemiology Kenneth Rothman. (Rothman, *JAMA*, 1993) For example, Rothman makes the provocative argument that when a journal editor asks for disclosure of "any relationships" that might be construed as causing a conflict of interest, authors might need to disclose their religious or sexual orientation along with any financial considerations. For Rothman, this smacks of McCarthyism.

Léon-Augustin Lhermitte, (1844–1925), *Claude Bernard and his Pupils.*

Rothman defined conflict of interest as "any situation in which an individual with responsibility to others might be influenced either consciously or unconsciously by financial or personal factors that involve self-interest." (Rothman, 1993) He noted that the term often has a pejorative connotation—what he called a "pernicious ambiguity." (Rothman, *J. Clin. Epidemiology*, 1991-supplement) The term, though, does not necessarily indicate any wrongdoing or even anything unethical. Kozlowski (*Science and Engineering Ethics*, 2016) defines it as "not being free to take an opposing position," i.e., "...not completely free to adopt opinions that would be unsupportive of the opinions of their agencies, superiors, close colleagues..." A conflict of interest refers to a *setting* in which one's conduct might be affected, and so could lead to a researcher's work being biased. (Rothman, 1991) Says Rothman, "...no scientist is objective...each of us is buffeted by the winds of unaccountable biases, some overt and some hidden." (Rothman, 1991)

Bias in research has been defined vaguely as any deviation from the truth. It includes systematic errors, as opposed to errors by chance, that can occur throughout the course of research and severely compromise the research's integrity. (Tripepi et al, *Kidney International*, 2008) David L. Sackett, one of the forefathers of evidence-based medicine, delineated 55 types of biases that he categorized by the stage of research involved, including conducting a literature review, choosing a population sample, and implementing the research's design, measurement, interpretation, and even its publication. (Sackett, *J. Chronic Diseases*, 1979.) More recently, others have described more than 70 subcategories of possible bias in scientific research. (Delgado-Rodriquez and Llorca, *J. Epidemiology and Community Health*, 2004)

Bias can result from a scientist's deliberate alteration, regardless of motivation, of his or her research findings, and that is considered overt fraud. When there is an unconscious "tilting" of study design or data collection, for example, says Rothman, "that is incompetence." (Rothman, 1991; Rothman, 1993.) Adds Rothman, "No one works in a vacuum…and no one is truly free of pressures that might distort intellectual endeavors," but those pressures notwithstanding, "The scientific work must be judged on its merits." (Rothman, 1993; Rothman, *J. Clinical Epidemiology*, 2001)

That is unquestionably the conclusion of obesity researcher and biostatistician David B. Allison, PhD,

Trouble Comes to the Alchemist, Dutch School, 17th century, artist unknown. Painting at Science History Institute, Philadelphia.

Wikimedia Commons/Public Domain.

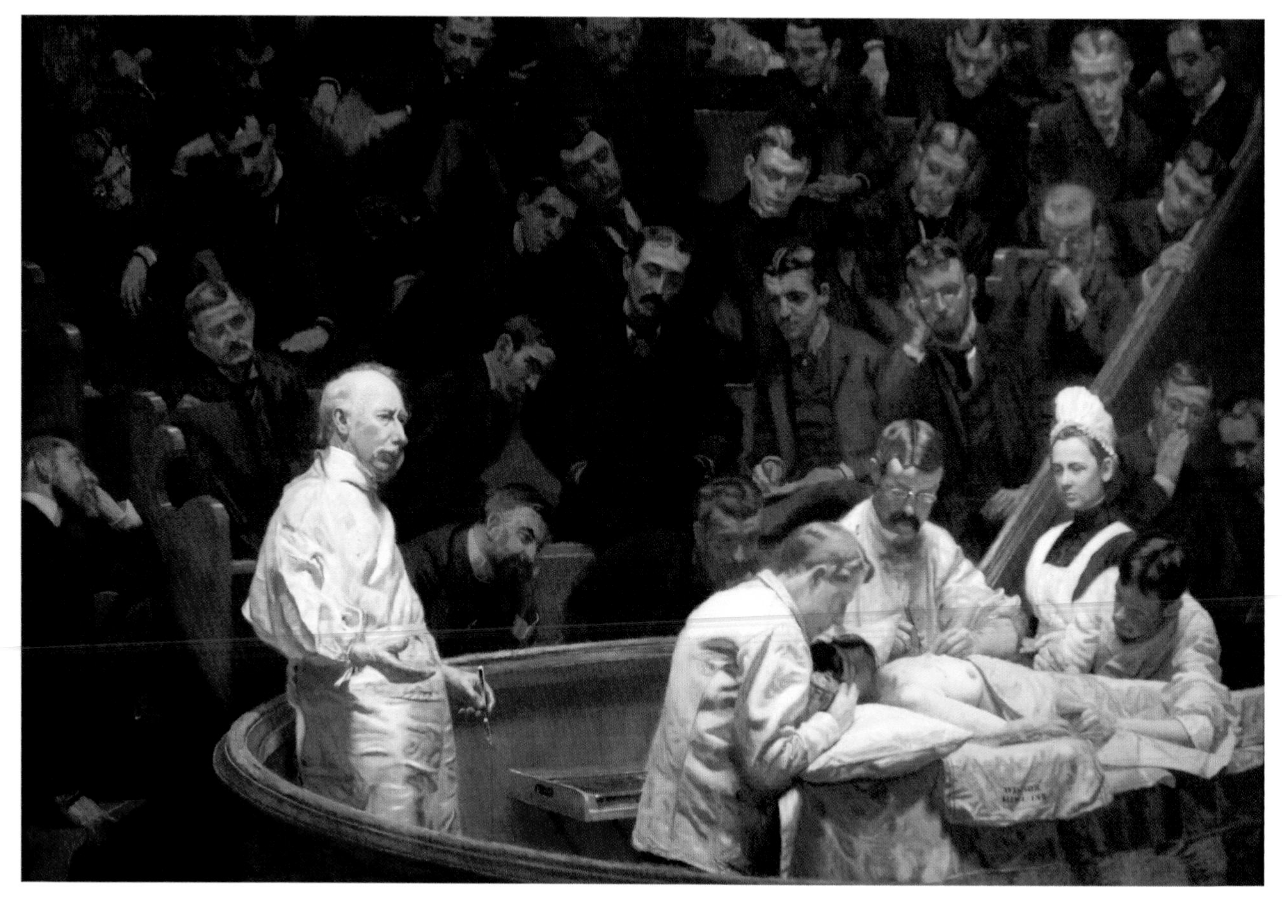

Thomas Eakins, *The Agnew Clinic*, 1889. Students from the University of Pennsylvania School of Medicine commissioned Eakins to make a portrait of their retiring professor of surgery, Dr. D. Hayes Agnew. Philadelphia Museum of Art.
Wikimedia Commons/Public Domain.

with whom I spoke, a Distinguished Professor of Public Health and Director of both the Office of Energetics and the Nutrition Obesity Research Center (NORC) at the University of Alabama at Birmingham. Says Allison, "We should reject *ad hominem* reasoning when we judge the quality of evidence, including judging research by its funding source. In science, three things are relevant: the data, the methods that generated the data…and the logic that connects the data to the conclusions. We should do everything we can to strengthen the rigor and transparency of those three things, and any focus on *ad hominem* attacks is not only uncivil, but also unscientific."

The label *conflict of interest,* though, has frequently been used as an *ad hominem* attack specifically with the intent to discredit someone personally and create *guilt by association.* Rothman emphasizes it is "… essential to the fabric of science" to avoid denigrating scientific research based on funding sources by *ad hominen* attacks, a practice that is clearly unethical itself. (Rothman, 1991; Rothman and Cann, *Epidemiology*, 1997)

It is important, though, to emphasize that funding for research has to come from somewhere and *any funding source*, whether industry, government, public, or private, can result in the appearance of potential conflicts of interest. (Rowe et al, *American J Clinical Nutrition*, 2009) Though perhaps sometimes unfairly (Thomas et al, *International Journal of Obesity*, 2008), industry funding has become particularly suspect and maligned. For example, a recent study published in the journal *JAMA Internal Medicine* uncovered correspondence that seemed to implicate the Sugar Industry's

Research and Experiment Department, Oxford, from the Collections of the Imperial War Museum, United Kingdom.
Rodney Joseph Burn, Wikimedia Commons/Public Domain.

clandestine ties to research in the 1960s and 1970s that sought to demonize fat consumption and minimize the effects of sugar intake on cardiovascular disease. (Kearns et al, 2016) Another study had demonstrated that industry-sponsored research was five times more likely to report inconclusive scientific evidence, in this case, for sugar-sweetened beverages and their relationship to weight gain, than studies independent of the food and beverage industries. The authors, though, acknowledged a limitation in their data: they could not rule out the possibility of publication bias in the studies that did not declare any conflicts of interest. (Bes-Rastrollo et al,*PLOS Medicine*, 2013) Publication bias occurs when a study is more likely to be published because of statistically significant results (i.e., its outcome) rather than its overall quality. (Allison et al, *International Journal of Obesity and Related Metabolic Disorders*, 1996)

Conflict of interest was the subject of a comprehensive report in 2009 by the Institute of Medicine (Lo and Field, eds., National Academies Press) that emphasized the "central goal" of any policy regarding conflicts of interest is "to protect the integrity of professional judgment and to preserve public trust." It also noted that conflicts of interest often involve degrees that are "not directly quantifiable" and depend on the context, i.e., the particular circumstances of the risk and not one personal decision. Furthermore, these circumstances are seen from the perspective of people who do not necessarily have all the information to evaluate the motives involved, and as such can be only "perceptions or appearances." Conflicts of interest, just like wicked problems, are not binary. In other words, they are "not simply present or absent, but rather more or less severe." Adams, in criticizing the "over-simplified perspective" whereby researchers are divided into "those willing to accept funding and those who are unwilling," describes a "continuum of moral jeopardy, stretching from those with minor involvements to those with unmanageable conflicts of interests." (Adams, *Addiction*, 2007)

Given the nature of wicked problems and given that as humans, we are all subject to biases—some conscious and some unconscious, are there ways to mitigate against the appearance of conflicts of interest? Most sources have suggested that full transparency is required and "adherence to the basic principles of good science like reproducibility, replicability, and reliable scientific reporting" is essential. (Binks, *International J Obesity*, 2014.) In other words, researchers should not only identify their funding sources with full disclosure (Rowe et al, 2009) but should share their data publicly. (Allison, *Science*, 2009.) Certainly, there is no place for *ad hominem* attacks that amount to professional bullying or what Sagner et al refer to as "a fallacy of relevance that undermines scientific progress." (*Progress in Cardiovascular Diseases*, 2016.) Only then is there the possibility that researchers can preserve their own professional judgment and maintain the public's confidence.

Nineteenth century French physiologist Claude Bernard, often considered the father of the scientific method in medicine, wrote in his 1865 *An Introduction to the Study of Experimental Medicine*, "A man of science should attend only to the opinion of men of science who understand him, and should derive rules of conduct only from his own conscience." Adds Allison, "We need to pursue truth through science not as a job but as a discipline, a vocation, and a privilege." ■

THE BRITTLE WORLD OF PEANUT ALLERGY

Navigating the daily potentially life-threatening challenges of food allergies

79 — Posted November 24, 2016

THE PEANUT HAS BEEN CULTIVATED worldwide for centuries and has been a decorative motif in art in civilizations as diverse as South America and China. There are, for example, images of peanuts on pottery that dates back from the 5th to 7th centuries A.D. from the Moche civilization of northern Peru. The Moche culture was known for its pyramid-like architectural mounds, ceramic stirrup-spouted pots, and metal works, as well as for an extensive irrigation system for agriculture. These people, though, were hardly peaceful farmers: we have anthropological evidence that they engaged in human sacrifices that involved mutilation of bodies. There are ceramics with residues of human blood as well as others indicating the earthenware had contained peanuts.

Technically, the peanut is not a nut but rather a legume that belongs to the bean or pea family *Fabaceae*. (van Erp et al, *Current Treatment Options in Allergy*, 2016) Peanuts can be eaten raw or after boiling or roasting; they can be used in recipes and made into solvents, oil, medications, textile materials, and of course, peanut butter. (Praticò and Leonardi, *Immunotherapy*, 2015) Peanut cultivation requires a southern climate, and today, Alabama and Georgia are the most common regions for its U.S. cultivation.

In the United States, peanuts were originally used to feed animals. Agriculturist, educator, and innovative researcher George Washington Carver, born a slave around the time of the Civil War, popularized their use for human consumption in the U.S. and laced his educational bulletins with his peanut recipes. Carver understood the need for crop rotation and the use of soil-enriching crops such as peanuts (and away from soil-depleting cotton.) (See the websites of Tuskegee Institute, Simmons College, and the National Park Service.)

U.S. Department of Agriculture. Photo : Alice Welch. Wikimedia Commons/ Public Domain.

In U.S. colloquial use for at least the past 150 years (*Oxford English Dictionary*), a "peanut" is a small or unimportant person, a small or insignificant amount of money, or even something that is trivial, worthless, or undersized. There is nothing trivial or insignificant, though, about peanut allergies that affect about 1 to 3% of children in westernized countries and have potentially deadly consequences in those particularly vulnerable. (Greenhawt, *Pediatric Clinics of North America*, 2015.)

Peanut allergies are not as prevalent as those to egg or milk, but they are less likely than milk, egg, soy, or wheat to be outgrown: (Syed, Kohli, Nadeau, *Immunotherapy*, 2013) fewer than 20% of cases resolve with age. (Greenhawt) Unintentional exposure is common enough since peanut products are found in many foods or processed in facilities where peanuts are found. Peanut proteins can be found in the dust in homes where peanuts have been consumed. (Benedé et al, *EBioMedicine*, 2016) As a result, a person's (and entire family's) quality of life can be severely compromised because there is the constant threat of exposure and the need for relentless watchfulness. (Antolín-Amérigo et al, *Clinical and Molecular Allergy*, 2016) Moreover, those affected may suffer from allergies to many other foods as well, and there is the phenomenon of the so-called *atopic march* (Bantz et al, (*Journal and Clinical and Cell Immunology*, 2014), i.e., a progression from skin sensitivity such as atopic dermatitis (e.g. eczema) to food allergies.

The development of a food allergy is characterized by two stages: *sensitization* where an allergic reaction is established on first exposure and *elicitation* where the immune system produces an inflammatory response when the person is re-exposed to that particular allergen. Only a small number of foods actually induce allergies. Why this happens is not known. (Syed, Kohli, Nadeau, 2013.)

Stirrup spout vessel depicting rows of peanuts. Moche pottery, northern coast of Peru, 250–500 A.D. Peanuts have been cultivated throughout the world for centuries.

Art Institute of Chicago, Gift of Nathan Cummings, 1958. Used with permission for scholarly work.

Ceramic stirrup spout bottle with peanut figure. Moche pottery, 5th to 7th century A.D., northern Peru,

Metropolitan Museum of Art, NYC, Gift of Mr. and Mrs. Nathan Cummings, 1964. Public Domain.

When there is a hypersensitivity reaction, though, patients may experience hives, wheezing, or vomiting that occurs after exposure to common allergens such as milk, peanut, or egg. (Benedé et al, 2016) For mild or moderate symptoms, anti-histamines are the first-line treatment. Anaphylaxis, though, is an acute, life-threatening, medical emergency that involves two or more organ systems and usually occurs within minutes of exposure. Symptoms include vomiting, skin rash, rapid and weak pulse, abdominal pain, swollen throat and lips, trouble breathing or swallowing, diarrhea, and chest tightness. (Benedé et al, 2016) The current standard of care, though, is strict avoidance of the culprit food and readily available intramuscular injectable epinephrine, the only treatment for the most serious, potentially lethal reactions that occur after unintentional exposure. (Yu et al, *International Archives of Allergy and Immunology,* 2012) It is no wonder there was such media uproar recently over the extraordinary price gouging from the manufacturer of the EpiPen. Nadeau and colleagues (Yu et al, 2012) note that peanut allergy is the leading cause of food-related fatal anaphylaxis in the U.S.

Food allergies affect roughly 15 million Americans (including 8% of U.S. children, Greenhawt, 2015) and 17 million Europeans, (Praticò and Leonardi, *Immunotherapy,* 2015) Anvari et al (*JAMA Pediatrics,* 2016) suggest that prevalence rates of reported peanut allergy have tripled in the ten years. The data on the epidemiology and natural course of food allergies, though, are not particularly accurate. (Savage and Johns, *Immunology and Allergy Clinics of North America,* 2015) Actual estimates vary widely because of differences in study methods, definitions of allergy, and different geographical areas. (Savage and Johns, 2015) Further, most statistical estimates are derived from self-or parent-report that can be notoriously inaccurate. Demographically, food allergies are more common in non-Hispanic blacks, Asians, and males; there is also a genetic predisposition,

Emergency injections of epinephrine can prevent death when anaphylactic shock due to peanut ingestion in those with severe peanut allergy.

Used with permission, istockphoto.com. Credit: MichelGuenette.

with specific genes likely involved. There are environmental factors: children with older siblings and those with pets are less likely, while those with vitamin D deficiency and those with atopy (e.g. asthma and atopic dermatitis), more likely. The most common age for presentation of peanut allergy is 18 months (Savage and Johns, 2015).

The mechanisms for food allergies are not well understood; it is not yet known whether these allergies represent pathological immune responses in allergic children or the absence of protective mechanisms normally found in healthy children. Current diagnostic techniques emphasize the importance of clinical history, family history, presence of other allergic conditions, and the timing of allergic reactions after ingestion. When allergic patients are exposed to a potential allergen by skin prick, they develop a skin reaction called a "wheal." While this test is usually safe, rapid, and highly sensitive, it does not provide specific information about severity, and extracts are often crude and unstandardized. (Syed, Kohli, Nadeau, 2013)

The gold standard for diagnosing a food, including peanut allergy, is the oral food challenge, but this has not been studied on a population level. (Savage and Johns, 2015) For increased diagnostic accuracy, rather than relying on "crude peanut extract," investigators use component-resolved diagnostics (CRD) to measure sensitization to purified or recombinant allergenic proteins within the peanut. Researchers can measure increases in levels of serum IgE antibodies when a patient is exposed to the specific peanut protein Ara h 2 that is pathognomonic for peanut allergy. (Syed, Kohli, Nadeau, 2013) CRD cannot be used to predict the risk of a severe allergic reaction. (van Erp et al 2016)

Why have rates of food allergies been increasing? The National Institute of Allergy and Infectious Diseases notes that there are insufficient data to suggest maternal diet, including during pregnancy, influences the development or course of food allergy in children. Studies from Israel where peanut allergies are significantly less common seem to indicate that early exposure to highly allergenic foods may even be preventive for food allergies. (Syed, Kohli, Nadeau, 2013)

The *hygiene hypothesis* suggests that changes in the pattern of the intestinal microbiome during infancy and decreased exposure to infectious agents in childhood are important factors in the development of allergic disease. (Syed, Kohli and Nadeau, 2013)

Prior to 2008, the American Academy of Pediatrics and others recommended delaying exposure to any peanut products in children before the age of three, especially for those at high risk (e.g. parent or sibling with family history or history of eczema in children) (Greenhawt, 2015) But international studies, including the LEAP Trial in England, confirmed that children at high risk of developing allergy to peanuts were actually more likely to develop a peanut allergy without any exposure than those exposed frequently as early as six months, who had a "clear protective benefit;" exposure, though, "did not unequivocally prevent peanut allergy from developing" in some exposed children. (Greenhawt, 2015) This study had limitations: it did not address different doses of peanut required to maintain tolerance, the minimal duration of therapy necessary to achieve tolerance, or the potential risks of premature discontinuation or the sporadic feeding of peanut. (Anvari et al, 2016)

DuToit et al (*NEJM*, 2016) conducted a follow-up to the LEAP study, called the LEAP-ON, for those high-risk children exposed to peanuts in the first year of life and then throughout their first five years. The study did demonstrate a persistent reduction in the prevalence of peanut allergy in a significant number of children, even after a one-year peanut hiatus.

Peanut specimen of George Washington Carver, born in the 1860s, was an educator, inventor, and agriculturist, who included many recipes for peanuts in his educational bulletins. Tuskegee Institute.

Photo taken by National Park Service employee. Wikimedia Commons/Public Domain.

Those who are allergic to peanuts must be constantly on guard for the possibility of exposure to peanuts or even peanut dust.

Licensed under Creative Commons Attribution Generic 2.0. Photographer: Dan4th Nicholas, Wikimedia Commons.

What about those who already have a potentially deadly food allergy such as to peanuts? There is no approved treatment for food allergies other than avoidance or emergency treatment when exposed. (Benedé et al, 2016) Clearly there is a need for "proactive therapies" since avoidance is hardly a foolproof, long-term solution. (Commins et al, *Current Allergy and Asthma Reports,* 2016)

Food allergy immunotherapy protocols have been tested in the past 25 years (Leung, Sampson, et al, *NEJM,* 2003) but to date, "it is not clear if this therapy is a myth or a reality." (Praticò and Leonardi 2015) The main goal of immunotherapy is a permanent oral tolerance of the offending food (even after a hiatus of exposure) by causing a decrease in IgE-specific antibodies and a significant increase in IgG4 antibodies. Different from tolerance, *desensitization* involves only *transitory* changes in certain intestinal cells and the need for continuous exposure to that food. Further, any factor that can affect the "intestinal barrier function" (e.g. exercise, gastroenteritis, stress) may cause a loss of protection to the previously tolerated dose, even after the end of immunotherapy, unless there is actual modulation of the pathological immune mechanisms. (Praticò and Leonardi, 2015)

Several techniques for food immunotherapy have been developed: sublingual immunotherapy (SLIT), oral immunotherapy (OIT), and epicutaneous (e.g. skin patch)(EPIT). (Benedé et al, 2016) A high percentage of those who achieve desensitization develop clinical reactivity when treatment is stopped. In all the protocols for immunotherapy, side effects

Peanuts and jujube dates, Qing Dynasty, 18th century, China. Metropolitan Museum of Art, NYC.

Gift of Heber R. Bishop, 1902. Public Domain.

are quite common. Initial phases are more likely to produce side effects, and rush protocols are usually more dangerous. A higher incidence of systemic reactions has been observed with OIT, while oral reactions are quite common with SLIT, and SLIT has been found too dangerous to use with peanut allergies. (Syed, Kohli, Nadeau, 2013) Patients may also experience cutaneous symptoms (e.g. itching) and gastrointestinal (e.g. nausea, vomiting, abdominal pain) and respiratory symptoms (e.g. asthma) that can be severe and usually linked to withdrawal from the study or poor adherence to the protocol. (Praticò and Leonardi, 2015)

The use of the anti-IgE (omalizumab, brand name Xolair), together with oral immunotherapy, has recently been reported to improve the safety of OIT, (Benedé et al, 2016) but note that this medication has, to date, been approved for use *only* with patients with asthma or chronic urticaria. (Bauer et al, *Journal of Allergy and Clinical Immunology*, 2015.) Further, its administration must be done under strict medical supervision in case of potentially life-threatening anaphylactic reactions. Clearly, much more research with standardization of protocols is warranted. (Syed, Kohli, Nadeau, 2013)

In their provocatively entitled, "Oral Immunotherapy for the Treatment of Peanut Allergy: Is It Ready for Prime Time?" Wood and Sampson (2014, *Journal of Allergy and Clinical Immunology: In Practice*) caution the use of OIT based on their own years of clinical experience and a review of the current literature. They believe, "Claims that OIT is safe are potentially misleading," and they strongly recommend that these experimental treatments be conducted under the oversight of institutional review boards and the U.S. Food and Drug Administration. Further, they note this treatment is not proven in either the short or longer-term, and patients may be living with a "false sense of security."

Other researchers have noted it is "easy to understand the trepidation" of practitioners and patients to oral immunotherapy, especially for a peanut allergy, since it is responsible for more anaphylactic deaths than other allergens. (Yu et al, 2012) More recently, Nadeau and her group (Ryan et al, *Proceedings of the National Academy of Sciences, USA*, 2016) have also found a way to measure OIT success by tracking and monitoring changes in specific cells involved in our immune system.

Bottom line: Food allergies, and particularly peanut allergies, can be lethal. There is currently no standard treatment other than avoidance of the specific food or emergency epinephrine in case of unintentional exposure. Immunotherapy shows promise but is still considered experimental; investigative trials have involved small, select populations and adverse effects have been common. ■

SUGAR BY ANY OTHER NAME? LOW CALORIE SWEETENERS

Why is there still confusion about sugar substitutes?

80 — Posted December 27, 2016

THERE IS A DIGITAL IMAGE currently on exhibit at the James A. Michener Art Museum in Doylestown, Pennsylvania by Seattle-based artist Chris Jordan. Jordan ingeniously uses Georges Seurat's iconic painting *A Sunday Afternoon on the Island of La Grande Jatte* (1884-1886) (immortalized by Stephen Sondheim's 1984 musical *Sunday in the Park with George*) as his template. Appropriating a version of pointillism, he has replaced all the thousands of Seurat's dots of color with 106,000 miniaturized aluminum cans of soda. For Jordan, who aptly calls his 2007 piece *Cans Seurat*, we are undoubtedly a culture of consumerism: 106,000 apparently represent the number of cans that U.S. consumers drink every thirty seconds.

There is nothing particularly more obesogenic about liquid calories, as for example, from our drinking cans of soda. (Allison, *British Journal of Nutrition*, 2014) After all, all calories count. There is, though, now scientific evidence from both experimental trials and observational studies that *suggest* that our consumption of sugar-sweetened sodas *may* cause excessive weight gain because of an excessive intake of calories and our often *incomplete* compensation for these excessive liquid calories at subsequent meals, at least in the short term. (Allison, 2014) As a result, sugar-sweetened sodas have been *associated* with an increased risk of type 2 diabetes and cardiovascular disease. (Malik and Hu, *Journal of the American College of Cardiology*, 2015).

Sugar was initially considered a "nutritional necessity" for health at the turn of the previous century, but "the positive health halo of sugar would not last." (Allison, 2014) In recent years, we have seen sugar demonized in both the scientific literature and popular press. So in an effort to lighten our caloric load and in response to an appeal to restrict full-calorie sugars (e.g. sucrose or table sugar; fructose, high fructose corn syrup, lactose), many of us have turned to low calorie sweetener substitutes.

What are low calorie sweeteners (LCS), the term preferred by some investigators? (Mattes, *Physiology & Behavior*, 2016) They are also called non-nutritive sweeteners (NNS), intense sweeteners, high potency sweeteners, sugar substitutes, artificial sweeteners, and non-caloric sweeteners, among others. (Mattes, 2016; Sylvetsky and Rother, *Physiology & Behavior*, 2016). By whatever name, they have a high sweetness potency compared to sugar on a weight-for-weight basis; only small amounts are required to provide sweetness without the calories. Sometimes these sugar substitutes are "blended together" for a synergistic effect. The sugar alcohols, such as sorbitol, mannitol, maltitol, typically found in sugar-free chewing gum, have two calories per gram and are not reviewed here.

Saccharin, 300 times sweeter than sucrose, the first low calorie sweetener, (Sweet'N Low; Sweet Twin), was discovered by accident in the late 1870s and remains on the market. (A. Roberts, *Physiology & Behavior*, 2016; Sylvetsky and Rother, 2016) **Cyclamate,** 30 times sweeter, was discovered much later in the 1930s but was subsequently banned from the U.S. market by 1970 after concern about the development of bladder cancer in rodents; though its carcinogen effects were later disproved, it remains off the U.S. market (though continues to be used internationally.) (A. Roberts, 2016) To date, Roberts notes that **acesulfame-K** (Sunett), 200 times sweeter than sucrose; **aspartame** (Nutrasweet; Equal), 200 times sweeter; **neotame,** (Sugar Twin; Newtame), 7,000-13,000 times sweeter; **sucralose,** (Splenda), 600 times sweeter, and most recently, **advantame,** (20,000 times sweeter), a chemical analogue of aspartame that does not have to contain that warning label for those with phenylketonuria (Sharma et al, *Indian*

Raphaelle Peale, *Still Life with Cake*, 1818. Metropolitan Museum of Art, NYC.

Maria DeWitt Jesup Fund, 1959. Public Domain.

Chris Jordan, *Cans Seurat*, an artistic rendering of Georges Seurat's iconic painting *A Sunday on La Grande Jatte*, 1884, with all the dots of color replaced by 106,000 cans of soda. Pigmented, ink-jet print, 2007.

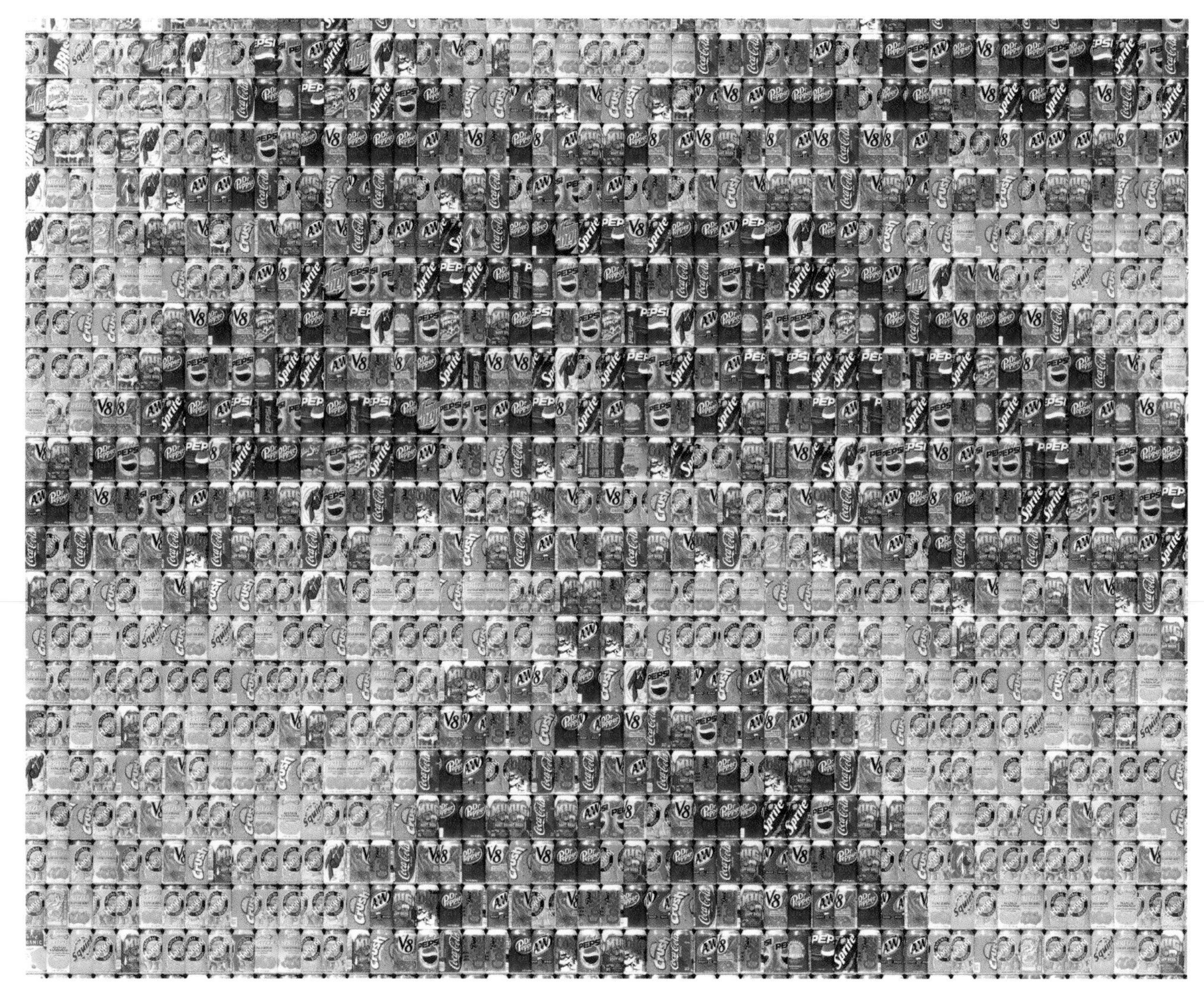

Detail from Chris Jordan, *Cans Seurat*, an artistic rendering of Georges Seurat's iconic painting *A Sunday on La Grande Jatte*, 1884, with all the dots of color replaced by 106,000 cans of soda. Pigmented, ink-jet print, 2007.

Journal of Pharmacology, 2016) are all now FDA-approved in the U.S. Rebaudioside A, 300 times sweeter (Stevia, Truvia), is a "highly purified product" from the stevia plant, is in a different category (as is lo han guo—monk fruit) and considered in the Generally Recognized As Safe—GRAS category, a much more streamlined approval process. (For a thorough discussion of the GRAS process, see the A. Roberts 2016 article.) Worldwide, low calorie sweeteners are expected to reach $2.2 billion by 2020, and while manufacturers must list the sweetener on the ingredient label, they are not required to disclose the quantity used. (Sylvetsky and Rother, 2016)

Those that are heat-stable can be used in cooking and baking; these products are ubiquitous in processed breads, cereals, granola bars, yoghurts, ice cream, jams, pancake syrup, and even in vitamins and mouthwash. Sucralose, with the most market-share, has now surpassed aspartame in the U.S. It is particularly stable at high temperatures and found in more than 4000 products in the U.S. and while most like table sugar in its "sensory profile," it does not lead to tooth decay. Further, sucralose been considered safe for those with diabetes because it maintains glucose homeostasis and does not increase insulin release; it has even been deemed safe during pregnancy and nursing.

For some, though, it may trigger migraines. (Shankar et al, *Nutrition*, 2013)

"Without question, low calorie sweeteners are some of the most extensively evaluated substances," and in the U.S., they have undergone thorough safety evaluations by manufacturers and the FDA toxicological testing program. (A. Roberts, 2016) Nevertheless, these low calorie sweeteners have been (and continue to be) the "subject of continuous acrimony" and their role "in a healthful diet has been contentious for over 135 years." (Mattes, 2016)

And as our usage of these low calorie sweeteners has increased over the past 35 years, so has the U.S. prevalence of obesity and overweight. Allison and his colleagues (Anderson et al, *Journal of Nutrition*, 2012) suggest this leads to the obvious question of whether there is any connection. The simple answer is the connection remains unclear: there is extraordinary complexity in determining all the elements that impact body weight, including psychobiological, cultural, and environmental factors that influence our decisions and motivations regarding food and health. (Anderson et al, 2012)

There are considerable methodological issues that make these studies particularly difficult. For one, dietary intake and physical activity are often not standardized, and there may be genetic differences in sweet taste receptors that can account for "clinically relevant individual differences," as well as a failure to account for other ingredients (e.g. carbonation, etc.) in these products that may have synergistic effects. (Sylvetsky et al, 2016, *Nutrition & Metabolism*) Further, most rely on self-reporting of diets that are notoriously inaccurate for many reasons, including faulty memory, overt misrepresentation, or even failure to assess portion size accurately. (Anderson et al, 2012) Many studies are observational, whether prospective or retrospective,

William Clark, *The Mill Yard : Grinding sugar cane in a windmill*, 1823. Image taken from *Ten Views in the Island of Antigua*. British Library, (National Library of the United Kingdom.)

Diego Rivera, *Sugar Cane.* 1930.

longitudinal or cross-sectional, but are not randomized and can tell us only about associations of variables and not causation. Observational studies are particularly subject to confounding where variables may seem directly related to each other but are not. For example, some observational studies have suggested an association between metabolic disorder and use of these products but when body weight (adiposity) is factored in, the relationship becomes much less clear or even non-existent. (Romo-Romo et al, *PLoS One*, 2016) Observational studies, though, may be the best alternative in situations where it is impossible, impractical, or unethical to assign randomization. (Anderson et al, 2012)

Establishing causation, in general, is difficult in these studies because those who consume these low calorie sweeteners may differ from those who don't in other ways. (Swithers, *Current Opinion in Behavioral Sciences*, 2016) For example, Azad et al (*JAMA Pediatrics*, 2016) found that those who used artificial sweeteners during pregnancy were more apt to stop breastfeeding and introduce solid foods earlier. Further, reverse causality may be a factor. (J.R. Roberts, *Current Gastroenterology Reports*, 2015) For example, do artificial sweeteners cause metabolic abnormalities or do those who consume these sweeteners already have increased weight and metabolic abnormalities? (Suez et al, *Gut Microbes*, 2015)

It is also often difficult to obtain information on metabolic dysfunction and use of these sweeteners because of the considerable time frames involved in assessing any effect on health. Many people do not even realize they are consuming these products, and it is even difficult to obtain a population that has not had some exposure to these sweeteners since they have virtually infiltrated so much of our food supply. (Shearer and Swithers, *Reviews in Endocrine and Metabolic Disorders*, 2016) Further, when some studies are crossover studies, they do not mention information regarding a so-called "washout" period.

Research on low calorie sweeteners is also challenging because they are all structurally different, with diverse physical properties, such as strength, mouth feel, aftertaste, duration of sweetness, stability, and how they affect the body. (A. Roberts, 2016) In other words, there may be methodological variations among reports because of the type of sweetener studied, its route of administration and duration of exposure, as well as an individual's body weight and prior exposure. (Burke and Small, *Physiology & Behavior*, 2015)

Canadian researchers, for example, as they correctly note, did not even distinguish among the different artificial sweeteners or account for their presence in solid foods when they conducted a large observational study with almost 2700 mother-dyad pairs at follow-up. Although they found that those mothers (over 29%) who consumed artificially-sweetened sodas or teas during pregnancy (by self-report) had babies with a higher BMI at one year of age, with the highest incidence of overweight seen in those born to mothers who reported daily consumption, they may have muddled their results by failure to distinguish among their sweeteners. (Azad et al, 2016)

Over the years, those studying low calorie sweeteners have focused on their effects on appetite, metabolism, energy intake, the microbiome, and body weight. Ironically, low calorie sweeteners have been used in animal feed to increase the animals' weight and decrease weaning time. (Shearer and Swithers, 2016) Rolls reviewed the effect of intense sweeteners on hunger, food intake, and body weight in humans (*American Journal of Clinical Nutrition*, 1991), particularly since some earlier studies had suggested that intense sweeteners had resulted in ratings of increased appetite—the so-called "paradoxical effect." Rolls made the point that increased feelings of hunger do not necessarily translate into increased food intake: much depends on a person's own motivation: if using low calorie sweeteners becomes an excuse to eat a high-calorie food, daily energy will remain unchanged. She found there are no data in humans that suggest intense sweeteners promote food intake or weight gain in dieters. Much more recently, Anderson et al, 2012, similarly found no evidence that these sweeteners cause increased body weight, and while data about these sweeteners "as a weight management tool

Much of our sugar comes from liquid calories in soda. Since all calories count, and many people do not fully compensate for these potentially additional calories by limiting intake of other foods, they may be prone to weight gain. Some people attempt to limit caloric intake by switching to sodas sweetened with low or no calorie sweeteners.

Used with permission/istockphoto.com. Credit: tibu.

Illustration from Beatrix Potter's book, *The Tailor of Gloucester:* Two mice stand next to a cup and a sugar bowl on a shelf overlooking a fireplace.

Ontario Institute for Studies in Education, the Internet Archive. Old Book Illustrations.

Stevia Rebaudiana plant, one of the newest low calorie sweeteners, considered in the Generally Recognized as Safe (GRAS) category. Though it is considered a "natural" product, nevertheless, it undergoes considerable processing prior to use.

Flyingbikie, 2012. Licensed under Creative Commons 1.0 Universal Public Domain, Wikipedia Commons.

are mixed, there is no evidence of harm or suggestion that they are counterproductive by leading to weight gain." (J.R. Roberts, 2015)

Many researchers have studied the effects of these sweeteners on hormone levels such as insulin and glucose. Tey et al (*International Journal of Obesity-London*, 2016), for example, investigated these effects in a small sample of lean, healthy males, in a randomized, crossover study with aspartame, monk-fruit, Stevia, and sucrose-sweetened beverages that occurred over the course of one day. They found that the use of low calorie sweeteners does not lead to overconsumption or increases in glucose or insulin levels compared to use of sucrose. Stevia may be particularly promising. Even rodents prefer its taste to aspartame and cyclamate, and Shankar et al (2013), in reviewing preliminary research on it, suggest it may lower insulin and postprandial glucose levels and even lead to lower total caloric intake.

Some researchers have investigated changes in the gut bacteria of our microbiome, even though these substances are essentially "inert," (i.e., excreted unchanged in the body.) (Suez et al, 2015) They noted that the composition of the microbiome of a person may create a need for a "personalized approach" to these sweeteners. They found that in mice and in a small subset of human subjects, some sweeteners could induce glucose intolerance by altering the microbiome. Shearer and Swithers (2016) found current "data do not present a clear consensus on the direction, type or magnitude of change on gut microbiome," and differences may be related to different sweeteners tested and other aspects of individual diets.

No prospective study to date has provided evidence that diet soda consumption *reduces* the risk of diabetes, heart disease, hypertension or stroke, and in some studies, may actually increase our risks. "Over the short-term, sugar-sweetened beverages do appear to be worse than those artificially-sweetened alternatives, but that does not mean that the 'diet' versions are healthy," says Swithers (2016). While "...scientific studies currently indicate that public health will be improved by reducing intake of all sweeteners, both caloric and non-caloric," she adds, "to be clear, excess sweetener consumption alone is not the only culprit" in our obesity epidemic. (Swithers, *Current Opinion in Behavioral Sciences*, 2016)

Note: For comparison, see blog 33, *Are We Sugar-coating Sugar Substitutes?* ■

BEHIND THE SMOKE-SCREEN OF VAPING: E-CIGARETTES

Condensing information from the nuance of vapor

81 — Posted February 3, 2017

THEY COME IN "Cherry Crush," "Snappin' Apple," "Chocolate Treat," coffee, mint, crème caramel, black cherry marshmallow, buttered popcorn, cotton candy, Fruit Loops, and over 7,700 other "unique flavors." (England et al, *American Journal of Preventive Medicine*, 2015; Hildick-Smith et al, *Journal of Adolescent Health*, 2015; Zhu et al, *Tobacco Control*, 2014.) Sound like gourmet specialty candy or ice cream? Hardly. They are among the flavor varieties available for electronic cigarettes.

Electronic cigarettes (e-cigarettes) are essentially electronic nicotine delivery systems, also referred to as ENDS. Nicotine, of course, is the major toxic chemical in traditional tobacco cigarettes, though traditional cigarettes contain 7,000 compounds and at least 70 "recognized carcinogens." (Hildick-Smith et al, 2015) Vapor from e-cigarettes contains lower levels of potentially toxic chemicals than traditional cigarettes, but users should not fall into the trap of thinking of them the way cigarette smokers had once thought of "lite" or "filtered" products that ultimately have proven no safer than traditional tobacco products. (Drummond and Upson, *Annals of the American Thoracic Society*, 2014)

It was in 1964 that the U.S. Surgeon General first issued a report linking smoking to lung cancer and chronic bronchitis. In attempts over the years to wean people from traditional cigarettes, manufacturers have created nicotine replacement therapies, such as nicotine transdermal patches, nasal spray, gum, and lozenges, all of which are FDA-approved. (Hildick-Smith et al, 2015). Nevertheless, traditional cigarette smoking continues "to cause a massive burden of avoidable disease and premature mortality" throughout the world. (Drummond and Upson, 2014) A January 2017 editorial in *The Lancet* notes that 6 million people die from tobacco each year, and there are projections that the number is expected to rise to 8 million by 2030; global health costs and loss of productivity are estimated at $1 trillion.

There have been two philosophies that have dominated efforts at controlling the use of tobacco—abstinence and harm reduction, and there has been "tension" between the two different camps. E-cigarettes have increased the tension because it is not clear whether they are able to "render combustion of tobacco obsolete," i.e., whether they "deliver promise or peril." (Abrams, *JAMA*, 2014)

The U.S. tobacco company Philip Morris had first begun working on electronic cigarette technology in the 1990s, ostensibly to develop a safer cigarette, but the company became concerned about possible FDA regulations on tobacco and tabled its product until around 2001. (Dutra et al, *Tobacco Control*, 2016). A Chinese pharmacist is credited with inventing e-cigarettes around 2003; the first device arrived in the U.S. market in 2007. (Orellana-Barrios et al, *American Journal of Medicine*, 2015) By 2014, there were over 450 brands on the market. (Zhu et al, 2014)

E-cigarettes are available in several different types: disposable cigarette-shaped; rechargeable cigarette-shaped;

Max Beckmann. *Quappi in Pink Jumper.* 1932–34. Matilde von Kaulbach (Quappi) was the second wife of Beckmann.

Smoking an E-cigarette.
Used with permission/istockphoto.com. Credit : mauro_grigollo.

E-cigarettes come in many sizes, shapes, and colors, all of which may be quite tempting to younger people.
Used with permission/istockphoto.com. Credit: kitiara65.

pen-shaped medium-sized rechargeable; and rechargeable vaporizers that are larger devices that apparently can deliver almost as much nicotine as traditional cigarettes. (Glasser et al, *American Journal of Preventive Medicine,* 2017; Grana et al, *Circulation,* 2014) There are also e-hookahs, e-cigars, and even e-pipes. (Zhu et al, 2014) These products have become big business and are expected to reach $10 billion globally in 2017 as many of the major tobacco companies have begun to purchase or develop this technology. (Tremblay et al, *BiomedCentral Medicine,* 2015; Zhu et al, 2014)

E-cigarettes have been called a "disruptive technology." (Glasser et al, 2017; Abrams, 2014; Etter, *Biomed Central Medicine,* 2015) First used to describe the Internet, a "disruptive technology" is one that can "radically alter market forces, profit expectations, and business models," as opposed to a "sustaining technology" that "reinforces the same markets and economic assumptions that existed previously." (Anderson, *Journal of the American Medical Informatics Association,* 2000.)

E-cigarettes have four component parts: a battery (usually lithium); a heating element that enables temperatures to rise high enough to produce an aerosol; a vaporizing chamber; and a solution cartridge that contains nicotine, flavoring, and either propylene glycol or glycerin. Batteries can be automatic or manual. In some models, batteries can be charged by using a USB port on a computer. Significantly, there have been reports of severe thermal burns, lacerations, and even explosions from overheated batteries. (Orellana-Barrios et al, 2015; Glasser et al, 2017)

Some e-cigarettes enable the user to adjust the amount of nicotine inhaled. Researchers note that the aerosol that is produced is technically not a vapor since it has a "particulate phase" and not just a gas phase typical of a vapor. Nevertheless, those who use e-cigarettes are called "vapers" and the process is calling "vaping." (Orellana-Barrios et al, 2015) These are indeed "very strange vapours" (Ben Jonson, *Bartholomew Fair,* 1614) and "confounding vapors." (G.K. Chesterton, *William Blake,*1910)

Essentially, e-cigarettes deliver nicotine to the lungs, but it is not clear where most of it is absorbed (Palazzolo, *Frontiers in Public Health,* 2013). An e-cigarette cartridge can deliver from 150 to 300 puffs; a traditional cigarette yields 10 to 15 puffs. Adverse reactions include mouth and throat irritation, vertigo, headaches, and nausea. Nicotine itself is not only toxic, but potentially lethal, especially for children.

Toxicity, i.e., the "toxicant fingerprint" (Orr, *Tobacco Control,* 2014) for e-cigarettes depends on the product's design and the amount of nicotine exposure. Because nicotine is delivered to the lungs and absorbed in the mouth, its absorption, metabolism, distribution, and excretion can be similar to traditional cigarettes. Both involve a respiratory route: e-cigarettes involve aerosol exposure while traditional cigarettes involve inhalation of smoke by the burning of tobacco. Apparently, it is difficult to extrapolate toxicity findings among the different brands that are commercially available, and there is no scientific consensus on "testing paradigms" to use to compare traditional cigarettes with e-cigarettes.

Many substances have been found in the solutions of e-cigarettes, including tobacco alkaloids, aldehydes, heavy metals (e.g., nickel, lead, cadmium, chromium), arsenic, and volatile organic compounds. (Orellana-Barrios et al, 2015) The vaping liquid products of glycerin and propylene glycol are of particular concern and potentially carcinogenic

Georgios Jakobides, *Smoke Rings (the Little Struggler)*, 1887. Cigarette smoking may lead to use of E-cigarettes, but sometimes E-cigarette use leads to smoking tobacco cigarettes.
Photograph by Sotheby's London, 2012. Wikimedia Commons/Public Domain.

Nikiforos Lytras, *Boy Rolling a Cigarette*, 1884. Adolescents may be quite susceptible to trying tobacco products, including E-cigarettes. National Gallery of Greece.
Donated by Argyris Chatziargyris. Wikimedia Commons/Public Domain.

as they are oxidized to formaldehyde and other toxic chemicals. (Callahan-Lyon, *Tobacco Control*, 2014) Other factors that may affect the inhalation of toxic chemicals include climate conditions, air flow, room size, number of other vapers, type and age of the system used, battery voltage, puff length and amount of suction on the device, interval between puffs (Callahan-Lyon, 2014; Evans and Hoffman, *Tobacco Control*, 2014), and experience of the user.

Further, there has been inaccurate product labeling such that products that supposedly do not contain nicotine have actually been found to contain it. (Palazzolo, 2013) There may also be inconsistent nicotine delivery because of a lack of industry standards in assessing the ingredients and methods used in manufacturing. (Callahan-Lyon, 2014)

Originally, there were no U.S. regulations on the manufacture, marketing, or sale of e-cigarettes, and many sales have been and continue to be transacted over the Internet. Significantly, Switzerland is the only country where the sale of nicotine-containing e-cigarettes is completely banned. (Greenhill et al, *Journal of Adolescent Health*, 2016) By mid-2014, 44 U.S. states had planned or already enacted legislative regulations on e-cigarettes, (Tremblay et al, 2015); it was only last May 2016 that the FDA finally decided to step in and legislate their sale as it does for other tobacco substances: once the legislation fully takes effect over the next few years, the sale of e-cigarettes will be banned to anyone under age 18, and a photo I.D. will be required. (FDA *Consumer Health Information*, June 2016)

The major focus and cause for concern now are on the growing use of e-cigarettes by adolescents. (Tremblay et al, 2015) There is considerable fear that vaping will encourage both teens and young adults to begin to use or continue using traditional cigarettes, i.e., that e-cigarettes may become a "gateway" to traditional cigarettes, especially in those who have never smoked previously. (England et al, 2015; Palazzolo, 2013) As tobacco smoking has fallen in these populations over recent years, from a high of 35% in the mid '90s to 9.2% in 2014 (Barrington-Trimis et al, *Pediatrics*, 2016), vaping has increased considerably: from 2014, in 21 cross-sectional studies involving adolescents, those reporting "ever use" rose from 6.4% to 31%. (Greenhill et al, 2016) In other words, the current use of e-cigarettes has, for the first time, surpassed cigarette smoking, and there is concern that the use "renormalizes" smoking. (Barrington-Trimis et al, 2016) Unfortunately, adolescents believe e-cigarettes are safer than traditional cigarettes, have a propensity for risky, thrill-seeking behaviors, and are particularly influenced by their peers. (Greenhill et al, 2016; Carroll Chapman and Wu, *Journal of Psychiatric Research*, 2014)

Studies have shown that nicotine exposure during "periods of developmental vulnerability" such as during adolescence, can impair neuronal development, brain circuitry that leads to changes in "brain architecture," and

Enrique Simonet, *Smoking Shisha at the Teashop*, 1892.
Often those who use E-cigarettes, especially adolescents, smoke in groups.
Wikimedia Commons/Public Domain.

neurobehavioral functioning. (Greenhill et al, 2016; England et al, 2015) Further, adolescents exposed to nicotine are more likely to become dependent on it. Nicotine exposure may have long-term effects on cognitive behavior, including reduced attention span and increased impulsivity in adulthood. (Yuan et al, *Journal of Physiology*, 2015) England et al (2015) acknowledge that ethical issues involving human experimentation "make it unlikely that there will ever be definitive human studies that fully quantify the effects of nicotine on the developing brain." They emphasize, though, that use of e-cigarettes "warrants extreme caution," including exposing the fetus to its effects during pregnancy.

Researchers also emphasize that since traditional cigarettes have not been advertised on television since the 1970s, heavy advertising of e-cigarettes is reaching a new generation not used to such marketing. (Grana et al, 2014) Even though as many as 85% of e-cigarette users report using the product to quit smoking, in some studies, e-users were found to be no more likely to quit one year later than non-users. (Grana et al, 2014) Further, since e-cigarettes are not approved for quitting, there are no "therapeutic instructions" for how to use them as replacements for traditional cigarettes.

Bottom line: Even though e-cigarettes technically don't send out smoke, we can think metaphorically that e-cigarettes may generate their own "smoke-screen." We don't know their long-term effects; further, they may not necessarily lead to decreased smoking, and they may even encourage "dual use" and the "renormalization" of smoking traditional cigarettes. There are now, though, enough data to warrant caution for everyone, but especially for vulnerable populations like adolescents and pregnant women.

Note: My subtitle was inspired by Neal Stephenson's 1992 disquieting, science fiction novel *Snow Crash* in which facial expressions convey information about what's going on in the mind of another: "Condense fact from the vapor of nuance." (p. 60)

For those who want the most complete bibliography on e-cigarettes (with 811 references), see the recent review by Glasser et al in *The American Journal of Preventive Medicine,* 2017) ■

SALT INTAKE: TAKING ADVICE WITH THAT PROVERBIAL GRAIN

Is there a salt set point for optimal health?

82—Posted March 9, 2017

Sixty-year-old Mahatma Gandhi, provoked by British regulations that prevented Indians from producing or selling their own local salt and that required them to purchase expensive, imported salt they could ill afford, led a group on a 240-mile march (*Salt Satyagraha*) in 1930 through a swath of western India to the Arabian Sea. Picking up "handfuls of salt along the shore," he and his followers thus "technically produced salt" and disobeyed Colonial law. (Pletcher, *Britannica.com*, 2015) This defiance, which led to his arrest and imprisonment, was considered Gandhi's first major step in his campaign of civil disobedience against British rule in India. (Pletcher, 2015)

History, from very ancient to modern times, is replete with examples of the importance of salt as a precious and versatile commodity. (For a breezy overview, see Kurlansky, *Salt: A World History*, 2002) For at least 5000 years, the Chinese appreciated the food preservative qualities of salt. (Ha, *Electrolytes & Blood Pressure*, 2014) In Roman times, it was used for trade: the Latin word "salarium," from which our word 'salary,' is derived was the "money allowed to Roman soldiers to purchase salt." (*Oxford English Dictionary*) And throughout history, people have been willing, like Gandhi, to take a stand against salt's unjust restriction. There are even some who believe the tax on salt (the *gabelle*), initiated centuries earlier, (Denton, *The Hunger for Salt*, p. 84, 1984) was one factor that ultimately led to the French Revolution. (Cirillo, *American Journal of Nephrology*, 1994)

Even the Bible has multiple references to salt. Perhaps most famous is The Old Testament's story of Lot's wife, who disobeys God's prohibition to look back on the fire and brimstone destruction of Sodom and is turned instantly into a pillar of salt. (Genesis 19: 15-26) (For those interested, see scientific articles on "The chemical death of Lot's wife" in a 1988 issue, *Journal of the Royal Society of Medicine.*) In The New Testament, Jesus says of his followers, "You are the salt of the earth" in his "Sermon on the Mount." (Matthew 5:13)

More recently, our English language is "peppered," if you will, with expressions involving salt. For example, being "worth one's salt" indicates competence or efficiency; "rub salt in one's wounds" means, "adding more pain to injury"; and "take with a grain of salt," suggests, "be skeptical." Even The Rolling Stones wrote the song "Salt of the Earth" for their 1968 album *Beggars Banquet.*

What then exactly is salt? For our purposes, common table salt is sodium chloride (40% sodium, 60% chloride) (Ha, 2014). Sodium is an "essential nutrient" and an important regulator of bodily fluids. Though Gandhi had used his "Salt March" for political purposes, he intuitively knew his physiology, "Next to air and water, salt is perhaps the greatest necessity of life," he wrote. (*Mahatma, Life of Gandhi*, Chapter 5, *The Epic March*, Commentary, Reel 11) Either too little (i.e., *hyponatremia)* or too much sodium (i.e., *hypernatremia*), such as can occur through excessive vomiting, sweating, dehydration, or diarrhea, can be potentially life threatening. About 90% of the sodium we eat daily is excreted in our urine. (O'Donnell et al, *Circulation Research*, 2015) The amount of sodium is tightly regulated, and the kidney is central to maintaining sodium levels by a complex system involving aldosterone, angiotension II, and renin, as well as the activation of the sympathetic nervous system. (Mancia et al, *European Heart Journal*, 2017)

Sodium can come from *discretionary* sources (e.g. salt added in cooking or by an individual during a meal) and *non-discretionary* sources (e.g. processed or pre-prepared

Salt March Statue, Delhi, commemorating 60-year-old Mahatma Gandhi's 240-mile *Salt Satyagraha* in 1930 to protest British Colonial regulations in India.

J. Miers, 2009. Creative Commons Attribution-Share Alike 4.0 International. Attribution: User: WT-shared) Jtesla 16 at wts wikivoyage.

Mount Sodom, Israel, showing so-called "Lot's Wife" who turned to a pillar of salt., Credit: Mark A. Wilson, 2007.

William Hamo Thornycroft, *Lot's Wife,* 1877, Victoria & Albert Museum, London.

foods.) In fact, about 80% of our daily sodium intake now comes from processed foods. (Taormina, *Critical Reviews in Food Science and Nutrition*, 2010) Further, there are many other compounds that contain sodium that enter our food supply, such as monosodium glutamate (MSG), sodium citrate, sodium nitrite, sodium benzoate, sodium bicarbonate, and sodium propionate, all of which have been used as food preservatives and have appeared as ingredients in processed foods. (Taormina, 2010)

In recent years, refrigeration helps preserve food, but for centuries, sodium chloride had been and remains one of the most effective agents against many bacteria (e.g. botulin toxin and E. coli) and even some viruses "and remains one of the most effective tools for the development of safe and wholesome food products." (Taormina, 2010) Salt is not only an important preservative, but it enhances flavor and can change the texture, consistency, and even moisture content of food (Taormina, 2010) For example, salt, by increasing the amount of water in meat, can increase its weight as much as 20%. (He and MacGregor, *Journal of Human Hypertension*, 2009) Salt can also aid in the process of fermentation.

Most people can handle substantial daily variations in sodium intake without any difficulty. Some, though, are more "salt-sensitive" than others. There is no consistent definition for salt-sensitivity. Blood pressure response to salt consumption varies among individuals: according to the response, some are considered *salt-sensitive* and some others are *salt-resistant*. In the total population, 51% of those patients with hypertension are salt-sensitive, and among

Researchers cannot always agree on how much salt is too much. Used with permission/istockphoto.com.
MichelGuenette.

Mosaic of Sodom. Another artistic rendering of the story of Lot, his wife who turns to a pillar of salt, and his daughters, who escape with their father. *Sodom Monreal.*

Dom von Monreale, Sizilien. Palermo, 12th century. Wikimedia Commons/Public Domain.

normotensives, 26% are salt-sensitive. Risk factors for salt-sensitivity include black race, intrinsic kidney disease, and aging. (Delahaye, *Archives of Cardiovascular Disease*, 2013) Rust and Ekmekcioglu (*Advances in Experimental Medicine and Biology*, 2016) note that salt-sensitivity in normotensive adults predicts future hypertension and has been associated with increased mortality in both normotensive and hypertensive persons. For those salt-sensitive, there is a recommendation to increase potassium consumption as well as reduce sodium intake. (Rust and Ekmekcioglu, 2016).

But despite that sodium chloride is "indispensible for life" for our body's homeostasis, salt has "attained the status of a villain, if not a poison" in the past century. (Drüeke, *Kidney International*, 2016) For example, in a recent study that received media attention, high sodium intake became "a key target" among the ten foods and nutrients that were studied and was implicated in the death of over 66,500 Americans in 2012—"the largest number of estimated diet-related cardiometabolic deaths." (Micha et al, *JAMA*, 2017) The authors did acknowledge they could not prove causality, and the researchers also appreciated that their "risk assessment does not prove that changes in dietary habits reduce disease risk." Further, this study did not actually measure sodium intake accurately.

The so-called "villainization" of salt grew out of research, both in animals and humans, from the 1940s and 1950s that implicated excessive sodium as a possible cause of hypertension. (Drüeke, 2016) Over the years, though, randomized controlled trials analyzing the role of lowering salt intake to decrease blood pressure have had conflicting results, often because of small samples and studies of short duration, often lasting less than six months. (Mancia et al, 2017) Furthermore, many of these studies could not separate out the effects of sodium specifically from other dietary factors, including potassium intake from fruits and vegetables, physical activity, and even excessive alcohol use that may also impact blood pressure. (Drüeke, 2016; Mancia et al, 2017) Further, obtaining sodium exposure from fresh fish is presumably different from exposure from highly processed "junk food."

In general, human and animal studies now support the concept that *excessive* salt intake is a major factor that increases blood pressure in the general population and reducing salt intake can lower blood pressure in many people. (He and MacGregor, *Journal of Human Hypertension*, 2009; Mancia et al, 2017) The mechanisms by which salt increases blood pressure are not completely elucidated: the kidneys, central nervous system, vasoactive substances, and neurohumoral factors are all involved. (Delahaye, 2013.)

Much of the support for the idea that a low sodium diet leads to lower blood pressure comes from the Dietary Approaches to Stop Hypertension (*DASH*) that enrolled over 400 people and randomly assigned them to different levels of sodium intake. The DASH diet, though, was significantly different from the control diet, and contained more fruits, vegetables, low-fat dairy, whole grains, less red meat and fewer sweets, etc. (DiNicolantonio et al, *American Journal of Medicine*, 2013) In another study (TOHP-II, with over 2300 subjects and follow-up of 36 months), different diets were also used, and the intervention cohort, for example, was advised to use more spices, use of which may have its own effects. (DiNicolantonio et al, 2013)

Researchers even note there is no conclusive evidence that a low sodium diet reduces the possibility of cardiovascular events in those patients who do not have hypertension, and they suggest that lowered sodium intake can even worsen those with congestive heart failure or type 2 diabetes. (DiNicolantonio et al 2013) Likewise, Mente et al (*The Lancet*, 2016) found there is a "U shaped curve, with evidence of increased risk of cardiovascular events with sodium excretion *lower* than 3 grams/day in those with and without hypertension, but the harm associated with high sodium consumption (greater than 6 grams/day) *seems to be confined to those individuals with hypertension*. These investigators believe "this argues against a population-wide recommendation to reduce sodium intake in most countries,"(Mente et al 2016) and it is "prudent" to recommend reduced sodium intake only in those with high sodium intake and hypertension. Further, the essential nutrient iodine is often added to table salt; lowering sodium intake may lead to depleted iodine reserves that can lead to thyroid disorders.

Researchers also found that there may be a diminishing effect of lowering sodium intake on blood pressure over time, and adherence to low sodium intake may not be sustainable over the long term. (O'Donnell et al, 2015) Further, investigators have found there is considerable variation in the amount of sodium intake worldwide: the INTERSALT study, for example, (involving over 10,000 subjects in 32 countries) found people in Northern China had the highest level of sodium excretion. (O'Donnell et al, 2015)

The gold standard for quantifying sodium is measuring its excretion in *repeated* 24-hour collections of urine. Day-to-day variations can be as much as 20%. (Kong et al, *Frontiers in Endocrinology*, 2016) Rarely do studies collect 24-hour urine more than once, and most studies rely on *self-report* (e.g. food frequency questionnaires or 24-hour food diaries) that are *notoriously inaccurate* in all nutrition research but particularly in these studies because there are so many hidden sources of sodium in food. (Mancia et al, 2017; He and MacGregor, 2009; Rust and Ekmekcioglu, 2009) Another complication is that there may be circadian rhythms involved in sodium storage not captured by these measurements. (O'Donnell et al, 2015)

Despite considerable clinical research, researchers cannot even agree on the range of sodium intake that is considered normal or adequate: clearly sodium is essential and beneficial, but how much is harmful? Rozin and colleagues (*Health Psychology*, 1996) note we use simplistic, categorical thinking in nutrition: foods are either "good or bad for health," and we have "dose insensitivity," i.e., "the belief that if something is harmful in high amounts, it is also harmful in low or trace amounts." They also found evidence in nutritional research of *monotonic thinking*, i.e., that we are reluctant to accept the idea that low and high doses may have opposite effects. "Salt is frequently considered 'bad' and in large quantities that is often true; but, just because a lot of salt is bad does not mean that a little salt is bad." (Brown and Bohan Brown, in *Nutrition in Lifestyle Medicine*, 2017, James M. Rippe, editor)

Lot's Flight from Sodom. This Bibical story has captured the imagination of artists for centuries. Woodcut from the *Nuremberg Chronicle*, 1493.

Michel Wolgemut, Wilhelm Pleydenwurff. Wikimedia Commons/Public Domain.

For sodium intake, the World Health Organization recommends less than 2000 mg/day; the American Heart Association, less than 1500 mg/day; and a 2013 evidenced-based report by the Institute of Medicine 2013 suggests there is "insufficient evidence to reduce sodium to below 2300 mg/day." The average intake of sodium hovers around 3400 mg/day or 8.5 g/salt in Canada, US, and UK. (DiNicolantonio and O'Keefe, *Progress in Cardiovascular Diseases*, 2016) (To obtain the amount of salt, multiply the amount of sodium in grams by 2.5; according to the American Heart Association, 1 teaspoon of salt equals 2300 mg sodium.)

Swedish researcher Folkow (*Lakartidningen*, 2003) believes there is actually a "*biologically determined set-point*" for salt intake—what he calls our "salt appetite" of 10 grams/day that is not reflective of "hedonistic abuse," when researchers consider both blood pressure and heart rate and this may be physiologically driven. He does support labeling

Piles of Salt in Bolivia. Photo by Luca Galuzzi, 2006.

Wikimedia Commons, Creative Commons Attribution-Share Alike 2.5 Generic.

the amount of salt in processed food and keeping it low, since "salt is easy to add but impossible to remove." Most other researchers recommend a considerably lower amount of 3-to-5 grams/day as optimal for cardiovascular health and believe that studies are too inconsistent to warrant decreases below 2 grams of sodium/day. (DiNicolantonio and O'Keefe, 2016; O'Donnell et al, 2015) Further, a meta-analysis of almost 170 studies found that sodium restriction lowers BP only by 1 to 3% in those without hypertension and 3.5 to 7% in those with hypertension, and some believe we may have "incriminated the wrong white crystals:" *sugar*, rather than salt, may be more detrimental and may "potentiate" the effects of salt, particularly in those who are salt-sensitive. (DiNicolantonio and O'Keefe, 2016)

Increased salt intake has been correlated with several pathological conditions, such as osteoporosis, obesity, and diabetes. For example, when salt intake is increased, there is an increased calcium loss in the urine, leading to a negative calcium balance that may predispose to greater bone loss. (Caudarella et al, *Journal of Endocrinological Investigation*, 2009; He and MacGregor, 2009) The relationship to obesity is complex and may be indirect: salt increases thirst that may then be quenched with highly caloric, sugary soft drinks. (He and MacGregor, *Pflügers Archiv, European Journal of Physiology*, 2015) Ma et al (*Hypertension*, 2015), though, did find a direct association between salt intake and obesity independent of energy intake: there is a suggestion that higher salt intake may lead to altered fat metabolism and increased fat deposition. Moosavian et al, (*International Journal of Food Sciences and Nutrition*, 2017) reviewed 18 cross-sectional studies and found an association between sodium intake and greater body mass index (BMI) and waist circumference, but conceded that they could not determine any "dose-response" association; further, some studies had not even measured sodium intake accurately, and unhealthy behaviors associated with obesity may be confounders. A recent study Radzeviciene and Ostrauskas (*Nutrients*, 2017) found a two-fold increase in diabetes in those who added salt to prepared meals, but used self-report measures.

Jaenke et al (*Critical Reviews in Food Science and Nutrition*, 2016) evaluated salt-reduced foods to assess the "extent to which salt could be reduced ...without jeopardizing consumer acceptability." These investigators found that salt could be reduced in breads by up to 37% and in processed meats up to 67%. When sodium is reduced, though, foods can become bitter. Often potassium is used as a substitute, but for example, when used in cheeses, where salt is used in the process of ripening, it can lead to a sour taste. (Rust and Ekmekcioglu, 2016) Our salt taste receptors on our tongue, though, can become more or less sensitive to lower salt exposure over time.

Bottom line: Guidelines for salt restriction are inconsistent, and studies are often methodologically flawed. There continues to be controversy and a lack of agreement about what level of salt intake is beneficial, and reducing salt intake may have its own harmful consequences. To date, there are *no randomized controlled trials that conclusively* demonstrate that moderate or lower reductions in salt intake actually reduce cardiovascular disease; (Mancia et al, 2017) blood pressure is, after all, only a "surrogate endpoint" for cardiovascular morbidity. (Kong et al, 2016) Weill Cornell Professor of Medicine Dr. Alan M. Weinstein adds, "If there were a clear risk for sodium intake to limit life expectancy, then good science should provide an estimate of that increased risk. I am not aware of any such estimate."

Note: For those interested, see Bisaccia et al's article, "The Symbolism of Salt in Paintings" in an issue of *The American Journal of Nephrology*, 1997. ■

OF MARCH AND MYTH: THE POLITICIZING OF SCIENCE

Scientific integrity, self-correction, and the public

83 — Posted April 19, 2017

TWENTIETH CENTURY Austrian-born British philosopher of science Karl Popper once wrote, “Science must begin with myths, and with the criticism of myths.” (*Conjectures and Refutations*, p. 66) For Popper, science is unique in its systematic approach to errors and its emphasis on self-correction. Popper was best known for his doctrine of falsifiability, i.e., the importance of the testability of a hypothesis. In other words, what distinguishes the scientific method from other methods of investigation is that it is a method of attempting to discover the weaknesses of a theory—to “refute or to falsify the theory.” (Popper, *All Life is Problem Solving*, p. 10) “Science has nothing to do with the quest for certainty or probability or reliability,” wrote Popper, “We are not interested in establishing scientific theories as secure, or certain, or probable…we are only interested in criticizing them and testing them, hoping to find out where we are mistaken…” (*Conjectures…*,p. 310)

To address some of the issues and difficulties within science, the National Academy of Sciences recently sponsored a three-day Arthur M. Sackler Colloquium *Reproducibility of Research: Issues and Proposed Remedies*, organized by Drs. David B. Allison, Richard Shiffrin, and Victoria Stodden, in Washington, DC. This colloquium brought together international leaders from multiple disciplines, including Nobel-Prize winning researcher Dr. Randy Schekman, its keynote speaker; 28 of these lectures can be retrieved on YouTube.com (Sackler channel.) (For a summary of these outstanding lectures, see Cynthia M. Kroeger's link on the internet. The subject is clearly opportune: *Rigor Mortis: How Sloppy Science Creates Worthless Cures, Crushes Hope, and Wastes Billions* by science journalist Richard Harris was released recently.

The “sloppy science” of which Harris writes, includes everything from misconduct such as falsifying, fabricating, or even plagiarizing data, to distorting the manner, by so-called “spin,” in which a study's results are reported. Says Schekman, “Sometimes, though, there is a fine line between sloppy science and overt misconduct and fraud.”

Sloppiness also occurs when imprecise language fails in “capturing the science” for the media, says colloquium speaker University of Pennsylvania Professor of Communications Kathleen Hall Jamieson. Without accuracy, scientists are “inviting misinterpretation” and what she calls “narrative infidelity.” Scientists, for example, generate confusion when they write of “herd immunity” as opposed to “community immunity” or when they create the “controversial frame” of a “three-parent baby” when they describe the technique of obtaining genetic material from a cell's mitochondria, the energy powerhouse of the cell, says Jamieson. She adds, “Mitochondria do not determine parenthood.”

Jamieson, though, is loath to call the existing scientific narrative a crisis. Rather than the accepting the narrative, “Science is broken,” Jamieson prefers to view science as

Sir Karl Popper, British philosopher of science, believed in the doctrine of falsifiability, namely, that hypotheses can be tested and refuted. That is the difference between the scientific method and other methods of inquiry.

Photo copyright by Lucinda Douglas-Menzies, National Portrait Gallery London, 1988. Used with permission.

"self-correcting," (Alberts et al, *Science*, 2015) just as Karl Popper had emphasized years earlier. Colloquium organizer Shiffrin also takes issue with the use of "inflammatory rhetoric" that merely increases the public's skepticism and undermines its trust in science. Shiffrin noted that since so much of science is exploratory, with a need for "successive refinement," it does not always lend itself to reproducibility. Researcher John Ioannidis, writing recently in *JAMA* (2017) acknowledges the importance of reproducibility and what we can learn when results cannot be replicated; he also appreciates, though, the complications that can arise from "unanticipated outcomes" in "complex and multifactorial" biological systems that can interfere with reproducibility.

One of the difficulties in reproducing results is that there are literally hundreds of kinds of bias—systematic errors as opposed to errors by chance— that can creep, either knowingly or not, into scientific studies. "Science, though, is a bias-reduction technique and the best method to come to objective knowledge about the world," says organizer Allison, who is a Distinguished Professor, biostatistician and Director of Nutrition Obesity Research Center (NORC) at the University of Alabama at Birmingham. "Its validity depends on its procedures but there is always room for improvement," he adds.

Joseph Wright of Derby, *An Experiment on a Bird in an Air Pump*, 1768. Research in the 18th century, while considerably more unsophisticated, was much less complicated than today. Experimenting on animals, though, cannot always lead to results applicable to humans. The National Gallery, London.

Wikimedia Commons/Public Domain.

William Blake, *Newton*, 1795. Tate, Britain.
William Blake Archive. Wikimedia Commons/Public Domain.

Another type of bias that can occur, as well, is what Allison and his colleague Mark Cope, in their 2010 papers in both the *International Journal of Obesity (London)* and *Acta Paediatrica*, call *white hat bias*, which they define as "bias leading to distortion of information in the service of what might be perceived as righteous ends." Examples of this kind of bias include misleading and inaccurate reporting of data from scientific studies by "exaggerating the strength of the evidence" or issuing media press reports that distort, misrepresent, or even fail to present the actual facts of the research or do not even mention any caveats or limitations. Cope and Allison note that *white hat bias* can be either intentional or unintentional and can 'demonize' or 'sanctify' research. Sumner et al (*PLOS One*, 2016) for example, note that press releases "routinely condense complex scientific findings and theories into digestible packets" that may produce "unintended subtle exaggerations" when they use simple language. Regardless of which way *white hat bias* leans, though, it can be "sufficient to misguide readers," say Cope and Allison.

Other difficulties that complicate research stem from the present reward system for scientific advancement that may inadvertently foster an unhealthy climate whereby "journal publications become the currency of science," writes Harris. One inherent difficulty in the scientific literature, for example, occurs when studies are more apt to be published if their results are perceived as statistically significant, unusually remarkable, or even improbable, particularly in what are called *high impact journals.* This is called *publication bias,* i.e., when publication depends more on a study's outcome rather than its overall quality. For Harris, this designation of *high impact* is "a measurement invented for commercial purposes to help sell ads and subscriptions." Statistically significant results, incidentally, may have considerably less genuine clinical significance. A medical trial for a cancer treatment, for example, may significantly improve patient survival but only by a few weeks—hardly clinically significant for an individual patient.

One of the procedures that scientists are beginning to recognize as a prerequisite for reproducibility and scientific integrity is the need for full transparency and a sharing of data as the default position. "Show your work and share, lessons we all learned in kindergarten," says colloquium speaker Brian Nosek, a psychology professor at the University of Virginia. Some researchers have now suggested those papers receive a badge, a kind of seal of approval, as an incentive for full disclosure. Cottrell, who supports the idea that peer review in science should no longer be anonymous and calls it an "historical anachronism" (*Research Ethics*, 2014) has described science as "struggling with a crisis of confidence." And since young researchers model themselves after the ethical conduct of their mentors, a laboratory's culture becomes exceptionally important.

Says Harris, "Getting biomedical research right means more than avoiding the obvious pitfalls…it's also critical to think about whether the underlying assumptions are correct." Getting it right, though, sometimes involves admitting errors and retracting papers whose results cannot be replicated, a fairly new concept that has grown only within the past fifteen years. Though the majority of journals now have retraction policies in place, actual paper retraction is a tedious and thankless endeavor, according to Allison. He found many journals were unwilling and overtly resistant to correcting blatant inaccuracies and spurious data that he and his colleagues have discovered while reviewing hundreds of papers weekly for their obesityandenergetics.org website that is freely available to over 80,000 of their subscribers. But as Harris says, "There is little funding and no glory involved in checking someone else's work."

Perhaps there should be. Mistakes in the literature, of course, are not just of academic interest: they can have far-reaching public health consequences. Many parents, for example, were wrongly discouraged from vaccinating their children because of the fraudulent connection between vaccinations and autism that had been published (and later retracted) in the reputable British journal *Lancet*. In recent years, there is now an organization *Retraction Watch* that reports on and encourages this self-policing practice.

While scientists are becoming more cognizant of their need to self-police, they are also aware of the need of calling attention publicly to the importance of science. Organizers

John White Alexander, portrait of Samuel L. Clemens (Mark Twain), American humorist, 1912 or 1913. Twain, back in the 1870s, attacked science when he spoofed the nonsensical extrapolation of data. Attacks on science are nothing new.
National Portrait Gallery, Washington. Wikimedia Commons, Public Domain.

of the April 22nd *March for Science*, in 500 cities worldwide, proclaim "Science, not Silence." While hundreds of thousands of scientists have registered, not all scientists believe the march will accomplish its goal. Coastal geologist Robert S. Young, for example, in a *New York Times* editorial back in January (1/31/17), thinks the march will politicize science even further and create a "mass spectacle." Said Young, "We need storytellers, not marchers." The main impetus for the upcoming march is the looming threat of a substantial reduction involving billions of dollars in public funding for scientific research in the proposed budget of the Trump Administration. How very different from other administrations. It was, after all, an Act of Congress, signed in 1863 by then President Abraham Lincoln that had first

Giovanni di Paolo, *The Creation of the World and the Expulsion from Paradise*, 1445. The view of Creation as depicted in the Bible conflicts with scientific thought on the evolution of our solar system and the development of life on Earth. Tempera and gold on wood. Metropolitan Museum of Art, NYC.

Robert Lehman Collection, 1975. Public Domain.

Paul Klee, *Adam and Little Eve,* 1921. The Biblical tale of Adam and Eve cannot be subjected to the scrutiny of scientific inquiry. Metropolitan Museum of Art, NYC.

The Berggruen Klee Collection, 1987. Public Domain.

Lorenzo Monaco (Piero di Giovanni), *Noah,* circa 1408–1410, tempera on wood, gold ground, Metropolitan Museum, NYC.

Gwynne Andrews Fund, and Gift of Paul Peralta Ramos, by exchange, 1965. Public Domain.

established the National Academy of Sciences, perhaps our nation's most prestigious assemblage of scientific scholars that now include almost 500 Nobel-Prize winners among its members. How has it now come to this, though, that science needs both its marchers and its storytellers?

The misguided policies of our current government notwithstanding, science itself, by its very complexity, has come under scrutiny and has become fair game for public assault. There has also been "a tendency to lump all opposition to science, despite topic, into a common 'anti-science' camp...with a 'them vs us narrative,'" write McClain and Neeley (*F1000Research*, 2014) Whether attacks are worse now or just apparently more prevalent due to increased social media exposure is not clear. There have, though, always been attacks. Mark Twain, for example, back in 1874, in his *Life on the Mississippi,* spoofed the nonsensical extrapolation of data and wrote, "There is something fascinating about science. One gets such wholesale returns of conjecture out of such a trifling investment of fact." Twain's quotation appeared in Darrell Huff's own 1954 best-selling spoof, *How to Lie with Statistics.* In more recent years, scientists themselves, as we have seen by the recent Sackler Colloquium, are now acknowledging the need to confront their own shortcomings. Storytellers, after all, can tell tales that are factual or fictitious.

Marches and stories are useful for generating awareness, but they are not enough. Scientists will convince the public and funding agencies of the importance of their science, though, only by their own unrelenting adherence to a culture that consistently fosters self-scrutiny and the value of self-correction. That is the science defined by Karl Popper over fifty years ago. ■

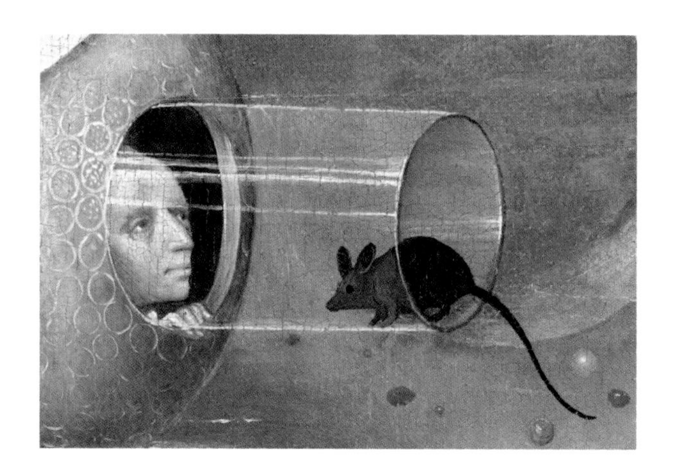

Hieronymus Bosch, *The Garden of Earthly Delights*, central panel: Detail of man with mouse, circa 1516. Prado Museum, Madrid.

Wikimedia Commons/Public Domain.

THE "FURRY TEST TUBES" OF OBESITY RESEARCH

The "intellectual leap" of translating science from mice to humans

84—Posted May 14, 2017

MICE FUNCTION WELL as stand-ins for humans in our allegories. Most notably, Spencer Johnson's bestseller, *Who Moved My Cheese?* (1998) recounts the tale of mice Sniff and Scurry to demonstrate the importance of anticipating change, recognizing the "handwriting on the wall," and of being flexible when evidence of change is inevitable. On a more profound level, Art Spiegelman wrote the extraordinary, Pulitzer-Prize winning graphic novel *Maus: A Survivor's Tale* (1973, 1986) that documents, using all cartoon figures of mice, his parents' experiences during the Holocaust. How well, though, do mice serve as substitutes for humans in scientific research? The answer is a complex one.

It was the renowned 19th century physiologist Claude Bernard (*Introduction to the Study of Experimental Medicine*, 1865) who was one of the first to defend the use of animals for scientific research. Bernard thought it was "wholly and absolutely" the right of humans to experiment on animals, "even though painful and dangerous," if that research benefited mankind. (Bernard, in Otis, *Literature and Science in the 19th Century*, 2002, p. 206) This belief differed with Bernard's view of *human experimentation*, "The principle of medical and surgical morality, therefore, consists in never performing on man an experiment which might be harmful to him to any extent, even though the result might be highly advantageous to science." (Otis, p. 205) The Nazi physicians of the 1940s, prosecuted and convicted at Nuremberg (Annas and Grodin, *The Nazi Doctors and the Nuremberg Code*, 1992), of course, in sharp contrast, engaged in practices that were "a frightening example of medicine gone wrong" (Annas and Grodin, p. 3) and subjected their human subjects "to conditions that would violate current animal research regulations, let alone human regulations." (p. 197)

As a result of the ethical considerations of using humans in research, scientists have focused throughout the years on creating animal models as substitutes. "The use of animals for research is a privilege granted to scientists with the explicit understanding that their use provides significant new knowledge without causing unnecessary harm." (Bailoo et al, *Institute for Laboratory Animal Research, ILAR Journal*, 2014) There is, though, "shockingly poor translation of basic animal work into human outcomes," writes Garner (*ILAR Journal*, 2014), who adds, "The secret to using animal models is to understand their limitations." Garner et al (*Lab Animal, NY*, 2017) have coined the term *therioepistemology* as a new discipline of how knowledge is gained from animal research. These researchers note that there can be considerable failure in this translation: for example, only one in nine drugs that are used in human trials will succeed despite that all of these drugs "worked" in an animal model. One devastating example occurred years ago with thalidomide, which caused only peripheral neuritis in animals but caused the severe birth defect of phocomelia (deformed limbs) in infants whose mothers were prescribed the drug during their pregnancy. (Rohra et al, *Journal of the College of Physicians and Surgeons*, Pakistan, 2005)

Sjoberg (*Behavioral and Brain Functions*, 2017) explains that researchers who use animal models may become susceptible to the fallacy of *false analogy*: assuming there are necessarily similarities between animals and humans and arrive at incorrect conclusions. Says Sjoberg, "Animal experiments are reconstructions, not replications, since animals cannot follow instructions." Results, though, can sometimes be extrapolated to humans.

Today, in obesity research, rodents, and in particular, mice, (i.e., murine models) have become the animal of choice for

Hatto, Archbishop of Mainz from the book, *The Nuremberg Chronicle* by Hartmann Schedel, a medical doctor (1440–1514). Published in Nuremberg in 1493. The book was considered a world history.

Wikimedia Commons/Public Domain.

Guardians of Day and Night, Han Dynasty. Paintings on ceramic tile, circa 200 A.D., anonymous Chinese artist.

Robert Temple's ***The Genius of China****, 1986, NY: Simon and Schuster, Wikimedia Commons/Public Domain.*

much of scientific investigation. Says Richard Harris (*Rigor Mortis, 2017*), "One reason everybody uses mice: everybody else uses mice…scientists have developed hundreds of inbred strains of mice…entire industries have grown up around the breeding, shipping, housing, feeding and care of mice." (p. 74) "The house mouse is an extraordinary animal …mice have followed us around the globe and colonized almost every environment humans ever created without the benefit of language, tools, or clothes." (Garner, 2014)

Mice are relatively easy to study because they have high reproductive rates, a short life cycle, and low maintenance costs relative to other mammals. (Nguyen et al, *Disease Models & Mechanisms*, 2015; Maloney et al, *Physiology*, Bethesda, 2014) But many variables can affect their functioning, including the environment of their cages, (e.g. bedding material and how often it is changed), the position of the cage from the ceiling (i.e., mice vary in their level of fear, abnormal behavior, and immune suppression according to their position in the cage rack), the gender of their handlers (e.g. more aggressive and stressed with male researchers); the strain of mice used, and how many mice are in each cage. (Garner et al, 2017; Slattery and Cryan, *Psychopharmacology*, Berlin, 2017) Because mice are nocturnal animals, with their inactive period during our daytime, even the amount of light to which they are subjected may be a factor. Further, the temperature at which mice are housed is important: most experiments on mice are conducted at temperatures well below where they are most comfortable (i.e., below their thermoneutral zone) so that they are under "chronic thermal stress" such that they increase their metabolism and food intake by as much as 50%, spend more time awake with sleep deprivation, and have an increased heart rate, all of which can impact research results and interfere with reproducibility. (Lodhi and Semenkovich, *Cell Metabolism*, 2009; Maloney et al, 2014)

Garner et al (2017) also emphasize that researchers often ignore or fail to measure aspects of an environment that do not matter to humans; for example, mice have different sensory ranges from us, including ultraviolet and ultrasound frequencies we don't detect. We also tend to "over-engineer" their environments—with "standard chow, standard barren environments…which in turn make them fundamentally abnormal" for mice. It then becomes "easy to fall into the trap of thinking of them as little more than *furry test tubes*

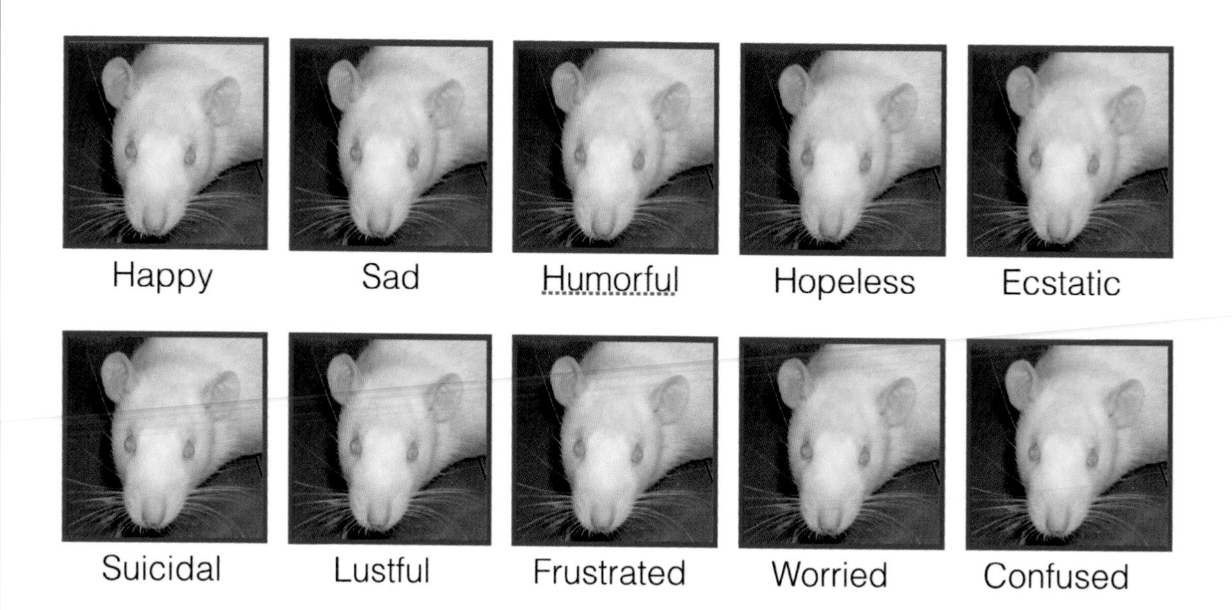

We can only guess at the psychology of the mouse, demonstrated beautifully by Dr. Nestler's slide. Many of the signs and symptoms of human disease cannot be ascertained from animal models.

Used with generous permission of Eric J. Nestler, MD, PhD. Mount Sinai Medical Center, NYC.

rather than animal patients." (Garner et al, 2017) Bailoo et al (2014) note that keeping everything constant—the *gold standard* of experimental design—works so well in the pure sciences like math or physics but may result in the *standardization fallacy* with mice: homogenization, "contrary to customary wisdom" may compromise the study.

And we tend to ignore human biology "through reductionism:" we create monogenic mutations when rarely are our human diseases the result of only one gene. (Garner et al, 2017) For example, Asrafuzzaman et al (*Biomedicine & Pharmacotherapy*, 2017) explain that type 2 diabetes is a human, not an animal disease, and develops from a combination of environmental factors and polygenic mutations. When we create an obese mouse with abnormal insulin levels and insulin resistance, we have altered one gene (e.g. ob/ob mouse with leptin deficiency.) These mice develop their abnormalities early on, whereas humans can develop type 2 diabetes at any time and often more likely with age. (Wang et al, *Current Diabetes Reviews*, 2014)

Further, many experiments done with mice fail to employ proper scientific technique. Ribaroff et al (*Obesity Reviews*, 2017) conducted a comprehensive meta-analysis of 171 papers that used animal models and found only 31% of those papers explicitly mentioned that the animals had been randomized, and none mentioned that the experimenters were blind to the interventions or treatment exposures. In marked

Das Spiel mit der Maus, (That Play with the Mouse), unknown artist, 1864.

Düsseldorfer Auktionshaus. Wikimedia Commons/Public Domain.

distinction to the inadequate design of many other investigators, Allison and his colleagues (Cope et al, *International Journal of Obesity*, London, 2005), exploring the link between atypical antipsychotic medication and weight gain, reported all the important experimental parameters (e.g. randomization, blinding, mouse strain, etc.) clearly and in great detail, including the importance of oral administration, rather than intraperitoneal or subcutaneous injections, to deliver medication less stressfully.

Bottom line: There is no perfect animal model for many of our human diseases. Using mice, for example, may require an *intellectual leap* or even an *anthropomorphic leap* when we cannot ascertain the animal's psychology, i.e., the *why* behind an animal's behavior, explain Nestler and Hyman, who study neuropsychiatric disorders like depression (*Nature Neuroscience*, 2010) They add "...it is hard to imagine significant progress in pathophysiology or therapeutics without good animal models" That is the challenge researchers face, but good experimental design obviates some of the inherent limitations and enables the possibility of valid and reproducible scientific results.

For those interested in an excellent summary of research involving mice, see Richard Harris's chapter, *Misled by Mice*, in his new book *Rigor Mort*is. ■

THE GAMBLER'S FALLACY IN RESEARCH

Down the "statistical garden path" for those participating in clinical studies

85 — Posted June 18, 2017

"It is impossible to approach the gambling table without becoming infected with superstition," writes Alexei Ivanovich, the Russian tutor in Dostoyevsky's powerful novella *The Gambler*. Even Alexei's "senseless and unseemly failure" earlier that day "has not left the slightest doubt" in him: he is still absolutely convinced that he will win.

The Gambler, set in the fictional German town of Roulettenburg, has the impact it does because Dostoyevsky was himself a compulsive gambler for many years who knew only too well the obsessive preoccupation and financial ruin that gambling can precipitate. Dostoyevsky's theme has appeared in many subsequent artistic adaptations, including a Prokofiev opera of the same name that was directly inspired by the novella. The subject of gambling is a frequent one in paintings, such as by Bokelmann, Cézanne, Rowlandson, and Caravaggio, among others.

Vasily Perov, portrait of Fyodor Dostoyevsky, 1872. Dostoyevsky was himself a compulsive gambler for many years and wrote the novella, *The Gambler*, which captures the mind of those afflicted with a gambling obsession. Tretyakov Gallery, Moscow.

Wikimedia Commons/Public Domain.

About the roulette wheel, Alexei explains, "...one morning, red will be followed by black and back again almost without any order, shifting every minute, so that it never turns up red or black for more than two or three strokes in succession. He continues, (p. 92) "Chance favors red, for instance, ten or even fifteen times in succession...Every one, of course, abandoned red at once, and ...scarcely anyone dared to stake on it..." Alexei even tries to warn elderly and infirm Granny, who too has become thoroughly absorbed by the game, "... zero has only just turned up, so now it won't turn up for a long time. You will lose a great deal; wait a little, anyway..." (p. 59)

Dostoyevsky's Alexei demonstrates the two aspects of the so-called *gambler's fallacy*: a statistical "misunderstanding" of the odds and unrealistic optimism that "supersedes statistical reasoning." (Swekoski and Barnbaum, *IRB: Ethics & Human Research*, 2013) The *gambler's fallacy* is "the belief that the odds for something with a fixed probability increase or decrease depending on recent occurrences," i.e., "so if red comes up four times in a row, on the fifth time, it is more likely to be black." (Wertheimer, *Rethinking the Ethics of Clinical Research*, 2011, note # 71, Chapter 3, p. 328) In other words, those who suffer from the gambler's fallacy will not accept that each turn is *independent* of another and has the same 50% probability of recurring; he or she *will instead* believe that the probability of another red, for example, must be much lower after a succession of previous reds. The other aspect of the gambler's fallacy, as evidenced in Alexei's words, is that "the odds are somehow suspended" and the odds of winning are "more certain." Even the "statistically sophisticated" can believe, "Tonight is my lucky night." (Swekoski and Barnbaum, 2013)

Though the gambler's fallacy exists in many contexts, it may occur in those who participate in randomized controlled trials, the gold standard of clinical research, in which an experimental treatment of unknown efficacy is compared with either placebo (i.e., an inactive treatment for the condition) or more likely, in recent years, a different

Caravaggio, *The Cardsharps*, circa 1594. Kimbell Art Museum, Forth Worth, Texas.
Wikimedia Commons/Public Domain.

therapy. (Wertheimer, 2011; Swekoski and Barnbaum, 2013) Says Wertheimer, "Gambling is an interesting analogue to participating in research because it may involve considerable risks." (p. 81) These investigators note that research subjects may likewise suffer from the same distorted reasoning and believe they, too, are "exempt from statistics" and have "undue optimism" about their chances of cure or treatment when they participate. (Swekoski and Barnbaum, 2013)

Subjects may volunteer for clinical research studies for different reasons, including a desire to help others, i.e., altruism; financial compensation given to participants; and variously other perceived personal benefits. (Detoc et al, *Expert Review of Vaccines*, 2017.) For example, participants may develop *therapeutic appropriation* (McDougall et al, *Journal of Medical Ethics*, 2016) whereby they may understand the research protocol but still want to join a study for individual therapeutic benefit, such as receiving additional tests and monitoring for their condition, increased access to hospital personnel or their own physician, and even other medical or social services. (McDougall et al, 2016) Henry Beecher, though, in his classic paper, (*NEJM*, 1966) has emphasized, "Ordinary patients will not knowingly risk their health or their life for the sake of 'science.'"

Research subjects may also suffer from a related concept, what Appelbaum and colleagues have called *therapeutic misconception* (Appelbaum et al, *International Journal of Law and Psychiatry*, 1982; Lidz et al, *Cambridge Quarterly of Healthcare Ethics*, 2015): they may believe they are special and unique and will receive *personal therapeutic benefit* from a study, even though they may realize (and have been told) that the goal of research is different from that of clinical care. *Therapeutic misconception* involves the *incorrect belief* that a person's *individual needs* will determine what treatment group (e.g., receiving medication rather than placebo) he or she will be assigned, *rather than by unbiased random* assignment. There are thus a misunderstanding and an unreasonable appraisal of both the process and goals of research.(Swekoski and Barnbaum, 2013)

Subjects who have a disease that is being investigated are "particularly vulnerable" to therapeutic misconception, and are often "desperate" to have access to an experimental treatment: they may "*overestimate the benefits* and *underestimate the risks* of research participation." (Wertheimer, 2011, p. 33; Henderson et al, *PLoS Medicine*, 2007) They may not appreciate the "possibility that they will receive no therapy or (even) sub-optimal therapy." (Charuvastra and Marder, *Journal of Medical Ethics*, 2008) Further, subjects may not even understand fully terms such as "double-blind," "placebo," or even "treatment" and "research." (Henderson et al, 2007) And their expectations may derive "from both cultural images of the physician-patient relationship and their previous experiences with medical caregivers;" they may believe that physicians would suggest research *only* when their participation would be in their own "medical best interests." In other words, they come to research with a "strong therapeutic bias" (Lidz and Appelbaum, *Medical Care*, 2002); unfortunately, though, "everything about a medical setting will evoke participants' expectations of personal care." (Lidz et al, 2015) Sometimes, therapeutic misconception "may be shared" by both subjects and researchers when researchers "may themselves believe they are acting in the best interests of the patient." (Charuvastra and Marder, 2008)

What is the goal of research? It is achieving *generalized knowledge* for the potential benefit of society, rather than that of clinical care whose goal is to *benefit the individual.* (Breault and Miceli, *Ochsner Journal*, 2016) In clinical care,

Roulette wheel with the "zero," black, and red.

Ralf Roletschek, 2013. Licensed under Creative Commons Attribution-Share Alike 3.0 Unported, Wikipedia.

Paul Cézanne, *Card Players*, between 1894–95.
Musée d'Orsay, Paris.
Wikimedia Commons/Public Domain.

physicians have the "fundamental ethical obligation" to "prioritize the interests" of their patients. (Lidz et al, 2015) While research studies must protect patients from undue harm, i.e., the *beneficence principle* (Lidz and Appelbaum, 2002) they do not "prioritize" their patients' interests above the study's interests: randomization is impersonal. (Lidz et al, 2015) Their primary purpose is "to answer a research question…" (Breault and Miceli, 2016) A researcher's *frame* is "independent of specific patient needs." (Lidz et al, 2015)

When researchers recruit their subjects for a study, they must convey to them five dimensions: (1) the *scientific purpose* is to produce generalizable knowledge and to answer questions about safety and efficacy; (2) *study procedures* are conducted for achieving scientific knowledge and not for their specific patient care; (3) there is *inevitable uncertainty* involved in both risks and benefits of the study; (4) there must be a *strict adherence* to the study's protocol that may involve an inability to receive their current medications, etc.; (5) the *clinician* involved in the study is *foremost an investigator*, not a personal physician, whose task it is to assess safety and efficacy and not administer treatment. (Henderson et al, 2007; Breault and Miceli, 2016)

Gustave Caillebotte, *The Bezique Game*, 1880–81,
Louvre Abu Dhabi.
Wikiart, Public Domain.

Patients, though, can compromise the well-meaning, but insensitive, intentions of clinicians, poignantly demonstrated in the film *Dallas Buyers Club* (2013), recipient of several Academy Awards. The film depicts the true story of Dallas electrician Ron Woodruff who contracted AIDS in the mid-1980s from unprotected heterosexual sex. It explores the desperation that Ron and his fellow sufferers experienced under the stringent controls of the FDA in the early days of the AIDS epidemic when clinical trials involved an inactive placebo control to assess the efficacy of AZT, the only potential medication then available in the U.S. Given a death sentence, Woodruff resorts to bribing a hospital staff member to ensure he receives active drug rather than placebo, and other study participants split their doses with those assigned to receive placebo. Ultimately, when his condition deteriorates, Ron resorts to smuggling in non-FDA approved medication from Mexico for himself and the other AIDS-ravaged patients in the hospital study, much to the exasperation of those physicians whose goal was to conduct unbiased research for generalized scientific knowledge.

Sometimes, though, patients may not necessarily be desperate, but just oppositional in their thwarting experimental protocol. British-born Sir Austin Bradford Hill (1897–1991), described as "a master of the methods by which arithmetic is made argumentative"[1] and considered by some to be the "greatest medical statistician" of the last century, was one of the first to appreciate the importance of designing and conducting randomized controlled trials on patients.

1. Doll's description of Hill was borrowed from British politician Sir John Simon, who had used those words to describe another British statistician, William Farr, one hundred years earlier. (Doll, 1993).

Christian Ludwig Bokelmann, *The Gambler*, circa 1873. Salford Museum and Art Gallery, Manchester, England.
Wikimedia Commons/Public Domain. Gift of Messrs. J. Johnson and Sons, 1881.

The Prodigal Son Gambles (One of eight scenes from the story of the Prodigal Son, German, 1532, Colorless glass, vitreous paint, and silver stain. Metropolitan Museum of Art, NYC.
Bequest of George Blumenthal, 1941, Public Domain.

(Doll, 1993, *Statistics in Medicine*). In his reminiscences, Hill describes the following as his personal favorite: "Doctor," said the young woman, 'Why have you changed my pills?' The doctor replied, "What makes you think I have changed your pills?" "Well," she replied, "last week, when I threw them down the loo, they floated, but this week they sank." (Hill, *British Medical Journal*, 1985)

Bottom line: Research participants may have a vastly different understanding of the nature of research from those who design and conduct it. It behooves investigators to sort out what patients expect so they can avoid having their studies compromised by patients' misconceptions, undue expectations, and failures to follow protocols.

Note: My subtitle is a quote from Sir Austin Bradford Hill (*NEJM*, 1952). For those interested in the many talents of Sir Austin Bradford Hill, see blog 66, *Toward a 'Knowledge of Causes…and All Things Possible'* and blog 75, *Advise and Consent.* ■

German artist Georg Flegel, painted *Snack with Fried Eggs* in the 1600s. Staatsgalerie Aschaffenburg, Germany.
Wikipedia Commons/Public Domain.

"CHOLESTEROL-PHOBIA" AND EGGS: WHAT DO WE KNOW?

Revisiting the "nutritional certainty" about restricting egg intake

86 — Posted Juyl 16, 2017

Dr. Seuss first published his *Green Eggs and Ham,* about young, oppositional Sam, in 1960. Had he written his classic in the late 1960s, when organizations like the American Heart Association issued a warning to the public about the dangers of cholesterol, Dr. Seuss might have run into controversy. As the main source of cholesterol in the American diet, the egg, with from 141 mg to 234 mg depending on its size, (Clayton et al, *Nutrition,* 2017) became the target of the food police. The public health mantra became, "Cholesterol in food equals cholesterol in the blood" and leads eventually to cardiovascular disease. (McNamara, *Proceedings of the Nutrition Society,* 2014.)

"Though there was no scientific rationale or justification" for 300 mg/day (and no more than 3 whole eggs per week), other than it was about half of what Americans were consuming at the time, it remains a "mystery" (McNamara, *Nutrients,* 2015) how that specific amount became synonymous with healthier eating. (McNamara, 2014). Adds McNamara (2014), "It has taken 50 years of research to undo the effects of those early condemnations and the *cholesterolphobia* much of the world suffered from for decades." The cholesterol saga "highlights" how researchers can overemphasize the role of a particular nutrient "without acknowledging limitations, uncertainties, and debates over this knowledge" such that information is presented as "nutritional certainty," (i.e., the "myth of nutritional precision.") (Scrinis, *Nutritionism,* p. 31, 2013) Cholesterol levels became an example of a "kind of biomarker reductionism." (p. 41)

What is cholesterol? The National Heart, Lung, and Blood Institute of NIH describes cholesterol as a soft, waxy substance that is one of the lipids (i.e. fats) in the body. It is found in all animal cells and is the precursor of steroid hormones and vitamin D. Further, it is a structural component of cell membranes, including nerve cells, and is found in gall stones. Our body makes its own cholesterol, primarily in the liver, and there is a complex homeostatic system in the body to regulate cholesterol levels when dietary intake is high.

Some people (and animals) are more sensitive to increased intake—called *hyper-responders*—than others—called *hypo-responders,* and it is not clear whether the hyper-responders are more efficient at cholesterol absorption or are not able to suppress their cholesterol pathway sufficiently enough, but there is a strong genetic component. (Beynen et al, *Advances in Lipid Research,* 1987) There is no specific test to determine if you are a hyper-responder other than taking multiple blood samples of cholesterol levels, and it is difficult to substantiate the concept because, unlike in animals, there are no 'inbred strains' and each human, except for identical twins, is genetically different. (Beynen et al 1987) Not only is our cholesterol level determined to some extent by dietary cholesterol but also by our weight, as well as our saturated fat and even fiber intake. Humans, though, are less sensitive to dietary cholesterol intake than some animals (e.g. rabbits or monkeys.) (Beynen et al, 1987) Further, says Ahrens, "...I cannot predict in advance how and by how much any given individual will respond to a given dietary challenge." (Ahrens, *The Lancet,* 1979.)

Cholesterol travels in blood plasma on lipoproteins, which also carry triglycerides and fat-soluble vitamins.

There are many different kinds of eggs, most of which are not edible by humans. This beautiful photograph had appeared on the June 23, 2017 cover of *Science*. It was taken by Frans Lanting.

Though there are five categories of lipoproteins, the most common are high-density lipoproteins (HDL, often called the 'good cholesterol') and low-density lipoproteins (LDL, often called the 'bad cholesterol.') The ratio of LDL to HDL is more important that total cholesterol.

Though loaded with cholesterol, a large whole egg is a major source of protein (over 6 grams and 18 amino acids); has only 4.76 grams of fat; 27 vitamins, including choline for fetal brain development, vitamin C, and folate; and 10 minerals, including calcium and magnesium. Some even believe eggs have a protective effect against inflammation and atherosclerosis because of their carotenoids and may even decrease the risk of developing cataracts and age-related macular degeneration because of their lutein. (McNamara, 2015; and see Clayton et al, 2017 for a full list of an egg's nutrients.) Further, a small study has suggested that an egg breakfast, as compared to cereal, leads to greater satiety and less food intake (Bailey et al, *FASEB Journal*, 2017) though a comprehensive meta-analysis by Allison and his colleagues has found that randomized controlled trials no longer support the general notion that eating (and not skipping) breakfast necessarily leads to greater weight control. (Bohan Brown et al, *FASEB Journal*, 2017; Milanes et al, *FASEB Journal*, 2016)

Given that eggs are so nutritious, how did they become vilified? In those early years, researchers found that both saturated fat and dietary cholesterol raised plasma cholesterol levels and led to heart disease but when dietary fat was eventually factored out, the relationship no longer held. It was Ahrens and his colleagues from Rockefeller University who emphasized that plasma cholesterol levels were more sensitive to saturated fat than dietary cholesterol. (McNamara et al, *Journal of Clinical Investigation*, 1987.) Some studies, as well, used animal models that did not mimic human responses and often used very high intake of dietary cholesterol so that compensatory mechanisms might have been overwhelmed. Further, they measured total cholesterol rather than LDL/HDL ratios. (McNamara, 2014) Even using different statistical methods led to inconsistencies and methodological flaws in the past. (Nicklas et al, *Journal of Nutrition*, 2015)

Theodor Geisel, popularly known as Dr. Seuss to children and adults worldwide, wrote his classic *Green Eggs and Ham* in 1960, before the cholesterol police clamped down on egg consumption.

Library of Congress Prints and Photographs Division. Photograph by Al Ravenna, 1957. Wikimedia Commons/Public Domain.

Glazed ceramic egg, late 17th- to mid-18th century, Armenian origin.

Metropolitan Museum of Art, NYC. Public Domain. Gift of Norma E. and Ralph D. Minasian, 2008.

Eat the parsley: this enormous ham and cheese omelet, loaded with saturated animal fat, is hardly healthy egg consumption.

Used with permission/istockphoto.com. Credit: JoeGough.

Allison and colleagues (Richardson et al, *International Journal of Obesity*, 2017), as well, describe how the success of the anti-smoking campaign led obesity efforts to focus more on public health approaches, "though these policies were not necessarily rigorously evaluated." They acknowledge, "Although we cannot always wait for perfect data to draw strong conclusions about causation," there is clearly a distinction between "*speculative* evidence that may lead to a *policy decision* and evidence required for a *scientific conclusion*." Back in the late 1970s, Ahrens had warned, "...if the public's diet is going to be decided by popularity polls and with diminishing regard for the scientific evidence, I fear that future generations will be left in ignorance of the real merits, as well as possible faults in any given dietary regimen." (*The Lancet*, 1979.)

Over the years, many researchers have now accepted that whole egg consumption (one a day) is *not associated* with a risk of cardiovascular disease and cardiac mortality in the general population. Frank Hu and colleagues conducted one of the most thorough prospective studies, including over 37,000 men and over 80,000 women followed for up to 14 years and reached that conclusion

Jean-Baptiste Greuze, French artist, *Broken Eggs,* 1756, Metropolitan Museum of Art, NYC.
Bequest of William K. Vanderbilt, 1920, Public Domain.

(*JAMA*, 1999), as have systematic reviews and meta-analyses by Kritchevsky and Kritchevsky, *Journal of the American College of Nutrition*, 2000; Shin et al, *American Journal of Clinical Nutrition, 2013*; ; Rong et al, *British Medical Journal*, 2013; Alexander et al, *Journal of the American College of Nutrition*, 2016, among others. Of note is that the 2015-2020 Dietary Guidelines for Americans no longer recommend limiting cholesterol intake to 300 mg/day, but add "individuals should eat as little dietary cholesterol as possible while consuming a healthy eating pattern." (For a discussion, see Clayton et al, 2017 and http://health.gov/dietaryguidelines/2015/guidelines.)

Bottom line: The scientific literature on the relationship between egg consumption, the major source of cholesterol in the U.S. diet, and the risk of cardiovascular disease, has been confusing and sometimes overtly inconsistent. Research has been complicated because almost all studies use notoriously inaccurate self-reports of dietary intake. Further, there are many confounding factors, such as smoking, weight, hypertension, and other dietary factors (e.g. too much saturated fat or too little fiber intake; size of the egg, its preparation, and many different patterns of use) that are not necessarily taken into account. What emerges from what we have learned is that eating a whole egg a day (genetic hyper-responders to cholesterol notwithstanding) does not lead to cardiovascular disease in the context of an otherwise healthy diet.

Note: For more on Dr. Gyorgy Scrinis's concept of the "myth of nutritional precision" in his book *Nutritionism*, see blog 44. ■

CROSSING THE THIN LINE TO STARVATION: CALORIC RESTRICTION

Beneficial or detrimental? The "dietary landscape" of caloric intake

87 — Posted August 17, 2017

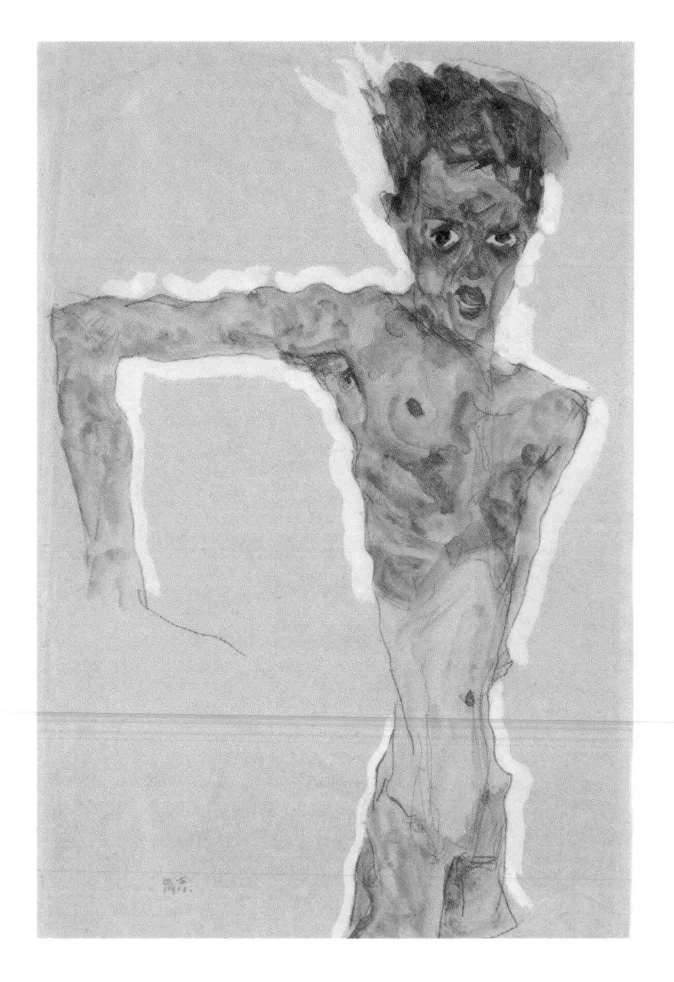

Egon Schiele, *Self-portrait,* 1911. Many artists, including Schiele, are known for drawing emaciated figures. There is often a fine line between restricting calories for health and emaciation. See the art of Giacometti, with his elongated, emaciated-looking figures.

Metropolitan Museum of Art, NYC. Bequest of Scofield Thayer, 1982, Public Domain.

In the creepy dystopian future of Max Ehrlich's *The Edict* (1971), the population has burgeoned out of control because there is no longer any cancer or cardiovascular disease. The average age in its "Senior City" was 100; many lived until 125 and some as long as 150 years because older people were "a patchwork of other people's parts." (p. 74) There was, however, not much food—only algae and plankton—and the average daily calorie allotment, scientifically calculated based on the ratio of births to deaths, was 652 calories. As a result, 90 percent of all deaths were due to "simple malnutrition."

For entertainment, the population went to titillating films called "Foodies," where they would watch vintage footage of old time supermarkets filled with fresh vegetables and fruits. The only part of the film this severely malnourished audience could not fathom was the diet section of the supermarket where there were shelves of low or no calorie items. But as the audience watched scenes of people eating real food like enormous pieces of roast beef or "a great slice of chocolate cake" with "a sticky bouquet of chocolate frosting," they would salivate and let out a collective moan, as if watching a pornographic movie, as their mouths "opened and closed in symbiotic union." (p. 133)

Throughout history, there have been many devastating famines, particularly in times of war and political conflict. During the Dutch famine of World War II, for example, the official rationing went from 1,800 calories a day in 1943 to 619 calories in the first quarter of 1945. (Keys et al, *Biology of Human Starvation,* Volume I, 1950, p. 25) It was even worse for the victims of the Lodz and Warsaw Ghettos (Weisz and Albury, *Israel Medical Association Journal,* 2013) and Nazi Concentration Camps: the official daily ration at Auschwitz was one liter of watery soup; 250 grams of bread; 20-25 grams of margarine or sausage or imitation honey. Explains Lucie Adelsberger, a Jewish physician imprisoned there, "These quantities became in time insufficient to support life, and the German camp doctors admitted that a prisoner could not hope to survive on them much longer than six months." (Adelsberger, *The Lancet,* 1946) In Belsen, technically a detention camp and not an organized extermination camp, where the daily caloric intake was under 800 calories a day, if that, (Lipscomb, *The Lancet,* 1945) the average male survivor weighed 44 kilos (97 pounds) and the average female survivor weighed 35.3 kilos (77.8 pounds) at the time of the camp's liberation, with an average loss of almost 40 percent of body weight. (Mollison, *British Medical Journal,* 1946.)

Most people can tolerate a weight loss of about 5 to 10 percent "with relatively little functional disorganization" but humans do not survive weight losses that are greater than 35 to 40 percent. (Keys et al, 1950, p.18, Vol. I) With war-ravaged Europe in mind, Ancel Keys and his colleagues at the University of Minnesota designed an experiment with a group of 36 conscientious objectors to assess the effects of caloric restriction that mimicked the malnourished diet (e.g. mostly potatoes, turnips, dark bread) of Europeans. The brochure proclaimed, "Will You Starve That They Be Better Fed?" The investigators' results were published in a meticulous two-volume tome, *Biology of Human Starvation (1950.)* Beginning in 1945, these men went from about 3,200 calories a day to about 1,800 calories, but continuously titrated down so that the men were to lose about 25 percent of their body weight (and walk 22 miles a week) in semi-starvation conditions over six months prior to a three-month rehabilitation period. The men became completely preoccupied with food, as they developed depression, nervousness, social withdrawal, anemia, fatigue, apathy, extreme weakness, irritability, neurological deficits, edema, loss of sexual interest, and inability to concentrate. Keys et al called the constellation of symptoms, a "semi-starvation neurosis." (p. 909, Volume II) Though the men found the experience grueling, when interviewed years later, they were proud to have participated. And unlike the conditions in Europe, said one participant, "We were starving under the best possible medical conditions. And most of all, we knew the exact day on which our torture would end." (Kalm and Semba, *Journal of Nutrition*, 2005.)

Singapore-The Cookhouse, Changi Gaol. British POWs Prepare their Main Meal. Typically, prisoners of war are deliberately starved, often to death.

Creator: Leslie Cole, 1946. From the Collection of the Imperial War Museum, UK, non-commercial license. Copyright Imperial War Museum (Art.IWM ART LD 5825) Used with permission.

Head of the Fasting Buddha, artist unknown., between circa 100 and 299 A.D., British Museum, London.
1907, given by Colonel F.G. Mainwaring. Wikipedia Commons/Public Domain.

Norwegian artist Christian Krohg, *Struggle for Existence*, 1889, National Gallery of Norway, Oslo.
Wikimedia Commons/Public Domain.

Keys et al noted that there is an important difference between a prolonged period of inadequate caloric intake and total fasting: With under-nutrition, the feelings of hunger get worse over time, whereas with total abstinence, hunger sensations dissipate in a few days. (Keys et al, p. 29, Vol. I)

What happens, though, when calories are restricted but there is adequate nutrition? Since the 1930s, researchers have questioned the value of restricting caloric intake, while maintaining adequate protein, fat, and carbohydrates, to benefit health and even to extend life. Caloric intake depends on our age, level of activity, and whether male or female. For example, adult men (before the age of 50) who are very active (e.g. walking quickly for more than three miles a day for more than 40 minutes, as well as usual activities) require about 2,400 to 2,800 calories a day; very active adult women may require about 2,200 calories a day. As we age, we require fewer calories. (Health and Human Services, NIH) The Institute of Medicine suggests the following ranges: 10 to 35 percent calories from protein; 20 to 35 percent from fat; and 45 to 65 percent from carbohydrates. (Dwyer, Chapter 95e, *Harrison's Textbook of Medicine*, 19th Edition, online.)

For years, researchers have found that many species (e.g. fruit flies, rodents), though not all (and even depending on strain of rodent) have not only substantially lengthened their lifespans through caloric restriction while maintaining adequate nutrition, but have also decreased their incidence of cancers as well as metabolic and immunological abnormalities. (Sohal and Forster, *Free Radical Biology & Medicine*, 2014) In an attempt to extrapolate their findings to humans,

Emaciated Horse and Rider, attributed to India, Deccan, Bijapur, circa 1625. Metropolitan Museum of Art, NYC.

Rogers Fund, 1944. Public Domain.

researchers have undertaken several major, long-term studies in non-human primates (e.g. rhesus monkeys) to assess the effect of caloric restriction (CR) on longevity. The results have been inconsistent: a study from the University of Wisconsin found a significant positive effect of CR on lifespan while a study from the National Institute on Aging did not find CR increased lifespan, though both groups found substantial health benefits. Allison and his colleagues, who conducted the statistical analysis for both studies, (Mattison et al, *Nature Communications*, 2017) explained the discrepancies in outcome: there were fundamental differences in each study's design and implementation. For example, there were different feeding schedules; different diet compositions including different amounts of protein and vastly different amounts of sugar in the diets; different age ranges for the initiation of the intervention of CR, and even different genetic strains of monkeys. Further, they acknowledged "the minimum degree of food restriction for maximum benefit has not been identified." And at a certain point, caloric restriction can go from being beneficial to being detrimental. (Roberts and Speakman, *Advances in Nutrition*, 2013)

Since non-human primates, though, share our "catalogue of pathologies" typical of our own aging, Allison and colleagues (Mattison et al, 2017) believe that there is a likely benefit of CR with adequate nutrition for lowering age-related morbidity in humans as well. This is the concept of *healthspan*-"maintaining full functioning as nearly as possible to the end of life." (Rowe and Kahn, *Science*, 1987) Explain Allison et al (Smith et al, *European Journal of Clinical Investigation*, 2010), "...the academic or esoteric question of whether lifespan can be truly extended by CR in humans may not be as important as the potential prolongation of healthspan." Further, say Kirkland and Peterson, (*Journal of Gerontology: Biological Sciences*,

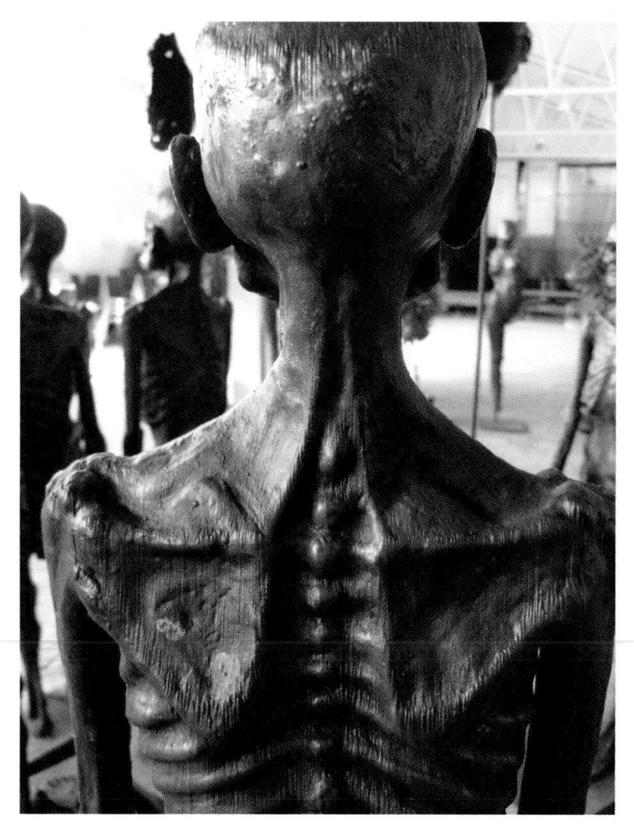

The Hunger March at the Forgotten Conflict Meeting in Copenhagen, 2003.

2009), who also emphasize the importance of healthspan, "It is not clear whether increasing lifespan will be associated with a pushing of morbidity out until near the end of life (compression of morbidity) or with increased disability and health care costs for society (expansion of morbidity.)

The problem is that CR may be too difficult to implement in humans, and as Keys et al had noted, hunger does not seem to dissipate with CR, and humans complain of being lethargic, tired, and cold. (Speakman and Hambly, *Journal of Nutrition*, 2007) The goal may be, say Balasubramanian et al, (*Ebiomedicine*, 2017) to understand its mechanism, rather than "promote it as a lifestyle." For example, there have been two human studies—the randomized controlled trials, CALERIE 1 and 2 (*Comprehensive Assessment of Long-term Effects of Reducing Intake of Energy*), to assess the physiological and psychological effects of CR on healthy, non-obese subjects. The researchers describe it as not exactly a clinical trial but rather "a model of a controlled experiment in free-living humans." (Stewart et al, *Contemporary Clinical Trials*, 2013) Initially the researchers had aimed for a 30 percent reduction in calories, but then realized that was not feasible and settled for 25 percent. By the end of the two years, though, CR was achieved at only 12 percent. (Das et al, *Molecular and Cellular Endocrinology*, 2017)

Bottom line: There are several theories about the mechanisms involved in caloric restriction with adequate nutrition, but there remain many unanswered questions, such as how much CR is beneficial; when should it first be implemented, including ethical issues of starting when someone is too young (where CR may affect growth, development, and even reproduction when CR may lead to amenorrhea in women) (Speakman and Hambly, 2007); and what are the most advantageous macronutrient percentages of fat, protein, and carbohydrate, i.e, the varying "dietary landscape." (Simpson et al, *Ageing Research Reviews*, 2017). Studies seem to demonstrate CR's efficacy in reducing age-related morbidities but for most, CR may not be feasible for the long term. Investigators are now studying compounds (e.g. resveratrol, rapamycin, metformin) to mimic the effects of CR. (Balasubramanian et al, 2017)

Note: In recent years, Ancel Keys has been unfairly maligned for his work on the connection between heart disease and fat intake in his major Seven Countries Study. For a comprehensive discussion, see the White Paper available online by researchers Pett, Kahn, Willett, and Katz, *Ancel Keys and the Seven Countries Study: An Evidenced-based Response to Revisionist Histories.* (2017) ■

Annibale Carracci, *The Bean Eater*, late 16th century, Palazzo Colonna, Rome.
Wikimedia Commons/Public Domain.

THE 'SOY-LING' OF OUR FOOD: THE VERSATILE U.S. SOYBEAN

Soy intake may be beneficial for health, but some still want more evidence

88 — Posted September 11, 2017

An unhappy little boy, obviously sick and swollen, stands with his mother, waiting in line for hours for their rationed water. Doctors have diagnosed him with "the kwash," the mother tells Shirl, one of the main characters in Harry Harrison's 1966 science fiction novel *Make Room! Make Room!* Shirl wonders if it is catching. No, explains Sol, because "'kwash' is short for 'kwashiorkor,' and it comes from not eating enough protein. They used to have it only in Africa but now they got it right across the whole U.S...there's no meat around, and lentils and soybeans cost too much," Sol adds. Some of the soybeans have even been poisoned by insecticide. The year is 1999, and there are massive shortages throughout the grossly overpopulated U.S. "I hope this proves to be a work of fiction," Harrison wrote in his introduction. In the much darker thriller film version *Soylent Green* (1973), soy as a protein source was not only expensive; it was no longer available.

Harrison's novel remains, at least for now, a work of fiction. Soybeans, considered the "crop of the century," are plentiful and "are likely to have unseated corn as the most widely sown crop in the U.S." (Meyer et al, June 20, 2017, *The Big Read: Agricultural Commodities*). Farmers throughout a vast swath of the Midwest (e.g. Ohio, Indiana, Illinois, Iowa, Nebraska, and now North Dakota) cultivate soybeans. (U.S. Department of Agriculture)

Soybeans can be black or yellow but when harvested prematurely, they are green and can be boiled in their pods and served with salt as edamame. The soybean is incredibly versatile. Soybeans can be non-fermented (e.g. soybean sprouts or soy milk, made through boiling or, when curdled, as tofu), or fermented, a process that can take months, to make soy sauce, miso, and tempeh. In Chinese cooking, soybeans with black hulls can be served as fermented black beans. There is also soybean oil, which contains both omega-6 and omega-3 fatty acids (Messina, *Nutrients*, 2016), and soy flour (finely ground roasted soybeans with the coat removed) that can be used in baking products. Soy flour washed with water can become soy protein, which is 70 percent protein by weight. (Barnes, *Lymphatic Research and Biology*, 2010). Even higher concentrations of protein (greater than 92 percent) can be created to form soy protein isolate. Both hydrolyzed vegetable protein and vegetable protein contain soy (Barnes, 2010). In general, soybean protein is higher than any other plant protein and similar to animal protein, and it is low in carbohydrate content. Further, it is a rich source of vitamins and minerals, including Vitamin K_2, calcium, and iron (Messina, 2016). Soy products can be used as "filler" for veggie burgers (i.e., as a meat alternative), processed into protein bars or shakes, or made into a nut butter similar to peanut butter. Soy can also be used in industrial products such as soaps, plastics, crayons, solvents, clothing, and even biodiesel fuel.

Further, much of the U.S. soybean crop is used for livestock feed, and there is even concern ("a clear and present danger") that people allergic to soy may have reactions to eggs and dairy foods that have soy residue from this feed given to cows, chickens, and fish without realizing this source. This "second-hand soy" has been called the "soy-ling of America," (Daniel, July 23, 2012, *The Weston A. Price Foundation*).

And due to genetic engineering and the new technology of CRISPR, we no longer have to worry about soybeans

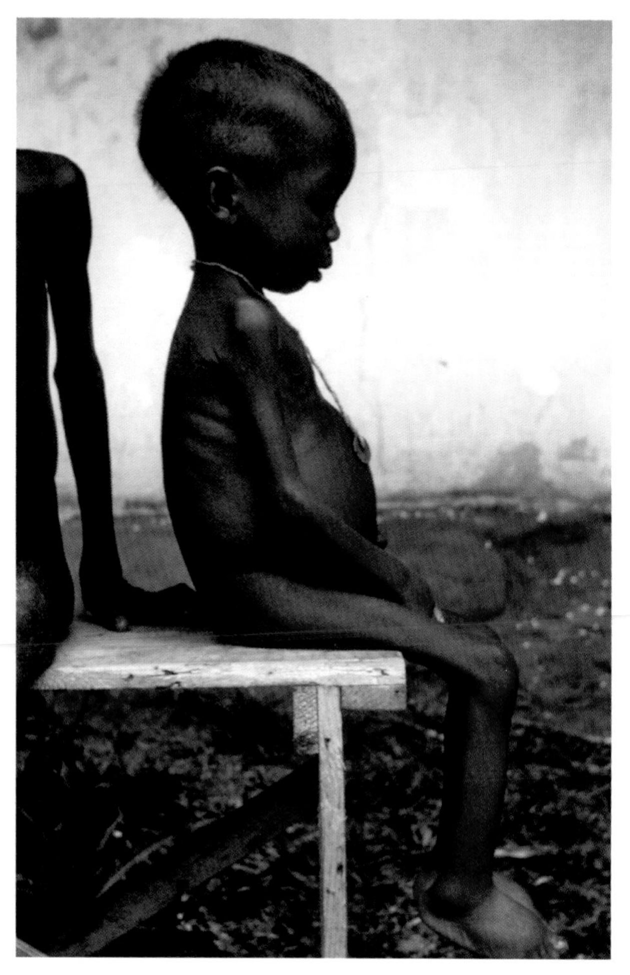

Starved girl, 1960s, with kwashiorkor, a protein deficiency disease. From the Nigerian-Biafran War.

Centers for Disease Control and Prevention, Atlanta. Credit: Dr. Lyle Conrad, Wikimedia Commons/Public Domain.

Soybean varieties.

Scott Bauer, 2004. Wikimedia Commons/Public Domain.

being poisoned by insecticides. CRISPR, an acronym for "clustered regularly interspaced short palindromic repeats," is the "holy grail of gene manipulation." It acts as a "pair of designer molecular scissors" that will eventually eliminate many genetic diseases, wrote Doudna and Sternberg, researchers instrumental in developing CRISPR technology, in their book *A Crack in Creation* (2017).

For years though, plant biologists have used CRISPR to edit genes in many crops, including soybeans (Jacobs et al, *BioMed Central Biotechnology*, 2015). For example, it has been used in corn, soybeans, and potatoes to enable these crops to develop natural resistance to insecticides. Further, by altering two soybean genes, researchers can produce a much healthier soybean oil with a "fat profile" that is similar to olive oil (Doudna and Sternberg, p.122).

There is some question regarding how long (and exactly where) soybeans (*Glycine max*) became domesticated from a wild variation (*Glycine soja*), but some archeological researchers believe it occurred from 3,000 to 9,000 years ago in several areas of East Asia, (Lee et al, *PLoS One*, 2011). Soybeans came to Europe and were first grown in the U.S. in Georgia in the late 18th century. They attract nitrogen-fixing bacteria so they can be used in crop rotation to restore nitrogen to the soil, (Barnes, 2010). Ironically, we are now shipping huge quantities of U.S. soybeans to China.

Soy is the main source of plant isoflavones, which are classified as both phytoestrogens and selective estrogen receptor modulators (SERMs). (Messina, 2016.) The two main isoflavones are genistein and daidzein and have a chemical structure similar to estrogen (Messina, 2016), but should not be equated with estrogens since they have 100 times *weaker* affinities than physiological estrogens, (Barnes, 2010). Initially, there was concern for soy exposure in postmenopausal women, but many (not all) believe there is no evidence of a negative effect on breast, thyroid, or uterus, (Messina, 2016).

Over the past 25 years, there have been overwhelming numbers of studies (2000 peer reviewed articles a year) that are "understandably challenging for health professionals to interpret" on the benefits of soy (Messina, 2016). Furthermore, animal studies have limited applicability to humans because many animals (rodents and even nonhuman primates) metabolize isoflavones very differently from the way we do, (Messina, 2016). Nevertheless, soy intake has been associated with better skin, decreased depression, and even less incidence of menopausal hot flashes, postmenopausal osteoporosis, as well as improved blood pressure, (especially fermented soy) (Nozue et al, *The Journal of Nutrition, Nutritional Epidemiology*, 2017), metabolic parameters (e.g. glucose, cholesterol), and overall cardiovascular health (Messina, 2016). Allison and colleagues (Cope et al, *Obesity Reviews*, 2008) conducted the most comprehensive evidence-based review of soy research and found many of the "proposed benefits" of soy "have yet to be clearly supported or refuted."

For some studies, the isoflavone component seems to be the most important part of soy, but biological efficacy and protective effects can vary, depending on the type of soy, the amount of soy used, the brand, and the processing technology. For example, tempeh has 3.1 mg of isoflavone per gram of soy protein, while tofu and some soy milk have 2 mg, (Kou et al, *Food & Function*, 2017). Isoflavones may help reduce blood pressure, for example, by causing vasodilation. For control of glucose, whole soy products like soy nuts may be more beneficial than textured soy protein, which suggested to researchers that other components in soy (e.g. fiber, polysaccharides) may be involved, (Ramdath et al, *Nutrients*, 2017). For reported cholesterol-lowering effects of soy, the gut microbiome may be a factor, (Ramdath et al, 2017).

Salvador Dalí (1904–1989). *Soft Construction with Boiled Beans (Premonition of Civil War)*, 1936.

The Louise and Walter Arensberg Collection, 1950. Photo credit: The Philadelphia Museum of Art/Art Resource, NY. Used with permission of ARS and Art Resource. Copyright 2018 Salvador Dalí, Fundació Gala-Salvador Dalí, Artists Rights Society.

Soybeans are often ground into meal for use in feeding livestock and even fish. Those allergic to soy may be getting second-hand exposure by eating animals or fish fed soy.

United Soybean Board. Licensed through Creative Commons Attribution 2.0 Generic, Wikimedia Commons.

Further, since breast cancer incidence is much lower in Asian women (and increases in prevalence once these women adopt a Western diet), researchers have suggested that soy intake may have a "chemopreventive effect." A recent prospective study (Baglia et al, *International Journal of Cancer,* 2016) of over 70,000 Chinese women (with follow-up of over 13 years) found that high intake during both adolescence and adulthood (greater than 16.4 grams a day) significantly reduced breast cancer risk. In those with breast cancer, a meta-analysis indicated soy consumption led to a statistically significant reduction in breast cancer recurrence and mortality and even may enhance tamoxifen treatment. These researchers caution that breast cancer is a "heterogeneous disease," and that factors such as hormonal and menopausal status, and the timing of exposure, can be important factors in influencing the relationship between soy exposure and breast cancer (and likewise affects the varying evidence in breast cancer studies). Furthermore, they note that other studies have not necessarily found soy as protective in Western populations, possibly due to considerably lower intake and a shorter period of exposure. (Baglia et al, 2016) Messina cautions, "The evidence that isoflavones reduce the risk of breast (as well as prostate cancer) is more preliminary." Some researchers recommend overt caution in using soy during or after breast cancer treatment, (Czuczwar et al, *Menopause Review,* 2017).

Allison and colleagues (Page et al, *Nutrition,* 2003), explain reasons for difficulties in studying the relationship of soy to breast cancer: should researchers study "breast tissue, the immune system, study humans, rats, or cell lines, look at short- or long-term exposure, and at what concentration? Or should researchers use whole soy, processed soy, or a component?"

What about the effect of soy on weight control and body composition? Some years ago, Allison and colleagues conducted two studies (Allison et al, *European Journal of Clinical Nutrition,* 2003; Fontaine et al, *Nutrition Journal,* 2003) and found soy replacement meals can lead to weight loss, though they are are not well-tolerated by some people, as evidenced by high drop-out rates (not uncommon in nutrition studies). More recently, Ramdath et al (2017) concluded: "soy protein does not seem to be an effective weight loss aid."

Bottom line: Soy products are an excellent source of protein and a healthy alternative to meat. There is no convincing evidence that soy protein is a better protein source than other proteins, though, or that it improves satiety or leads to weight loss when compared to other diets of comparable caloric restriction. There are countless reports that suggest soy intake may be beneficial, but researchers continue to recommend further studies. ■

TIME PRESENT AND TIME PAST: OBESITY AND CHRONOBIOLOGY

Setting limits and confronting our chronically fed state in a 24/7 environment

89 — Posted October 19, 2017

Ambrogio Lorenzetti (1285–1348), Detail from *Temperance*, 1338–1344. Palazzo Pubblico, Siena, Italy.

Does it matter whether we eat three large meals, six smaller meals, or whether we eat irregularly throughout the day? Or if we eat most of our food in the earlier part of the day or the most at night?

Researchers have been asking these questions as they note the importance, in "every piece of human physiology," including food intake, of our biological clocks.

"It is hard to find things that are not rhythmically fluctuating," says Rockefeller University's Dr. Michael Young, one of three scientists to win this year's Nobel Prize in Physiology and Medicine for work on identifying genes involved in the "inner workings" of our circadian rhythms. (Burki, *The Lancet*, 2017) Young adds, "If you have feeding cycles that are occurring with a daily rhythmic pattern, then you want organs that are dealing with the incoming food to be best aligned with those changes in the food supply. That is exactly what you see: genes switching on and off according to what the organism expects will be the pattern of food coming into the system." (Burki, 2017)

The suprachiasmatic nucleus (SCN) in the anterior hypothalamus is the "master regulator" of our circadian rhythms, and it is synchronized (i.e., "entrained") by the light/dark 24-hour cycle of the sun's rotation. There are also so-called "peripheral" clocks in almost every cell in our body, and these cellular clocks can be entrained on a daily basis by *zeitgebers*, ("time givers") i.e., other environmental cues such as food intake, noise, or exercise. (Bray and Young, *Current Obesity Reports*, 2012)

The suprachiasmatic nucleus (SCN) in the anterior hypothalamus is the "master regulator" of our circadian rhythms, and it is synchronized (i.e., "entrained") by the light/dark 24-hour cycle of the sun's rotation. There are also so-called "peripheral" clocks in almost every cell in our body, and these cellular clocks can be entrained on a daily basis by *zeitgebers*, ("time givers") i.e., other environmental cues such as food intake, noise, or exercise. (Bray and Young, *Current Obesity Reports*, 2012)

The timing of eating, though, is thought to be one of the most powerful zeitgebers. (Bray and Young, *Obesity Reviews*, 2007) It is as if our tissues can sense time. (Kohsaka and Bass, *Cell Metabolism*, 2007.) When our peripheral clocks become desynchronized from the central clock, we have *chronodisruption* (Garaulet and Gómez -Abellán, *Physiology & Behavior*, 2014) and the development of metabolic disorders. (Engin, *Advances in Experimental Medicine and Biology*, 2017)

Some researchers, though, believe the distinction between light and food as "entrainable circadian oscillators" is "convenient" but "ultimately a false dichotomy." (Mistlberger, *Physiology & Behavior*, 2011) For a complete discussion of circadian rhythms, see Karasu and Karasu, *The Gravity of Weight*, 2010, pp. 297–334.

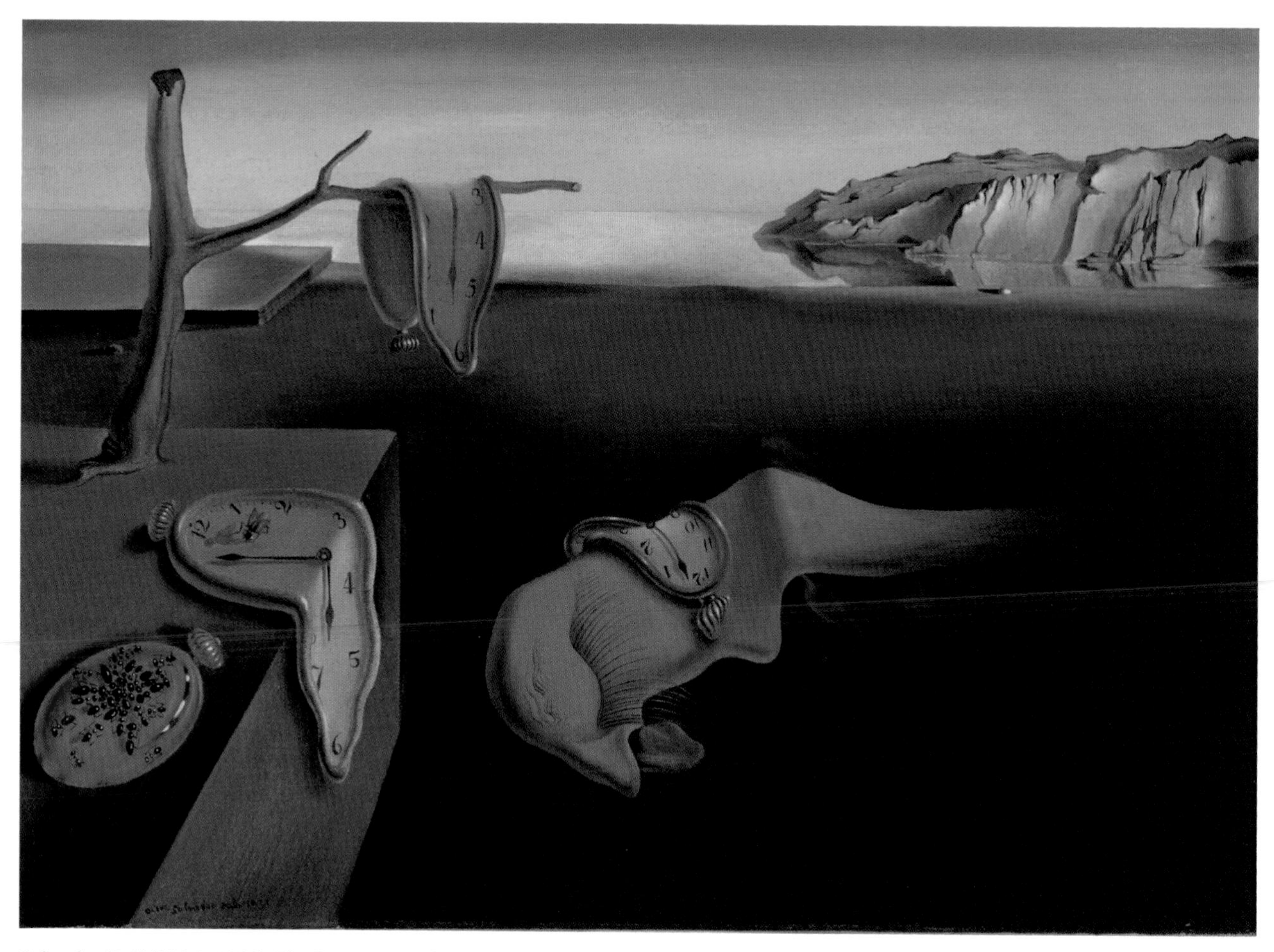

Salvador Dalí (1904–1989), *The Persistence of Memory,* 1931. It is as if our tissues can "sense time" and have a memory of when we had our previous meal.

Many even believe that obesity is a chronobiological disease. (Beccuti et al, *Pharmacological Research*, 2017) Given the importance of the timing of eating for our biological clocks, do we need what may be called "increased dietary structure" (Kulovitz et al, *Nutrition*, 2014) when we consider the timing and frequency of our meals? Some researchers believe so. For example, Allison and his colleagues (Mattson et al, *Proceedings of the National Academy of Sciences*, USA, 2014) explain that our industrial modern lifestyle has "perturbed" our circadian rhythms in three ways: shift work (reversal of the day/night pattern); prolonged exposure to artificial light of our increasingly 24/7 day; and erratic eating patterns where in many parts of the world, food is continuously available. From an evolutionary perspective, eating three meals and additional snacks is abnormal and leads to "daily overconsumption." (Mattson et al, 2014.)

In other words, we all tend to be in a "chronically fed state." (Cronise et al, *Metabolic Syndrome and Related Disorders*, 2017.) And we may need to consider "appropriate circadian timing" because of increased prevalence of obesity and diabetes and challenges to our circadian hygiene inherent in our 24/7 lifestyle. (Jiang and Turek, *American Journal of Physiology, Endocrinology, and Metabolism*, 2017)

Studying meal pattern, though, is difficult, including whether to conduct controlled research trials or to use "free-living" adults within a community setting. One of the major limitations is the "predominance of observational cross-sectional studies," rather than randomized

English artist Johann Zoffany, *Self-portrait (with Hourglass and Skull)*, circa 1776. Uffizi Gallery, Florence.

Wikimedia Commons/Public Domain.

Clio in the Car of History, in the Old Senate Chamber in the U.S. Capitol (in National Statuary Hall.)

Image taken by Capitol employee. Wikimedia Commons/Public Domain.

controlled trials, that are often short-term. (Kulovitz et al, 2014) Further, many trials fail to hold calories constant, and calories may be even more important than meal frequency in affecting metabolism and body composition. (Alencar et al, *Nutrition Research*, 2015.) There is also a lack of standardized terminology, such as what constitutes a meal (e.g. defining what constitutes "breakfast") or a snack or *ad libitum*. Typically, most studies rely on notoriously inaccurate self-report data, and many do not address the issue of exercise or that those who eat smaller meals may be engaging in other health-protective behaviors. And there is always the question of compliance and adherence to the research protocol. (Kulovitz et al, 2014) Many other studies have small sample sizes "that consequently lack statistical power." (Schoenfeld et al, *Nutrition Reviews*, 2015) Despite the difficulties, researchers have studied different patterns of food intake. There is caloric restriction (20 to 40 percent), with meal frequency kept constant; intermittent energy restriction, involving either fasting or reducing intake (e.g. 500 or 600 calories per day) on two non-consecutive days and eating "regularly" on the other 5 days (5:2 plan); and time-restricted eating, involving eating food within a 4-to-6 hour window per day. (Mattson et al, 2014.) Intermittent energy restriction, for example, works by the principle of *hormesis*, namely that exposure to a mild stress (e.g. restricted eating) results in adaptive behaviors that protect from more major stresses. There is a suggestion that this pattern, (athough there is considerable variation among studies), can lead to increased insulin sensitivity and other favorable metabolic parameters and even inhibit certain cancers. (Mattson et al, 2014) A recent small study has demonstrated that a 5:2 diet can be successful for some but not necessarily superior to other approaches. (Conley et al, *Nutrition & Dietetics*, 2017) When reached for comment about 5:2 energy restriction and its relationship to circadian rhythms, though, both Dr. Michael Young and Dr. Molly Bray each believe that this pattern does not make much sense. Dr. Bray added, "Such a regimen might be more tolerable, and thus, adherence might be better."

What about eating more calories later in the day? Allison and colleagues (Casazza et al, *Critical Reviews in Food Science and Nutrition*, 2015) note that the common saying, "Eat breakfast like a king, lunch like a prince, and dinner like a pauper" may be questionable because few studies have looked at calorie counts specifically. They find there is "little direct evidence to support or refute a unique obesogenic effect of calories consumed in the evening" but they cannot rule out a chronobiological effect on weight. They do acknowledge that those who work late shifts tend to be obese and are more prone to metabolic disturbances, but there may be other factors involved such as doing less exercise, eating unhealthy foods, or sleeping fewer hours. We do know from considerable research in rodents, for example, that when fed during their inactive time or fed throughout day and night, rodents tend to get obese and develop metabolic abnormalities. (Garaulet and Gómez-Abelán, 2014) There are few randomized controlled trials

Philippe de Champaigne, *Still-Life with a Skull*, circa 1671, Tessé Museum, Le Mans, France.
Wikimedia Commons/Public Domain.

Clock ornament, 18th century, French. Regardless how elaborate our external clock, our bodies have their own internal sense of time. Metropolitan Museum of Art, NYC.
Gift of J. Pierpont Morgan, 1906. Public Domain.

for intermittent energy restriction or time-restricted eating in humans. "Prescriptions" for meal frequency or timing remain to be "developed, validated, and implemented" (Mattson et al, 2014), and short-term benefits may depend more on the type of diet, calorie restriction, body composition, and genetics.

Bottom line: "There is *no* shortage of information available to the public" on different patterns of intermittent fasting, but there is "a shortage of evidence-based support." (Patterson and Sears, *Annual Review of Nutrition*, 2017) When a dieting strategy *may* help and fits with your schedule and your willingness to comply, "from a purely practical perspective," it may be worth trying. (Kersick et al, *Journal of the International Society of Sports Nutrition*, 2017) Whether changing meal frequency leads to increased satiety, appetite suppression, or metabolic benefit is not yet known conclusively. Circadian rhythms affect all aspects of our physiology. It seems reasonable, therefore, that meal patterns (e.g. timing and frequency) may be relevant for maintaining health. But for patterns of fasting and food restriction to be "more than a weight-loss fad," there needs to be "greater scientific rigor than our current studies provide." (Horne et al, *American Journal of Clinical Nutrition*, 2015)

Note: My title is from T.S. Eliot's "Burnt Norton," from *The Four Quartets.* ■

FROZEN: WHAT DO WE KNOW ABOUT CRYOLIPOLYSIS?

Let it go: Removing adipose tissue through controlled cooling

90 — Posted November 20, 2017

Elsa, in the 2013 Walt Disney 3D-animated film, *Frozen*, loosely based on a Hans Christian Andersen fairy tale, *The Snow Queen*, possesses cryogenic magical powers: Everything she touches turns to ice. The story follows the trials and tribulations of the relationship between Elsa and her younger sister Anna as they encounter a deceitful prince intent on plotting to seize the throne in their kingdom of Arendelle. This extraordinarily successful film garnered two Academy Awards, including for its iconic song *Let It Go*, and is generating *Frozen-The Musical*, to be released on Broadway this coming Spring, and *Frozen 2*, the film's sequel, in 2019.

While we see gorgeous animation of ice-covered landscapes, villages, wildlife, and people completely encapsulated in ice, we, of course, hear nothing of the actual impact of ice on the tissues of the human body. The negative effects of extreme cold are well known: for example, physical evidence of frostbite, i.e., injury to tissues that can lead to gangrene and ultimate amputation of the affected extremities, was seen in a pre-Colombian mummy dating back 5000 years, and Napoleon's Surgeon-in-Chief wrote a definitive treatise on frostbite injuries seen in soldiers during the failed 1812-1813 winter invasion of Russia. (Handford et al, *Emergency Medicine Clinics of North America*, 2017)

Less severe and often transient reactions to "an acute freezing reaction" when exposed to cold, though, had been reported in the literature by the early 20th century, and Haxthausen (*British Journal of Dermatology*, 1941) describes fat necrosis (with "firm infiltrations") due to prolonged, intense cold exposure in the "particularly conspicuous" cheeks which reminded him of a "Dutch cherub" of four children and one adolescent. He called this syndrome "adiponecrosis e frigore," and noted these infiltrations gradually "melt away" and present a clearly different clinical picture from frostbite; they are "extremely benign," with a tendency to spontaneous recovery. (Haxthausen,1941) In the 1950s, there is mention that subcutaneous fat necrosis developed in infants during the use of general hypothermia in cardiac surgery. (Adams et al, *Surgical Forum*, 1955) Epstein and Oren labeled a transient syndrome "popsicle panniculitis" (i.e., inflammation of adipose tissue) with erythematous nodules they had seen in the cheeks of several infants given popsicles. (*NEJM*, 1970) Further, Beacham et al described another similar reversible reaction "equestrian panniculitis" in the upper lateral thighs of young women, wearing non-insulated, tight-fitting pants, who rode horses for hours in very cold weather. (*Archives of Dermatology*, 1980.)

The Snow Queen, a fairy tale by Danish author Hans Christian Andersen, is the inspiration for the immensely popular Disney film *Frozen.*

Depicted here by Elena Ringo, 1998, licensed under Creative Commons Attribution 3.0 Unported, Wikimedia Commons.

Are there therapeutic uses, though, for extreme cold on the body and, in particular, on adipose tissue? *Cryotherapy*, from the Greek for "frost" or "icy cold," has a history that dates back to the early part of the 20th century (*Oxford English Dictionary*: reference in *Lancet*, 1909, mentioned, but not defined, as one of the seven sections of treatment to be discussed at the 1910 International Congress of Physiotherapy.)

Throughout the years, as well, dermatologists have used "cold-based" therapies to destroy actinic keratoses and superficial skin tumors (Jalian and Avram, *Seminars in Cutaneous Medicine and Surgery*, 2013), but it was not until the mid-2000s that Manstein and his colleagues reported on a "novel" and noninvasive method, which they called "selective cryolipolysis," an intentional procedure that selectively damaged adipose tissue by "controlled cooling." (Manstein et al, *Lasers in Surgery and Medicine*, 2008) These researchers noted that all cell types are susceptible to damage by

Lev Lagorio, *Transport of Ice* (Winter view of the former wine town on the Basil Island in St. Petersburg on the Black River, 1849). Irkutsk Regional Art Museum, Russia.
Wikimedia Commons/Public Domain.

"conventional cryosurgery," with the extent of injury related to freezing rate, temperature used, duration of the exposure, and the thawing time, but they noted that adipose tissue was "preferentially sensitive" to cold exposure. Their initial experiments were conducted on Yucatan pigs. (Manstein et al, 2008) Initially, there is an early inflammatory phase but after several months, there is a subsequent loss of fat tissue, without damage to the epidermis, dermis, or muscle—and without regeneration of the fat tissue or a rise in circulating lipid levels. Their speculation was that fat cells crystallize at temperatures well above the freezing point of water and "lipid ice" forms "at much higher temperatures than water ice." (Manstein et al, 2008)

The interest in noninvasive controlled cooling for "body-contouring" grew out of the "significant risks" and complications, such as pain, infection, scarring, edema, prolonged recovery time, and even more serious deep vein thrombosis and pulmonary embolism associated with liposuction. (Kennedy et al, *Journal of the European Academy of Dermatology and Venereology*, 2015)

The procedure of "body sculpting" or "body contouring" (Ho and Jagdeo *Journal of Drugs in Dermatology*, 2017) is considered one of those "lunchtime procedures." (Krueger et al, *Clinical, Cosmetic, and Investigational Dermatology*, 2014) It is done by a special machine without any anesthetic or even mild analgesics. Cooling panels with gel are applied to the area to be treated, and vacuum suction is used to draw the adipose tissue slightly away from the body. Depending on the area, the suction may be more uncomfortable than the cooling intensity. The treatments

German paintner Caspar David Friedrich (1774–1840), *The Sea of Ice, Polar Sea,* also known as *The Wreck of Hope,* 1823–24, Kunsthalle Hamburg, Germany.
Wikimedia Commons/Public Domain.

vary according to the examiner and can range from 30 to 120 minutes each; multiple sites can be treated simultaneously without any effect on serum lipid levels. (Derrick et al, *Aesthetic Surgery Journal,* 2015; Klein et al, *Lasers in Surgery and Medicine,* 2017) Most sites require more than one treatment, often spaced from two weeks to two months apart, again depending on the clinician's own protocol. After each treatment, there is a recommendation to do two minutes of manual massage (though no study delineated the science behind that specific two-minute time recommendation), and some researchers question whether post-treatment manual massage and repeat treatments increase the procedure's efficacy. (Ingargiola et al, *Plastic and Reconstructive Surgery,* 2015) Transient reactions include erythema, bruising, edema, numbness, and mild pain after the procedure. There is speculation that a "more-pronounced inflammatory response" may lead to a more-pronounced treatment response. (Dierickx et al, *Dermatologic Surgery,* 2013)

The most common complication, although rare, is a temporary decreased sensation in the treatment area that can last several weeks. The most troublesome, but very rare complication, is the development of paradoxical hyperplasia (incidence of one in 4,000 cases), possibly (although the etiology is not known) caused by a reactive fibrosis secondary to the damaged fat cells, that occurs three to nine months after the treatment. There may be a higher predisposition in men and a possible genetic susceptibility (in one study, all four cases were Hispanic men) (Kelly et al, *Plastic and Reconstructive Surgery,* 2016) as well as more likely seen with large applicators. (Ho and Jagdeo, 2017)Clinicians have evaluated cool sculpting in several ways, including measurement by calipers, photographic comparisons, evaluation of before and after photos by investigators who are blind to the procedures, and patient satisfaction surveys. Some use objective measurement by ultrasound. (Derrick et al, 2015) On average, patients lost about 23 percent of the fat in the treated area. In one study of 513 patients (73 percent female), 73 percent reported being "extremely satisfied" or "satisfied," and 82 percent said they would recommend the procedure to a friend. (Dierickx et al, 2013) Sometimes, those dissatisfied had gained a "significant" amount of weight at six months of follow-up. (Wanitphakdeedecha et al *Lasers in Medicine and Surgery,* 2015)

Patient selection and patient counseling, though, are crucial. Controlled cooling is not a treatment for obesity nor a substitute for weight control through diet and exercise. Potential patients must have "realistic expectations" (Stevens et al, *Aesthetic Surgery Journal,* 2013) and appreciate this is a "fine-tuning" cosmetic treatment that results in only a modest reduction of subcutaneous (and not visceral) adipose tissue in selected areas. Its results cannot be seen for months, and it is not as effective as liposuction. (Jalian and Avram, 2013) Further, certain areas of the body are apparently more amenable (e.g. abdominal area and "love handles" of the flank) than others, (e.g. inner thighs and knees), even with multiple treatments.(Stevens et al, 2013; Dierickx et al, 2013)

Despite heavy marketing targeted to professionals as well as patients (e.g. television and magazine advertisements), there are many unanswered questions about cryolipolysis. For one, adipose tissue is "not just a fat storage depot"

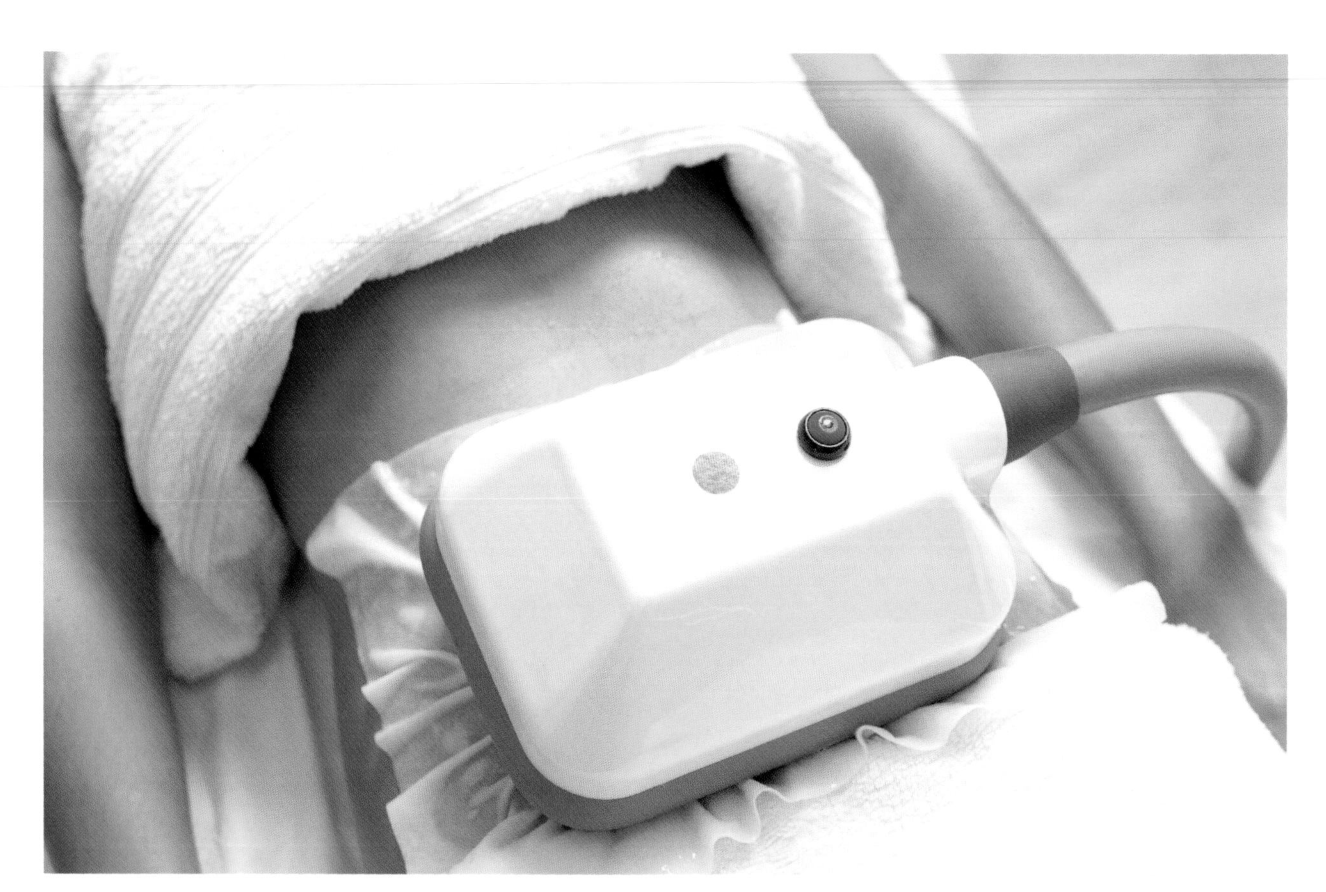

Technique of cryolipolysis. Note the cooling pad (with gel) and the vacuum machine. Patients require no anesthesia and can even work on their computers during the procedure. Discomfort from the cooling and suction is mild, and there is no recovery time, but do not expect miracles. Fat reduction is fairly minimal and may take months to be apparent.

Used with permission/istockphoto.com. Credit : Tutye.

Caspar David Friedrich, *Winter Landscape,* 1811. Staatliches Museum, Schwerin, Germany.
Wikimedia Commons/Public Domain.

(Henry et al, *The International Journal of Biochemistry & Cell Biology*, 2012) but rather is a fascinating and complex endocrine organ that secretes many compounds called "adipokines" (e.g. leptin, adiponectin) and occurs in depots throughout the body. There are differences among these depots, with the notion that these adipose depots exist as individual "mini-organs." (Cleal et al, *Adipocyte*, 2017) To what extent selective destruction of even a few grams of subcutaneous adipose tissue in different areas of the body does over the long-term is not completely known. It is not even known what determines a person's total number of adipocytes or total adipose volume. (Henry et al, 2012) Nor is it known "how the local microenvironment determines adipose tissue function and its impact on systemic metabolism." (Vegiopoulos et al, *The EMBO Journal*, 2017) Further, adipose tissue can be white, beige, or brown, and I could not find any reference to what effect controlled cooling has on brown fat (brown fat is "stimulated" by cold temperatures) (Hansen et al, *Experimental Cell Research*, 2017), nor could I find any reference to whether the timing of adipose tissue removal makes any difference. Adipocytes, like all cells in the body, are affected by circadian rhythms, and circadian clocks within these fat cells may be sensitive to environmental stimuli at different times of the day. (van der Spek et al, *Progress in Brain Research*, 2012)

To date, there is only one long-term follow-up of more than a few months (Bernstein, *Journal of Cosmetic Dermatology*, 2016) in two male patients followed for from 6 to 9 years without a recurrence of adipose tissue. As a result, the long-term duration of the procedure's effects

American painter John Singer Sargent, *Mannikin in the Snow,* circa 1891–93. Metropolitan Museum of Art, NYC.

Gift of Mrs. Francis Ormond, 1950. Public Domain.

has yet to be established, though it is not likely that the fat removed will regenerate in that area. (Krueger et al, 2014) In an editorial in *Aesthetic Surgery Journal* (2015), Nahai wonders why some physicians and surgeons are "so willing to jump on the bandwagon of new, relatively unproven innovations," especially when there is a "lack of high-level evidence."

For a thorough review of the noninvasive devices (e.g. those using radiofrequency, ultrasound, low-level light laser, mechanical suction, or cryolipolysis) for selective adipose tissue removal, see comprehensive articles by Nassab, *Aesthetic Surgery Journal*, 2015; Kennedy et al, 2015; or Alizadeh et al, *International Journal of Endocrinology and Metabolism*, 2016). Nassab emphasizes that there are currently no randomized controlled or comparative trials of cryolipolysis devices to prove their effectiveness and notes many of the studies have been industry-sponsored or have failed to consider patient variables such as weight loss over time that are due to changes in diet or exercise that may impact results. In published studies, average fat reduction ranges from 14 to 25.5 percent, depending on the area involved (i.e., on average, about one to two cm.)

Bottom line: Cryolipolysis is one of several treatments currently available for selective destruction of small, localized deposits of unwanted adipose tissue. It has now been FDA-approved for many areas of the body, (earliest FDA approval in 2010) including the abdomen, knees, back, flank, chin, and inner thighs, though there are no randomized controlled trials or long-term follow-up. Most studies are retrospective and observational. (Naouri, *Journal of the European Academy of Dermatology and Venerology*, 2017) It is a purely cosmetic procedure with modest effects for those with a normal or slightly overweight BMI and clearly not a treatment for obesity. Though not as effective as liposuction in removing quantities of adipose tissue, cryolipolysis is noninvasive, well-tolerated, with mild discomfort during the procedure, no recovery time, few transient side effects, and almost no adverse effects. Further, patient satisfaction, at least according to short-term follow-up in many of the studies, is high. ■

THE SO-CALLED "Seven Sleepers of Ephesus" were Christian men who attempt to escape Roman persecution in the 3rd century A.D. by hiding in a cave. Their persecutors seal off the cave, but eventually, by chance, the entrance is opened, and these seven men emerge. When they send one of their group to obtain some food, he finds everything has changed, and Christian crosses now appear on buildings. What's more, he realizes his coins are ancient and no longer in circulation. Apparently, so this apocryphal tale goes, according to Dutch professor of religion, Pieter W. van der Horst, the men had been sleeping in the cave for over 300 years. A version of their story, depicted in illuminated manuscripts and classical paintings, can be found in Greek, Jewish, Christian, and Muslim sources. (van der Horst, 2011)

19th century German votive painting of *The Holy Seven Sleepers*, artist unknown. The theme of the *seven sleepers* captured the imagination of many artists.

THE LONG AND THE SHORT OF IT: SLEEP DURATION AND HEALTH

What we know about the metabolic consequences of too little or too much sleep

91 — Posted December 21, 2017

Then there is the more familiar story of Rip van Winkle, who is "one of those happy mortals, of foolish, well-oiled dispositions, who take the world easy...and would rather starve on a penny than work for a pound." This does not suit his wife, "a terrible virago," who has no tolerance for Rip's lazy and careless disregard for her demands. One day, in an effort to have some brief respite from Dame van Winkle, hen-pecked Rip goes walking with his dog and gun in the mountains of the Hudson Valley. What happens to him is not entirely clear, only that he meets up with some mountain characters that get him drunk. When he finally awakens, he finds his dog is no longer there, his beard has grown over a foot, and his gun is now rusted. And when Rip returns to his village, he recognizes no one, and no one initially recognizes him for it is some 20 years since he had disappeared.

In these fantastical tales, the so-called "long sleepers" have no idea how long they have been asleep, and we hear essentially nothing of the physiological effects of their long sleep on the body. Fictional accounts seem to be captivated by those who can awaken after a very long sleep without any dire consequences. Rip is apparently no worse for wear; the Seven Sleepers of Ephesus, though, do die shortly after they emerge from their cave.

What do we know about our need for sleep? Sleep occurs universally in all animals, and it is "infinitely more complex, profoundly more interesting, and alarmingly more health-relevant," says UCLA Berkeley Professor Matthew Walker, in his new book *Why We Sleep*. The exact function of sleep remains "controversial" but researchers acknowledge that had sleep not been essential, over the years of evolution, there would have been "natural selection pressure" to eliminate it, particularly since all animals are more vulnerable to predators and external threats while asleep. (Ogilvie and Patel, *Sleep Health*, 2017.)

There are several theories about our human requirement for sleep, including that sleep allows an opportunity for

▲ *The Seven Sleepers,* Menologion of Basil II, circa 985, anonymous artist. The Menologion of Basil II is an illuminated manuscript that was designed as a church calendar or a service book for the Eastern Orthodox Church.

Wikimedia Commons, Public Domain.

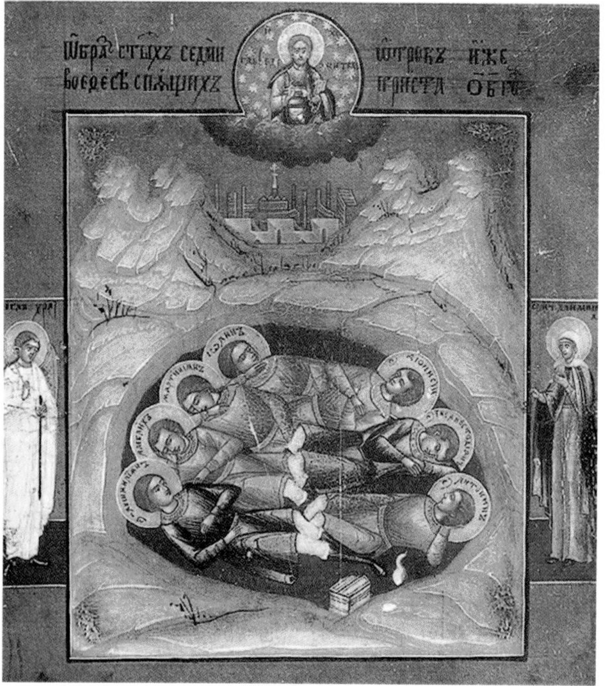

◀ *The Seven Sleepers,* 17th to 19th century, anonymous artist, Russian icon.

Wikimedia Commons, Public Domain.

synaptic pruning in our brain (e.g. to facilitate learning and memory consolidation) or allows for the clearance of neural waste products, as well as increases immune functioning. (Ogilvie and Patel, 2017) In other words, sleep is a homeostatic function that is important for health, and we have known for years from animal studies that complete sleep deprivation can lead to death within a few weeks. (Ogilvie and Patel, 2017) Ironically, in classical Greek mythology, Hypnos, the God of Sleep is the brother of Thanatos, Death, depicted in John William Waterhouse's classic 19th century painting, "Sleep and his Half-Brother Death."

How much sleep do we need? Researchers are beginning to appreciate there is a therapeutic window for the amount we require, i.e., too much or too little may be associated with poor health. Experts, though, cannot agree on that exact amount: Consensus from the American Academy of Sleep Medicine and the Sleep Research Society now recommends at least 7 hours for adults (Watson et al, *Sleep*, 2015); the National Sleep Foundation recommends 7 to 9 hours for adults (and not more than 10 hours), and 7-8 hours for those older than 65 (and not more than 9 hours.) (Hirshkowitz et al, *Sleep Health*, 2015) There is, though, a growing concern that a huge percentage—almost 1/3, at least by self-report, are getting 6 or fewer hours a night (Ford et al, *Sleep*, 2015), often due to our 24/7 increased use of artificial lighting (e.g. television, smart phones, and computer screens) and longer working hours. Furthermore, as we tend to sleep less, obesity rates have doubled among adults and tripled among children and adolescents over the past thirty or so years, and many researchers question whether there may be some connection. (Capers et al, *Obesity Reviews*, 2015)

In their comprehensive and now classic review, Allison and over 20 of his colleagues (McAllister et al, *Critical Reviews in Food Science and Nutrition*, 2009) explain that reasons for this increase in obesity prevalence are

Illustration of Rip van Winkle from *Uncle Sam's Panorama of Rip van Winkle and Yankee Doodle,* circa 1875, illustrated by Thomas Nast, toy published by McLoughlin Bros., Inc.

General Collection, Beinecke Rare Book & Manuscript Library, Yale University. Wikimedia Commons/Public Domain.

John William Waterhouse, *Sleep and his Half-brother Death*, 1874. Private Collection.
Wikimedia Commons/Public Domain.

"incompletely understood," but they do include *sleep debt* as one of their ten "putative contributors." Before World War I, Americans reportedly averaged up to 9 hours a night. Evidence from animal and human studies suggests that sleep deprivation "is consistently associated" with "profound effects" on levels of hormones (e.g. leptin, ghrelin) and peptides (e.g. galanin) that result in increased food intake, and in turn, potentially lead to an increased risk of diabetes, heart disease, increased body mass index, and even increased mortality. (McAllister et al, 2009) Further, fatigue due to sleep deprivation may lead to decreased activity and subsequent weight gain.

Allison and his colleagues (Davis et al, *Obesity*, 2018) revisit the topic of contributors to the obesity epidemic in their just-published discussion; they note the importance of considering "complementary hypotheses" and a "multifactorial approach," including the "behavioral factor" of sleep deprivation. Sleep debt, incidentally, may have an even greater effect on body weight in children and adolescents. (McAllister et al, 2009)

Studying sleep duration in children is complicated because children have different sleep needs as they get older. Recommendations from the National Sleep Foundation (Hirshkowitz et al, 2015) include 10-13 hours for preschoolers; 9–12 hours for school-aged children; and 8–10 hours for adolescents. Li et al (*Journal of Paediatrics and Child Health*, 2017) found a 45% increased risk of obesity in those who are considered "short sleepers" in their review of studies, including over 44,000 children from the U.S., Canada, Australia, and the UK. Most studies relied on self-report (or parent report) questionnaires or diaries to assess the duration of sleep. Questionnaires, often not even validated, rely on retrospective recall and may be less accurate than sleep diaries that are filled out nightly. (Tan et al, *Sleep Medicine Reviews*, 2018) Both methods may be significantly less accurate than objective measurement of sleep duration by actigraphy, which is rarely used in studies.

One study of over 380 male and female adolescents found a dose-response relationship: for every hour of sleep reduction, there was an 80% increase in obesity. (Gupta et al, *American Journal of Human Biology*, 2002) "Overall, the published literature supports the presence of an association between sleep duration and weight" but there were considerable differences in the definition of "normal" and "short sleep" duration. (Patel and Hu, *Obesity*, 2008) "So that a six year old with 9 hours of sleep could be classified as having

American artist John Singer Sargent's painting, *Repose*, 1911, National Gallery of Art, Washington, DC.

Wikimedia Commons/Public Domain.

Vincent van Gogh, *Noon-Rest from Work (The Siesta)* (after Millet), 1890–91. Musée d'Orsay, Paris. Many studies of sleep duration do not consider napping during the day in their assessments of sleep duration.

Wikimedia Commons/Public Domain.

short, intermediate, or even long sleep duration, depending on the study." (Patel and Hu, 2008)While many large epidemiological studies have found a significant relationship between short sleep and obesity, Allison and his colleagues (Capers et al, 2015) note that an actual "causal pathway" is far from clear, particularly because of differences in patient populations, small sample sizes, study design, and studies that are short-term. In their meta-analysis of 16 randomized controlled trials involving the relationship between sleep duration and body composition and energy balance, for example, they had to adjust their inclusion criteria (from a duration of at least four weeks to those that lasted only at least 24 hours) because only two studies had met that initial criterion. (Capers et al, 2015)

More recently, Itani et al (*Sleep Medicine*, 2017) reported on 153 prospective cohort studies in their systematic review and meta-analysis to assess the relationship between short sleep and multiple health data of over 5,100,000 people. The definition of "short sleep" varies with cultures and ethnicity. Another confusion is that some of their studies list "hours per day" and some list "hours per night." They found, though, that short sleep (defined by fewer than 6

Philippe Laurent Roland, French sculptor, *Sleeping Boy*, circa 1774, terracotta, painted white. Metropolitan Museum of Art, NYC.

Wrightsman Fund, 1990, Public Domain.

hours) was associated with increases in mortality, diabetes, hypertension, cardiovascular disease, and obesity, but these researchers, too, noted the mechanisms for these increases "do not seem straightforward." Furthermore, they emphasize the need for caution from these community studies since "there is no rigorous evidence that lengthening sleep duration can lead to smaller frequency of these outcomes," and the role of individual differences in sleep duration "is still uncertain."

In addition, many studies don't ask about napping during the daytime and so may underestimate total 24-hour sleep duration. The relationship of napping to sleep debt warrants further investigation. (Faraut et al, *Sleep Medicine Reviews*, 2017) There may also be large night-to-night variability (including differences between weekday and weekend sleeping duration.) Researchers also question the possibility of reverse causation, i.e., obesity can increase the risk of medical conditions such as osteoarthritis, GI reflux, asthma, and heart failure as well as obstructive sleep apnea, all of which can affect sleep duration. And there may be residual confounding, such as when psychiatric disorders such as depression, or use of medications, can affect both sleep and weight bidirectionally. (Krittanawong et al, *European Heart Journal Acute Cardiovascular Care*, 2017; Patel and Hu, *Obesity*, 2008)

"Long sleepers," often defined as greater than 9 hours a day, are also at risk for obesity and diabetes, at least from a review of observational studies. (Tan et al, *Sleep Medicine Reviews*, 2018) These researchers, though, note differences in results among studies, including how "long sleep" is defined. (Tan et al, 2018) They further wondered whether some long sleepers are more apt to use medication because they have a poor quality of sleep and may also have a more sedentary lifestyle, be less likely to exercise, and make unhealthy diet choices, as well as have a habitually later bedtime, i.e. what has been called an unhealthy "nocturnal lifestyle." (Knutson et al, *Sleep*, 2017)

Bottom line: Since sleep deprivation has become so common in recent years, any causal association between short sleep duration and obesity would have substantial public health ramifications. (Patel and Hu, 2008) Though we cannot prove causality, (and we must consider the possibility of reverse causation, namely that obesity and its related disorders lead to changes in sleep duration), research suggests there are strong associations between the amount of sleep we get each night and metabolic health. Too much or too little sleep has each been linked to an increased risk of obesity, type 2 diabetes, hypertension, cardiovascular disease, and even increased mortality in adults and increased obesity in children and adolescents. Proposed mechanisms include increased hunger due to changes in hormone and peptide levels; additional time for increased food intake; increased pleasure from food, i.e., an "up-regulation" of food's salience and reward value (St-Onge, *Obesity Reviews*, 2017); decreased physical activity due to fatigue; altered thermoregulation due to circadian rhythm misalignment; and even neuro-cognitive changes that lead to impaired judgment and decision-making regarding food choices. (St-Onge et al, *Circulation*, 2016) ■

THE HANDWRITING ON THE WALL: MENU LABELING

Is posting nutritional information in restaurants "public health paternalism"?

92 — Posted January 18, 2018

Preparing for a great palace feast, King Belshazzar, an ungodly man, commanded his underlings to fetch the golden vessels that his father Nebuchadnezzar had confiscated from the temple in Jerusalem. He and a thousand of his lords, so the *Book of Daniel* tells us, drank from these holy goblets and praised "the gods of silver, gold, brass, iron, wood and stone." Suddenly, there appeared "fingers of a man's hand" that inscribed the famous four words on the palace wall.

Alarmed and much shaken, Belshazzar assembled his soothsayers and astrologers, none of whom could interpret the strange words. The Queen, though, suggested they send for righteous Daniel, who interpreted this "handwriting on the wall" as a sign from God: "God hath numbered the days of your kingdom and brought it to an end; you have been weighed on the balances and found wanting; thy kingdom is divided..." That night Belshazzar was killed. (*Oxford Annotated Bible, Daniel,* 5:1–30) This story has been appropriated by musicians (e.g. George Frideric Handel's oratorio), artists (e.g. paintings by Rembrandt and John Martin), and poets (e.g. Lord Byron and Heine); the famous words are even the title of a short story by John Cheever. (*The New Yorker,* 4/27/63)

While hardly as menacing or alarming as those Biblical words, menu labeling—the clear and conspicuous posting of nutritional information in restaurants—may require some of its own interpretation as well.

Menu labeling has had different meanings. Most commonly, it refers to information on calories specifically, but it can refer to other kinds of nutritional information (e.g. fat, sugar, salt content.) Sometimes, it refers to the so-called "traffic light system," whereby there is not merely information but an *evaluative judgment* presented (e.g. "green" for healthy; "red" for unhealthy.) (Fernandes et al, *Nutrition Reviews,* 2016) And it even can include the amount of exercise (e.g. walking) required to expend the calories eaten. Further, there is the suggestion that calorie counts alone are not as effective unless consumers are given some guidelines about how many calories they should be consuming in a day or at one meal. (Shiv and Fedorikhin, *Journal of Consumer Research,* 1999.) This, of course, adds to the complexity because caloric needs for a particular individual are quite varied.

Mandatory labeling of packaged food began in the early 1990s, with the implementation of the *Nutrition Labeling and Education Act.* The plan to offer information on calories in restaurants grew out of the concern that obesity rates have continued to increase throughout our country over the past thirty years, and people have frequented restaurants (and particularly fast food establishments) much more commonly than years ago. Stunkard and colleagues, in the late 1970s, for example, noted that people ate most of their meals at home. (Coll et al, *Archives of General Psychiatry,* 1979.) More recently, Urban et al (JAMA, 2011) reported that 35% of our *daily intake* in the U.S. now comes from food purchased outside of the home. VanEpps et al (*Current Obesity Reports,* 2016) reported that half of all food dollars are spent on "away from home foods." This Act, though, did not lead to a decrease in obesity. (Bernell, *Food & Drug Law Journal,* 2010.)

Over the years, as well, portion sizes have grown considerably: in 1955, a McDonald's hamburger came in one 1.6-ounce size; more recently, people can choose from several sizes, including an 8-ounce burger. (Young and Nestle, *Journal of Public Health Policy,* 2007.) Bassett et al (*American Journal of Public Health,* 2008) surveyed over

John Martin's painting *Belshazzar's Feast,* 1820, Yale Center for British Art.

Wikipedia Commons/Public Domain

Claude Monet's painting, *La Grenouillère,*1869. This restaurant is not exactly what menu posting laws had in mind.

Metropolitan Museum of Art, NYC,. H.O. Havemeyer Collection, Bequest of Mrs. H.O. Havemeyer, 1929, Public Domain.

Vintage decorative menu, part of Cooper Hewitt Collection, NYC, gift of unknown donor.

Cooper Hewitt Museum (Smithsonian Design Museum). Public Domain.

7300 customers from 11 fast food chains and found that people purchased a mean of 827 calories a meal, with 34% purchasing over 1000 calories, and 15% over 1250 calories.

Furthermore, people are notoriously inaccurate in assessing calorie counts, called by the NYC Department of Health, "the calorie information gap." (Farley et al, *Health Affairs,* 2009.) A now classic survey, for example, of professional nutrition experts found even they significantly underestimated calorie and fat content of common foods; when it came to a Porterhouse steak and onion rings, for example, the respondents were more than 600 calories off. (Bankstrand et al, 1997, www.portionteller.com/pdf/cspistudy97.pdf.) More recently, Burton et al (*American Journal of Public Health,* 2006) likewise found consumers significantly underestimated calories and fat in food as well.

New York City was the first to propose laws for posting of calories in restaurants, but the New York State Restaurant Association filed suit against the city, claiming, among other things, that the law violated the First Amendment (e.g. calorie posting was seen as a form of "compelled speech," and restaurants should have a right not to have to say something.) David B. Allison, PhD., now Dean of the School of Public Health at Indiana University at Bloomington, was the expert witness for the Restaurant Association. In a logically argued and carefully worded affidavit (2007), Allison questioned whether posting calorie counts in one setting would *necessarily* lead to changes in calorie intake in other settings. He maintained there was not sufficient evidence that calorie postings would necessarily reduce obesity in individuals or in the general population and might be ineffective or even possibly have unforeseen consequences. (For a summary of Allison's affidavit, see either Banker, *Food & Drug Law Journal,* 2010 or Bernell, 2010.) Ultimately, the judge, while acknowledging Dr. Allison's argument that there was not adequate evidence one way or the other, nevertheless ruled in favor of NYC and upheld the law that any restaurant with 15 or more establishments had to post calories. The law, incidentally, excluded condiments, daily specials, or custom orders. (Banker, 2010)

Subsequently, other states followed NYC and ultimately led to the passage of a federal law as part of the *Affordable Care Act* that any restaurant with 20 or more facilities had to post their calories. (fully effective by May 2018.) Now, over ten years later, with several systematic reviews published (Downs et al, *American Journal of Public Health,* 2013; VanEpps et al, 2016; Bleich et al, *Obesity,* 2017) (including data before and after the law has taken effect in certain cities), results are decidedly mixed on the effect of calorie postings, with some research finding no effect at all.

Rembrandt's *Belshazzar's Feast,* circa 1636–1638, National Gallery in London, based on the Biblical story in the *Book of Daniel.*

Wikipedia Commons/Public Domain

Most of the studies are observational and cannot prove causation. When studies did find a reduction in calories in response to postings, people *purchased* (assessed by receipts) only minor decreases in their numbers of calories (e.g. 38 calories; 22 calories, etc.) per order. Cafeteria settings may be more responsive, possibly because people eat there more regularly and don't think of that setting as a treat or because they attract a more educated, health-conscious population (e.g. hospital or university setting.) (Bleich et al, 2017) To date, there are 11 studies involving the influence of menu labeling on children and adolescents; those conducted in the "real world" as opposed to an artificial lab situation, are less supportive of an effect and overall, many are of weak quality. (Sacco et al, *Perspectives in Public Health,* 2017)

Menu labeling is an example of a "nudge," first elaborated by Thaler and Sunstein in their book of that name. (2008) It is an approach that (often for their own good) "steers people in a particular direction, but also allows them to go their own way." (Sunstein, *Behavioural Public Policy,* 2017) For some, nudges are considered an example of paternalism, especially when there is a sense that people need influencing to make the right decision. Sometimes, though, nudges are not effective: they can be counterproductive if people have strong contrary preferences or if they are confusing or present too much information. "Attention is a scarce resource," says Sunstein (2017.) Nudges may also have only a short-term effect, as for example with repeated exposure, the information can become more "like background noise." And they can produce a "rebound effect" whereby people will compensate for the behavior originally produced by the nudge. All of these issues are potentially relevant to menu posting.

What, though, constitutes scientific evidence and how much should we require in obesity studies, especially when considering public health policies? Allison and colleagues (Richardson et al, *International Journal of Obesity*, 2017) raise these provocative questions. They remind us that there is a distinction between evidence to reach a decision about public policy and evidence to reach a scientific conclusion. A scientist, says Allison (*International Journal of Obesity*, 2011) is concerned with truth while the "well-meaning public health advocate" asks, "...given what we know today, is it prudent to implement one plan in the

Renoir's *A Waitress at Duval's Restaurant*, circa 1875, Metropolitan Museum of Art, NYC.

Bequest of Stephen C. Clark, 1960, Public Domain

The Hand-Writing upon the Wall, published by Hannah Humphrey, 1803, hand-colored etching, by James Gillray, English. Napoleon Bonaparte is warned by God that his kingdom will fall.

Courtesy of the Warden and Scholars of New College, Oxford/ Bridgeman Images, used with permission.

hopes that it will create a certain response?" Along those lines, he is concerned with "public health paternalism" that "may focus more on changing the unhealthy eating habits of the less powerful social classes." (Allison, 2011.)

Sandro Galea, in his new book *Healthier: Fifty Thoughts on the Foundations of Population Health* (2018), cautions about the unintended consequences of oversimplifying complex systems in public health. He believes that action "does not need to follow causal certainty;" we can and sometimes should act "even when we don't know all the answers" as long as we recognize the uncertainties and are willing to adjust our course "when some of our ignorance falls away." (pp. 206–7)

Bottom line: Just like Belshazzar, menu labeling has been, as it were, "weighed...and found wanting." Some people (myself included) find menu labeling helpful for regulating their caloric intake; others, especially over time, may ignore the information as background noise and still others may resent these paternalistic nudges and in opposition, increase their intake.

Most studies are of weak or only of moderate quality, and none have sufficiently demonstrated that calorie labeling actually leads to reduced overweight and obesity on a population level, nor even to reduced intake from one meal to another, nor even distinguishes calories purchased from calories consumed. (Allison, 2011) Further, studies usually don't discuss the process by which calorie counts are obtained or even how accurate they are. Several studies have found significant differences between reported calories and those subsequently tested, often due to portion size discrepancies. (Urban et al, 2011; Feldman et al, *Appetite,* 2015) Nor do studies differentiate thin customers from obese ones by measured BMI. There is a suggestion that menu labeling has led some food establishments to offer lower calorie options and healthier choices, (one of the original goals of posting), but this has not yet become widespread. Further, a focus on calories exclusively, without considering other aspects of nutrition, may be misleading. (Lucan and DiNicolantonio, *Public Health Nutrition,* 2015) Essentially, then, menu labeling is one public health strategy but "it is wrong to expect too much from menu labels alone." (Carter, *Public Health Ethics,* 2015) ■

"Well," said Sam, "of all the cool boys ever I set eyes on, this here young gen'lm'n is the coolest. 'Come, vake up, young dropsy!"

Pickwick Papers.

SLEEPERS AWAKE! OBSTRUCTIVE SLEEP APNEA

Poetic license in the study of sleep-related breathing disorders

93—Posted Mar 08, 2018

THERE IS A TALE in 19th century European folklore about the supernatural water sprite, Ondine, who falls in love with the knight-errant, mortal Hans. In one account, Ondine is warned that Hans will ultimately betray her and return to his betrothed Bertha, who had sent him out on his quest. A curse is put on Hans: if he ever deceives Ondine, he will die. Torn between his love for Ondine and his love for Bertha, Hans ultimately chooses Bertha, and the curse takes effect: when he falls asleep, he "forgets" to breathe and dies. The story has been immortalized in a 1930s play by Jean Giraudoux (Olry and Haines, *Journal of the History of the Neurosciences,* 2017) and a 19th century painting by John William Waterhouse. The moral: don't transgress the bounds of nature. Is the Academy Award-winning film, *The Shape of Water,* a somewhat more sanguine, science fiction version of this story?

Joe, "the fat boy," in Charles Dickens' *The Posthumous Papers of the Pickwick Club*, (1837) had obesity, snoring, and fell asleep repeatedly throughout the day.

Walker Art Gallery, Alamy stock photo, used with permission.

Ondine's curse is also the eponym given to one of the "strangest diseases in the history of neurology," i.e., the central ventilation syndrome. (Olry and Haines, 2017) It is one of the sleep-related breathing disorders, extremely rare (1/200,000 births) and potentially deadly, usually caused by an autosomal dominant genetic mutation, but can also be caused by severe trauma or tumor to the brainstem. Patients have impaired autonomic control of breathing, with a loss of sensitivity to carbon dioxide chemo-receptors so that they actually do "forget to breathe" during sleep but maintain normal control of their breathing while awake. (Zaidi et, *Autonomic Neuroscience,* 2017) With newer, advanced techniques that can monitor sleep, those who manifest symptoms in infancy can now survive into adulthood.(Suslo et al, *Advances in Experimental Medicine and Biology,* 2015) Anesthesiologists have sometimes used the eponym, *Ondine's curse*, as well, to refer to an unexplained and unexpected unconsciousness and hypoventilation that occurs in patients after they had initially awakened from anesthesia. (Suslo et al, 2015).

Another sleep-related breathing disorder also has an eponym, the *Pickwickian syndrome*, named for a character in Charles Dickens', *The Posthumous Papers of the Pickwick Club* (1837.) Though references to obesity and disordered breathing had occurred from the time of Hippocrates (Kryger, *Archives of Internal Medicine*, 1983), it was Burwell and his colleagues, in a classic 1956 paper, who labeled a syndrome of extreme obesity associated with hypoventilation the *Pickwickian syndrome* (reprinted, *Obesity Research*, 1994.) These patients, just like Joe, Dickens' "fat and red-face boy," who snored loudly and could not seem to stay awake, have an "extraordinary degree of somnolence in which sleep may overcome them while they are sitting up or even engaged in conversation." (Burwell et al, 1994, reprinted.) Clinical features for these researchers included marked obesity, somnolence, twitching, cyanosis, periodic respiration, secondary polycythemia, right ventricular

John William Waterhouse, *Undine,* 1872, the water sprite who puts a curse on Hans, her mortal beloved, who betrays her. Waterhouse spells her name with a "U" but most others call her *Ondine.*

Wikimedia Commons/Public Domain

Charles Bargue, *A Footman Sleeping*, 1871, from the collection of the Metropolitan Museum of Art, NY. Those with sleep-related breathing disorders, such as obstructive sleep apnea, may have excessive daytime sleepiness and fall asleep at inappropriate times and places.

Bequest of Stephen Whitney Phoenix, 1881. Metropolitan Museum of Art, Public Domain.

hypertrophy, and right ventricular failure, with a resulting high blood level of carbon dioxide (hypercapnia) and low levels of oxygen.

Not everyone has agreed with Burwell et al's diagnosis of the Dickens' character, and clinically, there has been controversy regarding what really constitutes the Pickwickian syndrome: some describe the term "as an erratic terminology of scientific expression and more justified by poetic license than ..medical precision." (Rössner, *Obesity Reviews*, 2012) Since Joe had a red face (possibly due to polycythemia) and dropsy (the old word for edema), he may have had *cor pulmonale*, but physicians over the years have considered endocrine or hypothalamic disorders and even the Prader-Willi syndrome. (Kryger, *Journal of Clinical Sleep Medicine*, 2012) Today, the Pickwickian syndrome is synonymous with the *obesity hypoventilation syndrome* (OHS), as manifested by obesity, chronic daytime hypercapnia, and sleep-disordered breathing, with snoring. (Liu et al, *Oncotarget*, 2017) It may present with acute hypercapnic respiratory failure and hence may be misdiagnosed as chronic obstructive pulmonary disease. (Liu et al, 2017) In patients with this syndrome, they may have increased morbidity, with the development of hypertension, diabetes, atrial fibrillation, or even heart failure. Its prevalence ranges from 9 to 20% in those referred obese patients. (Pierce and Brown, *Current Opinion in Pulmonary Medicine*, 2015) Their upper airway may collapse not only while reclining in sleep, but also when they are sitting. (Liu et al, 2017) OHS is more common in postmenopausal obese women, who have more "deranged blood gases" and more comorbidities than men, and may go unrecognized. (Piper et al, *Sleep Medicine Clinics*, 2017) Delivering positive air pressure via mask will alleviate the sleep disordered breathing and increased blood carbon dioxide but

Johannes Vermeer, *A Maid Asleep*, circa 1656–1657. Perhaps this maid, who fell asleep on her job, may have had obstructive sleep apnea, though OSA is more common in women after menopause.

Bequest of Benjamin Altman, 1913. Metropolitan Museum of Art, NY/ Public Domain.

patients are strongly encouraged to undergo weight loss and increase physical exercise. (Piper et al, 2017).

The most common sleep-related breathing disorder, though, is *obstructive sleep apnea (OSA),* which sometimes coexists with the obesity hypoventilation syndrome. The vast majority of patients with OHS will have comorbid obstructive sleep apnea. When OSA occurs independently, there are impaired episodes of breathing only during sleep, when the patient's upper airway collapses, and not during the day. (Liu et al, 2017)

Estimates of the prevalence of OSA vary due to differences in what constitutes severity, but reportedly 9 to 49% of middle-age men and 3 to 23% of women will have "moderate to severe" episodes of apnea, as defined by more than 15 apneic episodes per hour during sleep. The *apnea-hypopnea index* (AHI) is the standard scale, though "an inadequate metric" for classifying OSA since it is "extremely limited" in predicting future clinical outcomes. (Sutherland et al, 2018, *Expert Review of Respiratory Medicine*, 2018) Greater than 30 episodes per hour is usually considered severe. By far, the major risk factor is obesity, but 20% of those diagnosed with OSA are not obese, and these patients can be more difficult to treat. (Gray et al, *Journal of Clinical Sleep Medicine*, 2017) Anatomical features, such as a thick neck, narrow pharyngeal airway, or even fat accumulation on the tongue, can predispose a person to OSA.(Osman et al, *Nature and Science of Sleep*, 2018) The diagnosis should be suspected in someone who has loud nightly snoring, daytime sleepiness, and obesity, but the "gold standard" for diagnosis is made in a sleep lab by polysomnography. OSA is also seen in children, particularly when they have obesity or enlarged tonsils and adenoids. (Andersen et al, *International Journal of Pediatric Otorhinolaryngology*, 2016)

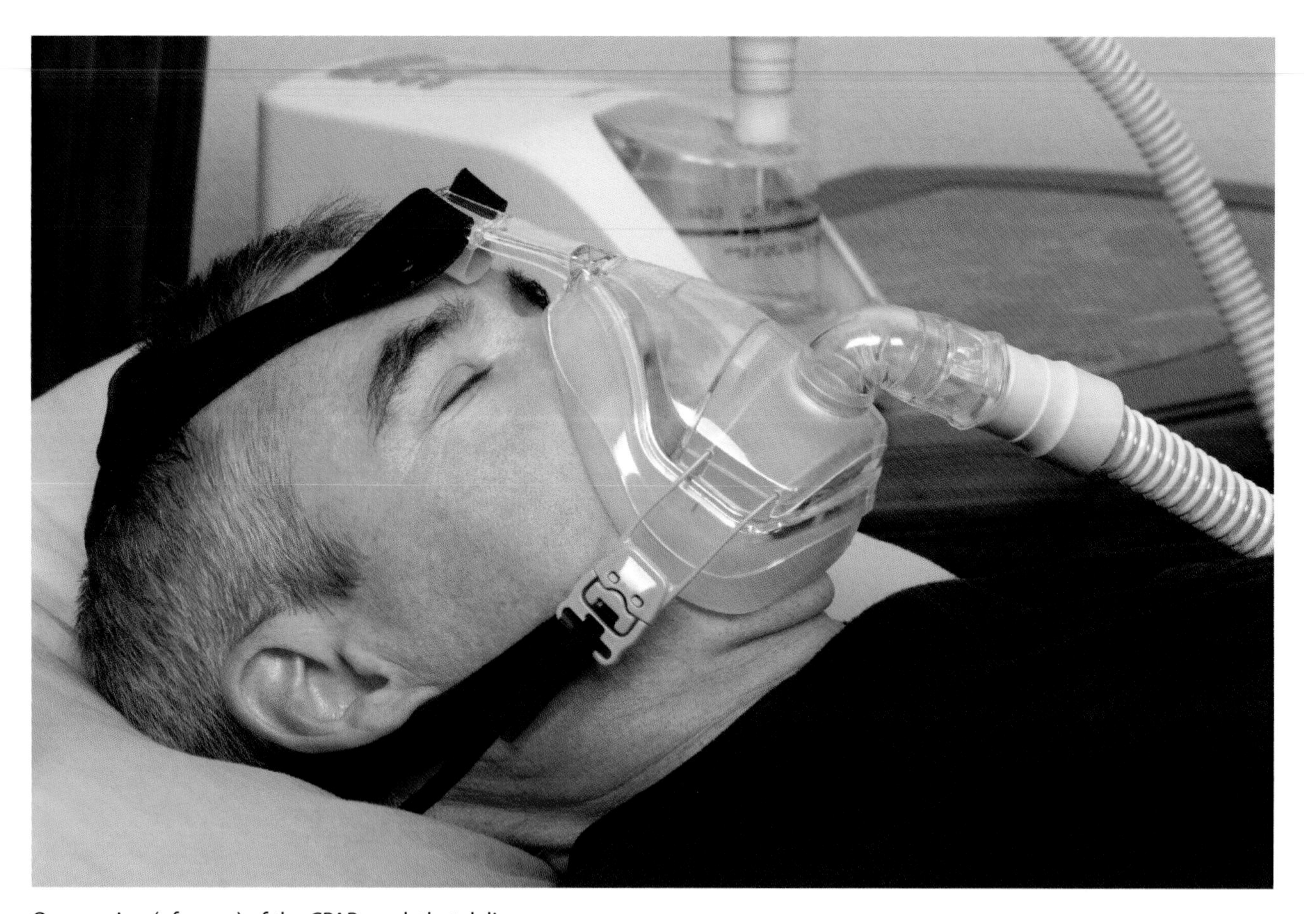

One version (of many) of the CPAP mask that delivers continuous airway pressure to those with OHS and/or OSA. Because the mask is cumbersome and even embarrassing, patients are often not compliant with treatment recommendations.

Panther Media GmbH/Alamy stock photo, used with permission.

The official White House portrait, 1911, by Anders L. Zorn, of William Howard Taft, our 27th President (1909–1913.) Taft was known as our most obese president: at the time of his inauguration, reportedly, he weighed between 300 and 330 pounds. Researchers have suggested he may have suffered from obstructive sleep apnea.
Wikimedia Commons/Public Domain.

There is perhaps fanciful speculation that some famous people, including Brahms (of Brahms *Lullaby*), (Margolis, *Chest,* 2000) and Queen Victoria (Conti et al, *Medical Hypotheses,* 2006) may have had undiagnosed OSA. President William Howard Taft may have had it as well. Taft reportedly could fall asleep in the middle of important conversations, had loud snoring, a thick neck, hypertension, cognitive "fogginess," and a body mass index (BMI) of 42, (Sotos, *Chest,* 2003) OSA is apparently not uncommon among some hefty players and retired players in the National Football League. (George et al, *NEJM,* 2003; Luyster et al, *Nature and Science of Sleep*, 2017)

Obstructive sleep apnea is characterized by the "periodic narrowing and obstruction of the pharyngeal airway" that can occur because the "human pharynx is unique in that it lacks a bony support" and hence is vulnerable to collapse during sleep. (Osman et al, 2018) It results in hypoxia and even surges in blood pressure during sleep, as well as excessive daytime sleepiness, sleep fragmentation and non-restful sleep, irritability, memory loss and cognitive impairment, morning headache, depression, and even an increased risk of car accidents. (Osman et al, 2018) It is associated with "downstream" morbidity, with increased risk of cardiovascular disease and metabolic abnormalities. (Sutherland et

Elias Gottlob Haussmann, *Johann Sebastian Bach*, original from1746. Bach was the composer of *Cantata 140, Sleepers Awake!*—the inspiration for my title.

Bach-Archiv, Leipzig, Germany. Bequeathed to the Bach-Archiv by William H. Scheide. Wikimedia Commons/ Public Domain.

al, 2018) Further, any patient being considered for bariatric surgery should be screened for OSA and OHS because of possible peri-operative complications, and pain medications that decrease respiration (e.g. opioids) should be avoided. (de Raaff et al, *Current Opinion in Anesthesiology*, 2018)

The primary treatment for both obstructive sleep apnea and obesity hypoventilation syndrome is continuous positive airway pressure (CPAP) delivered by a mask, but difficulty with the "mask interface" is common and leads to non-compliance with treatment recommendations of using it long term, every night, for the entire night. Even those "compliant" sometimes use it for only four hours a night. (Dibra et al, *Sleep Medicine Clinics*, 2017) Some patients feel claustrophobic and others fail to comply out of embarrassment. There are many choices of mask, and patients are encouraged to try different ones. If the mask does not fit properly, patients can get eye and skin irritation, and air leakage.

Bottom line: There are many breathing disorders associated with sleep. The most common is obstructive sleep apnea (OSA). The diagnosis should be considered whenever a patient has obesity, daytime sleepiness, heavy snoring at night, and observed (by partner) apneic episodes during sleep, though it can occur in the non-obese and in children. CPAP, if used as prescribed, is helpful for the breathing difficulties, but most clinicians recommend interventions for weight loss in those patients who are obese. OSA can lead to cardiovascular disease and metabolic abnormalities. Once cardiovascular disease develops, though, studies suggest that CPAP will not necessarily prevent future "cardiovascular events." (McEvoy et al, *NEJM*, 2016)

The musical scholars among you will know my title comes from Bach's *Cantata 140, Sleepers Awake!* ■

CHOCOLATE: GLORIFY OR DEMONIZE?

A bittersweet look at this most commonly craved food

94—Posted April 13, 2018

Pierre Auguste Renoir, *Cup of Chocolate*, 1914.

The Barnes Foundation, Philadelphia, Pennsylvania, USA/Bridgeman Images, used with permission.

Willy Wonka tells the children who are fortunate enough to tour his factory, in Roald Dahl's classic *Charlie and the Chocolate Factory*, that he has *Supervitamin Chocolate* that contains all the vitamins from A to Z ("except for vitamin S, which makes you sick, and vitamin H, which makes you grow horns") and the most magical vitamin of them all—vitamin Wonka." Clearly Willy Wonka has manufactured his own fantasied concoctions, but what do we know about the origins of chocolate, how chocolate is made, and whether it is as healthy as some would have us believe?

Cultivation of the cocoa bean originated in Mesoamerica, possibly 1000 years B.C., (Rusconi and Conti, *Pharmacological Research*, 2010), and was used later by both the Mayan and Aztec civilizations. According to one myth, cocoa came from the blood of an Aztec princess, who chose to die rather than betray her kingdom's wealth. (Gianfredi et al, *Nutrition*, 2018) In another, it was discovered by the gods in the mountains. (Dillinger et al, *Journal of Nutrition*, Supplement, 2000.)

One of the first Europeans to mention a bitter drink made from the cocoa bean was Hernando Cortés, who landed on the eastern area of Mexico and described how the Aztec emperor Montezuma used it as an aphrodisiac. (Dillinger et al, 2000; Lippi, *Nutrition*, 2009) It was the Swedish naturalist Linnaeus, in 1753, who used a Mayan word in his scientific description and called cocoa, *Theobroma cacao*, "Food of the gods." This early Mayan drink was made by dissolving dried cocoa beans in water with cinnamon and pepper. (Verna, *Malaysian Journal of Pathology*, 2013). Others described cocoa as having "as much nourishment as a pound of beef"(de Quélus, 1719) or even that it was a "universal medicine," (Lavedan, 1796) which was literally used to treat hundreds of ailments, from wasting diseases to hypochondria and even hemorrhoids. (see Dillinger et al, 2000)

Though anecdotal evidence of the healing powers of cocoa has existed for centuries, it was not until later in the 20th century that researchers appreciated that cocoa beans are one of the richest sources of polyphenols, which are antioxidants that trap dangerous free radicals in the body and prevent them from destroying cells and tissue. (Oracz et al, 2015, *Critical Reviews in Food Science and Nutrition.*) Cocoa beans, whether fresh or processed, contain more polyphenols (and give the beans their bitter, pungent taste) than coffee, black or green tea, or wine. Flavonoids, the largest group of polyphenols, are believed to have anti-inflammatory, anti-allergic, and anti-bacterial properties, among others. (Oracz et al, 2015) Cocoa beans are the seeds of the tree, and each seed contains 40 to 50% fat as cocoa butter (Rusconi and Conti, 2010). These seeds are "embedded in mucilaginous pulp," within pods that come from cocoa trees. The trees grow in moist, hot regions in a belt around the Equator. (Kongor et al, *Food Research International*, 2016) There are three main varieties of cocoa trees, with the most common being Forastero. The concentration, though, of polyphenols, depends on the genetics of the bean, but also environmental conditions, such as soil, sun exposure, rainfall and even storage time. (Oracz et al, 2015)

Aztec woman pouring chocolate to generate foam, circa 1553, from the Tudela Codex, a pictorial religious document, now in Madrid, at the Museo de America. Creator: Mexican School, 16th century.

Bridgeman Images, used with permission.

Woman Chocolate Vendor, by Paul Gavarni, between 1855–57, Walters Art Museum, Baltimore, Maryland. Water color with graphite underdrawing and white heightening on cream.

Walters Art Museum, Baltimore, Maryland. Wikipedia Commons, Public Domain.

Chocolate manufacturing. Mixing room, French School, late 19th century-early 20th century, French educational card chromolithograph, Private Collection.

Bridgeman Images, used with permission.

Cocoa beans, from which our chocolate products originate, looked almost like almonds to 16th century Europeans.

Wikimedia Commons/Public Domain. GNU Free Documentation. Author: SuperManu, 2008.

Today, 70% of our cocoa beans comes from Western and Central Africa, with Ghana, the primary producer of high-quality cocoa. (Oracz et al, 2015) Cocoa is a major crop and is grown by 5 to 6 million farmers throughout the world, (Kongor et al, 2016) with the demand for cocoa growing. There is even a prediction that by 2020, there may be a world shortage. (Wickramasuriya and Dunwell, *Plant Biotechnology Journal,* 2018.)

Chocolate, from the cocoa bean, is a highly processed substance. Cocoa liquor, which contains non-fat cocoa solids and cocoa butter, is a paste that comes from the bean. Cocoa butter contains both monounsaturated (mostly oleic acid) and saturated fatty acids (stearic and palmitic). Cocoa powder results when some of the cocoa butter is extracted from the liquor. Chocolate is a combination of cocoa liquor, with cocoa butter and added sugar. (Magrone et al, *Frontiers in Immunology,* 2017) Nibs are cocoa beans without the outer shell. (Di Mattia et al, *Frontiers in Immunology,* 2017) Milk chocolate contains not less than 20 to 25% cocoa. (Verna, 2013) White chocolate contains cocoa butter, sugar, milk powder, and vanilla. (Verna, 2013) When chocolate develops a whitish and opaque "lacy" appearance, there is nothing wrong with it: part of the cocoa butter has solidified or recrystallized and

An 1896 Art Nouveau poster for the hot chocolate drink from Pepinster in Belgium.
Alamy.com, used with permission.

come to the surface, a phenomenon known as "chocolate bloom," and can be prevented by storing it in a cool place. (Aguilera, *Edible Structures: The Basic Science of What We Eat*, pp. 126–7, 2017)

The processing of chocolate involves many steps. The first, which can take from 5 to 10 days and reduces the bitterness, is fermentation of the pulp surrounding the cocoa beans. Next, the beans are sun-dried, following which they are roasted, which can give the beans their typical color, aroma, and taste and texture. Then there is conching, in which the beans undergo an agitation process at high temperatures, and finally, tempering. At every step, there is considerable loss of the polyphenol content. (Di Mattia et al, 2017) For example, after 8 days of fermentation, polyphenol levels drop by as much as 58%. (Oracz et al, 2015) For the most thorough general discussion of the effects of processing on our food supply, see Gyorgy Scrinis' book *Nutritionism* (2013), as well as his recent article on ultra-processed foods, (Scrinis and Monteiro, *Public Health Nutrition*, 2017.)

Despite the substantial loss of polyphenols during the multi-steps involved in processing, cocoa has been touted in recent years as a food that has almost magical powers. Chocolate, and particularly dark chocolate, with a cocoa content of at least 70%, has gone from being blamed for obesity and type II diabetes only a few years ago, to a food that is considered actually beneficial to health. (Verna, 2013) There are positive effects even on mood, behavior, and cognition, (Tuenter et al, *Planta Medica*, 2018) as well as "an extensive range of benefits" on blood pressure, insulin resistance, cardiovascular disease, and even body weight.

Chocolate production. Taking the cocoa beans to be dried. Liebig collectors' card, 1929. There is a bitter reality to chocolate cultivation/production, particularly in Western Africa in the countries of Ghana and Côte d'Ivoire, where most of our cocoa for the major chocolate companies comes. Because of their own poverty, cocoa farmers have resorted to exploiting children and even using them as slave labor.

Bridgeman Images, used with permission.

(Kord-Varkaneh et al, *Critical Reviews in Food Science and Nutrition*, 2018.) The exact mechanisms for these positive benefits are not completely understood as, for example, why cocoa decreases platelet aggregation and reduces platelet adhesion. Cocoa may cause vasodilation of blood vessels by increasing levels of nitrous oxide, which, in turn, may affect mitochondrial functioning. Increased levels of uric acid, detrimental and painful to those suffering from gout, may also have a role here. (Latif, 2013; Ludovici et al, *Frontiers in Nutrition*, 2017) Cocoa may also improve the barrier function in the gut by changing the gut's microbiome. (Strat et al, *Journal of Nutritional Biochemistry*, 2016)

Scientific investigation, though, on cocoa suffers from many of the same difficulties as seen in much of nutritional research, and results are often inconsistent regarding actual benefits. For example, subjects are often not blinded because it is hard to mask the typical characteristics of chocolate. Further, there are often methodological differences among studies that make meta-analyses challenging: participants may vary in terms of their BMI, age, initial health status, and kind of intervention (e.g. what type of cocoa was used), whether it was in liquid or solid form, or even how long or how much cocoa was given. Further, when cocoa is mixed with other substances—the so-called "food matrix"—results may vary. (Ellinger and Stehle, *Nutrients*, 2016; Di Mattia et al, 2017) For example, in some but not all studies, the addition of milk interferes with the absorption of the antioxidants and may negate any potential health benefits. (Lotito and Frei, *Free Radical Biology & Medicine*, 2006) Often, as typical of many nutrition studies,

Typical 17th century scene that demonstrates the preparation of chocolate.

Wikimedia Commons/Public Domain. Creative Commons Attribution 3.0 unported.

researchers use food frequency (i.e., self-reporting) questionnaires that can be inaccurate or not even distinguish among different kinds of chocolate. (Latif, *The Journal of Medicine*, 2013) Further, sometimes studies are sponsored by chocolate manufacturers so that conflicts of interest must, at least, be considered. (Latif, 2013)

There is, though, a bitter reality to chocolate cultivation, particularly in Western Africa in the countries of Ghana and Côte d'Ivoire, where most of the cocoa for the major companies (e.g. Mars, Nestlé, Cadbury, Hershey) comes. Apparently, because cocoa farmers live in poverty, they have exploited children, often resorting to overt slavery, to maintain competitive prices. (*Food Empowerment Project*, Chocolate Industry, 2014) Conditions for many of these children are unsanitary and unsafe: at least until recently, most are not allowed to attend school. They are fed poorly, work from sunrise to sunset, climb the high cocoa trees, cut the cocoa bean pods with a machete, and are exposed to toxic chemicals used to control insect infestation and disease. One report noted that "virtually" all children have scars all over their bodies from accidents with the machetes. Rarely, if ever, have these children even tasted any chocolate products. (*Food Empowerment Project*, 2014). For a discussion of some recent attempts to protect these children and mandate school attendance, see the article in *Fortune Magazine* by Brian O'Keefe, March 1, 2016.

Bottom line: Throughout the years, chocolate and the cocoa from which it is made, have been both glorified and demonized. In recent years, cocoa has been seen as having many health benefits, particularly because of its antioxidants, but study results are not always consistent. And all researchers acknowledge the dangers to health of overeating highly caloric chocolate concoctions laden with sugar and other food additives. ■

NUTS WITH BENEFITS

What are some of the plausible effects on our health?

95—Posted May 21, 2018

"A boy puts his hand into a jar of filberts," so the Aesop's fable goes, "and grasps as many as his fist could possibly hold. But when he tried to pull it out again, he found he couldn't do so, for the neck of the jar was too small to allow the passage of so large a handful." A bystander chides the boy for being too greedy and suggests that he will be able to remove his hand if he can be satisfied with only half the amount. The story is a familiar one, with the moral, "Do not attempt too much at once."

For some, though, grabbing as many nuts as possible may not reflect greed: it may indicate a superior knowledge of the health benefits of nuts. What do we know about how nuts fit into our diet?

Nuts, particularly pistachios and almonds, say Allison and his colleagues, have been consumed since biblical times, and references appear in *Genesis* (Lewis et al, *American Journal of Clinical Nutrition*, 2014) and throughout the Bible. (https://bible.knowing-jesus.com/topics/Almonds) Though eaten for centuries, they were much maligned and considered "undesirable" due to their high-fat content from the mid-1950s until the mid-1990s. (Lewis et al, 2014) In recent years, though, we have begun to appreciate that not all fat is equal, and nuts may have plausible health benefits. Tree nuts include walnuts, hazelnuts (filberts), pecans, almonds, pistachios, cashews, Brazil nuts, and macadamia nuts. Peanuts are technically not nuts, but legumes. (For more on peanuts and their allergic potential specifically, see blog 79, *The Brittle World of Peanut Allergy.*)

Nuts are all "nutrient dense," and depending on the particular nut, contain different levels of healthy monounsaturated (mostly oleic) and polyunsaturated fatty acids (mostly linoleic), and low levels of saturated fats, as well as protein, soluble and insoluble fiber, vitamins E and K, folate, thiamine, minerals such as magnesium, copper, potassium, and selenium, and antioxidants. (De Souza et al, *Nutrients*, 2017) For example, walnuts have the highest level of polyunsaturated fatty acids. (Kim et al, *Nutrients*, 2017); almonds have the highest fiber of the tree nuts, and peanuts have the most protein and fiber. (De Souza et al, 2017) Even cashews, which initially had been exempt from health claims made by the FDA in the early 2000s because they "exceeded the disqualifying amount of saturated fatty acids," have been exonerated because a third of their saturated fat comes from stearic acid, now considered "relatively neutral" on blood lipids. And the majority of fat (60%) in cashews is monounsaturated (primarily oleic acid.) (Mah et al, *American Journal of Clinical Nutrition*, 2017) (For a comprehensive review of the nutrient contents of different raw nuts, see Kim et al, 2017)

Further, because of their high fiber content, nuts take more work to chew and digest (i.e., may enhance calorie expenditure) and may even act as an appetite suppressant (i.e., decrease food intake from other sources due to their satiating quality.) A review of 21 studies demonstrated no weight gain from consumption of various nuts, and "under most circumstances," nuts can be added (or at least substituted for other foods) in a diet without significant weight change. (Kim et al, 2017) The way, though, that nuts are processed can change the number of calories absorbed. For example, there are fewer calories in natural, roasted, or chopped almonds than in almond butter, and whole, natural almonds (without additional oil or salt) are the healthiest. (Gebauer et al, *Food & Function*, 2016) Further, not only can processing "differentially impact" the number of calories available, but studies now show that the degree of almond processing can change the ratios of bacteria in our gastrointestinal microbiome. (Holscher et al, *Nutrients*,

A jar of filberts, also called hazelnuts. Perhaps the boy, whose hand got stuck in a jar in the classic Aesop's fable, appreciated the benefits of eating nuts.

Grant Heilman Photography. Source: used with permission, Alamy Stock Photo. Wikimedia Commons/Public Domain.

2018) Even without processing, in one small study, 42 grams (about one and one-half ounces) of walnut halves changed these ratios. (Holscher et al, *Journal of Nutrition*, 2018)

The notion that nuts can be part of a healthy diet grew out of a belief that a Mediterranean-style diet, which incorporates fresh fish, vegetables and fruits, grains, olive oil, wine, legumes, and nuts, may be cardio-protective. Researchers are especially interested to find connections between diet and cardiovascular disease (CVD) since CVD is the "primary global cause" of death, with 17 million deaths attributed to CVD each year. (Mattioli et al, *Journal of Cardiovascular Medicine*, 2017.) This type of diet had been systematically studied by Ancel Keys in the late 1950s, in his now classic *Seven Countries Study.* Keys' methodology had been criticized in the media in recent years, but researchers do believe Keys' work remains a "seminal study" and was a "first step" in exploring the important relationship of diet to cardiovascular health. (Menotti and Puddu, *Current Opinion in Lipidology,* 2018; see Pett, Kahn, Willett, and Katz, *Ancel Keys and the Seven Countries Study: An Evidenced-based Response to Revisionist Histories* (2017) for a thorough discussion refuting criticisms of Keys' and his *Seven Countries Study.*)

Power figure: Male (Nkisi), 19th- to mid-20th century from the Democratic Republic of Congo, made of many materials, including nuts. Nuts have been eaten since biblical times, but also used as a medium in art.

Michael C. Rockefeller Memorial Collection. Bequest of Nelson A. Rockefeller, 1979. Metropolitan Museum of Art, NYC. Image copyright, The Metropolitan Museum of Art. Image source: Art Resource, NY, used with permission.

Umberto Moggioli, *Still Life*, 1918.

Private Collection/Mondadori Portfolio/Walter Mori/Bridgeman Images. Used with permission.

British artist William Henry Hunt, drawing, *An Apple, Grapes and a Hazelnut on a Mossy Bank*, 19th century.

Metropolitan Museum of Art, NYC, Public Domain

Photo of Ancel Keys, 1958, who did some of his groundbreaking research on human starvation during World War II at the University of Minnesota. Years later, Keys performed "seminal" research linking diet to cardiovascular health, though some consider his work controversial.

Archives of the University of Minnesota. Credit: Courtesy of University of Minnesota Archives, University of Minnesota-Twin Cities. Used with permission.

1950s doll from Bermuda. This doll is made of banana leaves and screw pine and has a walnut for head. Nuts are not just for eating but can be used in crafts as well.

Author's private collection.

Since Keys' original study, there have been countless scientific reports on the relationship of diet, including eating nuts specifically, to health, but research in this area can be problematic. For example, when considering the effect of diet on the heart, there are such "long induction periods" for cardiovascular disease to manifest itself that short-term randomized controlled studies cannot necessarily capture a "dietary etiology." (Satija and Hu, *Trends in Cardiovascular Medicine,* 2018). Casazza and Allison (*Clinical Obesity,* 2012) emphasize that researchers must give "careful consideration" to *what is and is not known,* and exactly what the "evidence base actually shows."

What we actually know can be difficult to assess. Not only is it important to consider and control for the so-called "background diet" (Neale et al, *British Medical Journal,* 2017) of the participants, namely what else they may be eating during the study, but also to isolate one component behavior, such as eating nuts, from other healthy behaviors a participant may engage in. For example, Hu and Willett (Bao et al, *NEJM,* 2013), in their study of over 76,000 nurses and over 42,000 male health professionals, exploring the relationship of mortality among men and women who eat nuts, found those who consumed nuts were more likely leaner, less likely to smoke, more likely to exercise, eat more fresh fruits and vegetables, take multivitamin supplements, and drink more alcohol. Further, they noted reverse causality had to be considered, namely, that those with chronic illness or in poorer health may refrain from eating nuts. Though their study

Giuseppe Arcimboldo, *Autumn*, 1573 (from *The Seasons*), Louvre Museum, Paris. Arcimboldo was famous for incorporating foods, animals, and household items into his paintings. This painting was commissioned by Emperor Maximilian II.

Bridgeman images, used with permission.

involved 30 years of follow-up and did indicate a 20% lower death rate in those consuming nuts seven or more times a week, they noted nut consumption, like so many studies of diet, was recorded by potentially (and notoriously) inaccurate self-report, and no data were collected on how nuts were prepared. And they acknowledged that their population was a specific one consisting of those in the health professions and may not generalize to other populations. (Bao et al, 2013)

In more recent years, Eslamparast et al (*International Journal of Epidemiology*, 2016) studied nut consumption and its possible relationship to mortality, but in a population, in northeastern Iran, "whose nut consumption does not track with a healthy lifestyle." They, as well, used self-report, with over 50,000 participants, and a median of 7 years of follow-up. Those women who ate nuts three or more times a week (28 grams, about an ounce) had a 51% lower risk of death, though men had only a 16% lower risk, but the researchers acknowledge that since their study was observational, they, as well, cannot assume causality.

Perhaps one of the most important (and reasonably consistent) findings about the benefit of nut consumption is its effect on cholesterol levels (total cholesterol and LDL, or so-called "bad" cholesterol) which, when elevated, are a major factor in the development of cardiovascular disease. A study by Berryman et al, *Journal of Nutrition*, 2017; and systematic reviews by Afshin et al, *American Journal of Clinical Nutrition*, 2014; Kim et al, 2017; and Del Gobbo et al, *American Journal of Clinical Nutrition*, 2015, have all found nut consumption can lower total cholesterol and LDL levels significantly. And it seems it is the *quantity of nut intake (i.e., dose-related) rather than any specific nut itself.* (Asgary et al, *Journal of the American College of Nutrition*, 2018) For

A leather Bible, printed in 1477 and bound in 1478, from Nuremberg, Germany. References to nuts, predominantly almonds, but also a reference to pistachios, appear throughout the Bible.

Fletcher Fund, 1924, Metropolitan Museum of Art. Source: Metropolitan Museum of Art, NYC, Public Domain.

example, one trial reviewed by Del Gobbo et al (2015) found those who ate 100 grams (slightly more than 3½ ounces) of nuts a day lowered their LDL cholesterol by up to 35 mg/dL.

Researchers have also examined the relationship between nut consumption and inflammatory and endothelial function (Neale et al, 2017), as well as with hypertension, stroke, and diabetes (Zhou et al, *American Journal of Clinical Nutrition,* 2014; Luo et al, *American Journal of Clinical Nutrition,* 2014), but these results are less conclusive. A very recently published prospective study of over 61,000 Swedish participants, with up to 17 years of follow-up, was also cautious in suggesting that nuts may "play a role" in reducing a risk for atrial fibrillation and "possibly" heart failure, but not with myocardial infarction, for example, once the researchers adjusted for "multiple risk factors" and other healthy behaviors of their subjects. Furthermore, this study also used self-report data. (Larsson et al, *Heart,* 2018)

Bottom line: In their editorial summarizing several of these studies, Allison and his colleagues (Lewis et al, 2014) concluded "with reasonable confidence" that nut consumption is associated with a reduced risk of coronary artery disease, but not with a risk of stroke or all-cause mortality, and from the studies they reviewed, they could not make conclusions about the relationship to diabetes. All in all, though conclusions about causation "would be premature," all nuts seem to have benefits, particularly for certain aspects of cardiovascular health, and it makes sense to incorporate them into our diet. (Lewis et al, 2014) Eating more nuts is associated with lower cholesterol levels and in general, nut consumption is not necessarily associated with weight gain, usually because it creates satiety and less intake of other foods. Incorporate nuts into your diet, but *not* at the expense of increasing your daily caloric intake. Everything in moderation, and like the moral of Aesop's fable, don't attempt too much at once! ■

Bird eating nuts, detail from a tablinium decorated with Egyptian-style paintings, Roman, 1st century B.C. (fresco), Villa dei Misteri, Pompeii, Italy.

Bridgeman images, used with permission

A POINT OF ORDER: NUTRITIONAL PRESCRIPTIONS AND FOOD SEQUENCE

The Complex "Foodscape" of Glycemic Control

96—Posted July 1, 2018

Count Alexander Ilyich Rostov, in Amor Towles novel, *A Gentleman in Moscow*, (2016) encounters the adventuresome, precocious nine-year-old Nina Kulikova in the Metropol Hotel to which the Count has been confined. In one scene, Nina, who will be having dinner with her own father somewhat later, "has taken the liberty of ordering herself an hors d'oeuvre—a small tower of ice creams." "Quite sensible," says the Count, as he watches her eat her ice cream "one flavor at a time, moving from the lightest to the darkest in shade." (p. 91)

Young Nina may be following the saying, "Life is uncertain; eat dessert first," but *this may not be the most sensible* strategy for weight control and particularly, for our long-term health, according to researchers Louis J. Aronne, MD, the Sanford I. Weill Professor of Metabolic Research, professor of clinical medicine, and Director of the Comprehensive Weight Control Center, Division of Endocrinology, Diabetes & Metabolism, and Alpana P. Shukla, MD, assistant professor, at Weill Cornell Medicine. In other words, the sequence in which we consume our food may have an impact, just as do the quantity, quality, and macronutrients (e.g., proportion of protein, fat, & carbohydrate) of our meals. (Shukla et al, *Diabetes Care*, 2018; Shukla et al, *BMJ, Open Diabetes Research & Care*, 2017.)

These researchers acknowledge that low calorie diets, especially those emphasizing intake of fewer carbohydrates, are difficult for people to adhere to, even "under the best of circumstances." "Most patients stray, to a similar degree, regardless of which diet they are on," and often tend to increase their carb intake substantially over time. (Orloff, Aronne, and Shukla, *American Journal of Clinical Nutrition, Letter to the Editor*, 2018) The key predictor to weight loss, not surprisingly, says Aronne, is "adherence to a diet." (Umashanker et al, *Current Athersclerosis Reports*, 2017) While increased caloric intake and decreased physical exercise are the "main drivers" of our overweight and obesity epidemic in the US (e.g. 69% Americans are overweight and 39.8% are obese), researchers are becoming increasingly aware that not just caloric intake per se, but dietary patterns—i.e., the *foodscape*—(Forouhi et al, *The BMJ*, 2018) may be promoting obesity and its metabolic consequences. (Stanhope et al, *Obesity Reviews*, 2018)

There is still controversy regarding the best proportion of macronutrients for a diet, (e.g. lower carb/higher fat; lower fat/higher carb) or even what constitutes "low carb" (Forouhi et al, 2018) because of a lack of long-term, well-controlled studies. (Stanhope et al, 2018) Generally, a "personalized nutrition" approach is warranted for most people. (Stanhope et al, 2018) While all macronutrients lead to increased plasma glucose levels and a subsequent increase in insulin after eating, carbohydrates have "the greatest effect" on plasma glucose and insulin. That is why carbohydrate restriction was the main prescription for control of diabetes prior to the discovery of insulin by Banting, Macleod, and Best in the 1920s. A diet higher in carbohydrates, especially those refined and without fiber, therefore, has more "potential" to increase fat gain by a cascade of metabolic effects that ultimately lead to increased hunger and greater food intake. (Stanhope et al, 2018)

In recent years, by developing *nutritional prescriptions*, investigators have explored whether food order could reduce postprandial glucose and insulin levels, improve long-term glycemic control, and avoid eventual metabolic complications. (Cavalot et al, *The Journal of Clinical Endocrinology & Metabolism*, 2006; Ma et al, *Diabetes Care*, 2009; Imai et al, *Journal of Clinical Biochemistry*

Italian female painter, Fede Galizia, *Maiolica Basket of Fruit*, 1610, Private Collection. Studies by Drs. Aronne, Shukla and colleagues suggest that eating carbohydrates last may be an effective strategy for glucose control.

Wikimedia Commons/Public Domain.

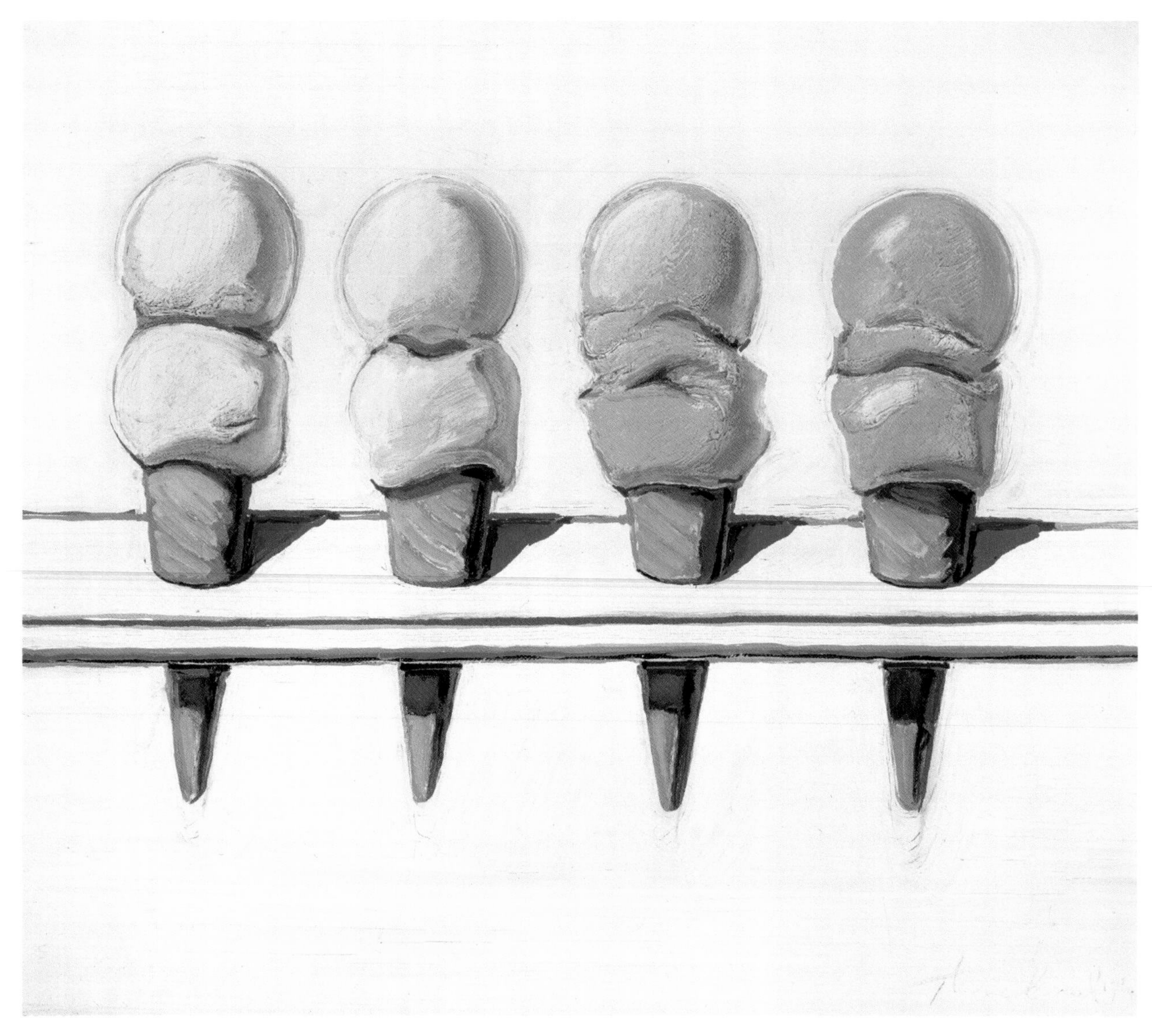

Wayne Thiebaud, *Four Ice Cream Cones,* 1964. In Amor Towles' *A Gentleman in Moscow* (2016), young Nina prefers to eat a dish of different flavors of ice cream as her hors d'oeuvre. She prefers to eat "one flavor at a time, moving from the lightest to the darkest in shade." (p. 91) Her scoop of lemon "perfectly matched her dress." Eating ice cream, though, at the beginning of a meal, especially without protein, may create a surge in plasma glucose levels and is not recommended by researchers.

▲ Joachim Beuckelaer, from Antwerp, Belgium, *Fish Market,* 1568, Musée des Beaux-Arts de Strasbourg. Studies suggest that eating protein, such as in fish, prior to carbohydrates at a meal, may lead to lower postprandial blood glucose levels.

Wikimedia Commons/Public Domain.

◀ Tibor Polya, Portrait of Dr. Frederick B. Banting, 1925, Library and Archives of Canada. Banting won the Nobel Prize, along with JJR Macleod, in 1923, for the discovery of insulin. Banting was age 32 at the time, and he shared his prize money with his colleague Dr. Charles Best.

Library and Archives of Canada, used with permission.

Gustave Caillebotte's *Fruit Displayed on a Stand,* circa 1881, Museum of Fine Arts, Boston.
Wikimedia Commons/Public Domain

and Nutrition, 2014; Alsalim et al, *Diabetes, Obesity, and Metabolism*, 2016; Kuwata et al, *Diabetologia*, 2016; Tricò et al, *Nutrition & Diabetes*, 2016; Faber et al, *Pediatric Diabetes*, 2018) These studies, albeit involving small numbers of subjects, have involved different populations worldwide (e.g. Japan, Italy, Australia, The Netherlands) and have included children with type 1 diabetes, as well as adults with type 2 diabetes. Consistently, they have found that food sequence, with carbohydrates eaten after protein and fat, leads to improved glycemic control.

For example, Aronne and Shukla conducted studies on their overweight and obese patients with type 2 diabetes (all treated with metformin) to assess the temporal effects of food sequence on glucose, insulin, glucagon-like peptide (GLP-1), and ghrelin. Their studies involved a sample of 16 subjects, with stringent and well-controlled protocols. Subjects consumed the same meal on 3 days in random order: either carbohydrates first (bread and orange juice), followed after a 10-minute interval by protein and vegetables; carbohydrates last; or altogether in a sandwich. (Shukla et al, 2018; Shukla et al, 2017) The researchers found a significant difference (lowest peaks) in levels of glucose when carbohydrates were consumed last and lower peaks when consumed all at once in a sandwich (e.g. chicken, bread, vegetables) as compared to carbohydrates first. Insulin levels were lower whereas GLP-1 (a gut hormone that slows gastric emptying) levels were higher in the carbohydrate-last meal as compared to the carbohydrate first

Refined carbohydrates, as found in many breads, increase plasma glucose levels and create a subsequent rise in plasma insulin. Research studies suggest that eating carbohydrates after protein and fat during a meal may have beneficial effects on these metabolic parameters and significantly decrease postprandial glucose levels. Unfortunately, many restaurants tend to bring baskets of tempting bread even before serving the appetizers or main meal.

Photo taken by 3268zauber, 2008. Source: Wikimedia Commons/licensed under the Creative Commons Attribution-Share Alike 3.0 Unported.

meal. Carbohydrates first, for example, led to a rebound in ghrelin, a hormone indicating hunger, that was similar to pre-prandial levels. Aronne and Shukla acknowledge they cannot yet generalize their findings since their sample was small, and they were studying a specific population for a short period of time with certain food choices. They speculated, though, that the carbohydrate last sequence delayed gastric emptying and led to a slower rate of its absorption, possibly also related to the presence of fiber in the vegetables. They intend to repeat their studies with patients who have type 1 diabetes, those with prediabetes, as well as those who are healthy, and they hope to determine the optimal timing for carbohydrate consumption.

What, though, is the potential relevance, of decreased plasma glucose levels (so-called *glucose excursions*) after a meal? Apparently, there is considerable relevance. For one thing, diabetes is a major *global* public health concern, with a prevalence estimated to increase worldwide to 629 million people by 2045. (Forouhi et al, 2018) According to the *National Diabetes Statistics Report* 2017, from the Centers for Disease Control and Prevention, in 2015, there were an estimated 30.3 million people in the US alone (9.4% of our population) with diabetes. (Type 2, most associated with obesity, accounts for 90 to 95% of these cases.) It is also

Thomas Gainsborough, Portrait of John Montagu, 4th Earl of Sandwich, 1783. National Maritime Museum, London. Reportedly Montagu asked his chef to prepare a food that he could eat while continuing to gamble—and hence the "sandwich" was created. Drs. Aronne and Shukla found that eating protein and vegetables, together with bread, decreased postprandial glucose levels, though not as much as eating carbohydrates last.

Wikimedia Commons/Public Domain.

Edward Hopper's *Table for Ladies,* 1930. Metropolitan Museum of Art.

George A. Hearn Fund, 1931. Copyright Metropolitan Museum of Art. Art Resource, NYC. Used with permission.

estimated that of that total, millions go undiagnosed, and an estimated 33.9% of US adults (older than age 18) had prediabetes in 2015. The percentage increases with age so that by age 65 years, nearly half of adults have prediabetes.

Many researchers acknowledge that patients with type 2 diabetes experience wide variations in their postprandial glucose levels, and a fasting plasma glucose level is a "poor indicator" of plasma glucose at other times of the day and does not even correlate with HbA1c levels. (Bonora, *International Journal of Clinical Practice*, Supplement, 2002.) Further, data indicate that postprandial levels are an independent risk factor for cardiovascular disease (Bonora, 2002) and the many other complications of diabetes. (Bell, *Southern Medical Journal*, 2001.) Acute increases in glucose levels may lead to the production of free radicals, endothelial dysfunction, and even a transient state of hypercoagulability. (Bell, 2001.)

Macrovascular complications include coronary artery disease, peripheral artery disease, and stroke; microvascular complications include retinopathy (e.g., a leading cause of blindness), nephropathy (e.g., chronic kidney disease requiring dialysis or kidney transplant) and neuropathies (e.g. neuropathic pain and burning, as well as orthostatic hypotension, tachycardia, female incontinence, and erectile dysfunction.) (Fowler, *Clinical Diabetes*, 2008; American Diabetes Association, 2018) Apparently many patients with

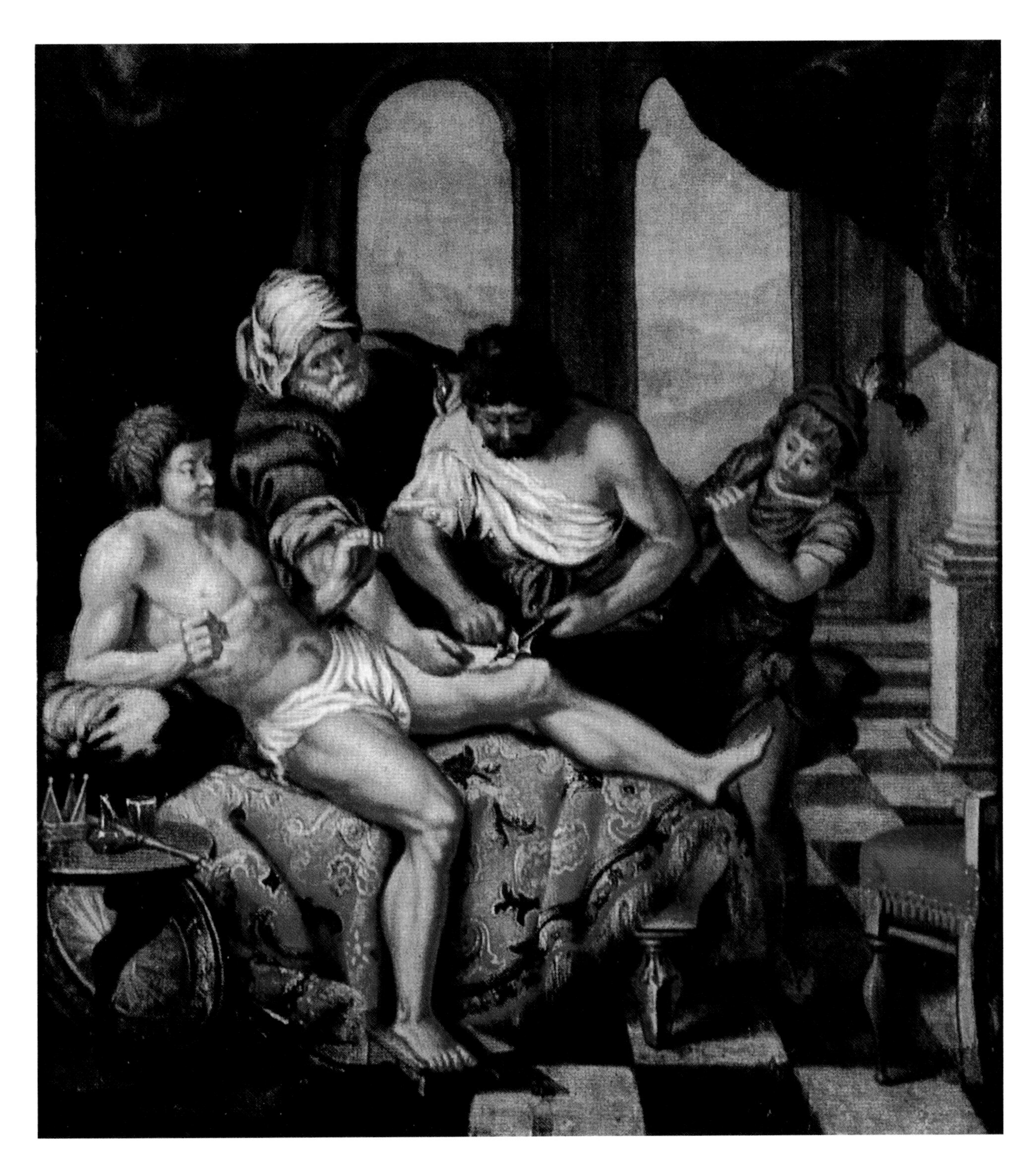

Treatment of Wound with Lance, unknown date and unknown artist, possibly Dutch. Those with diabetes, both types 1 and 2, develop many complications over time, including poor wound healing due to microvascular compromise, that may even require a limb amputation.

József Antall, 1981, Wikimedia Commons/Public Domain.

Georg Flegel, *Still Life with Bread and Confectionary,* first half of 17th century, Städel Museum, Frankfurt, Germany. Researchers recommend against eating sugary confections and white flour bread initially at a meal to avoid strong surges of plasma glucose levels.

Wikimedia Commons/Public Domain.

type 2 diabetes still develop certain neuropathies despite eventual adequate glucose control because they have had asymptomatic hyperglycemia for years prior to their diagnosis. (Pop-Busui et al, *Diabetes Care,* 2017) And since there are no adequate treatments for nerve damage once it develops, prevention, through a focus on glucose control and especially postprandial levels becomes essential. (Pop-Busui et al, 2017)

That diabetic complications persist even after postprandial levels stabilize is known as *metabolic memory,* (Mobbs, *Frontiers in Endocrinology,* 2018) first described in 1973 by Szepesi et al (*Proceedings of the Society for Experimental Biology and Medicine.)* The term was again used by Cahill (*NEJM,* 1980) to describe how a diet can "result in metabolic patterns that may persist…a type of chronic adaptation or *metabolic memory.*"

Bottom line: In his new book *The Order of Time,* (2018) Carlo Rovelli writes, "…the past leaves traces of itself in the present." (p. 166) "…things change one in respect to the others," he says (p. 120) Perhaps this is a poetic way of looking at the long-term pathological effects of uncontrolled postprandial hyperglycemia. In those without diabetes, postprandial hyperglycemia is transient; in those with either type 1 or type 2 diabetes, these glycemic excursions may have serious repercussions. (Madsbad, *Journal of Diabetes and Its Complications,* 2016) Since type 2 diabetes, in particular, has become an epidemic itself among the US population and is expected to worsen over the next years, (and because so many cases remain undiagnosed), it seems worthwhile for all to consider ways of curtailing hyperglycemia. Attention to our food sequence, with eating carbohydrates *after* protein, non-starchy vegetables, and fat, is a potential effective first-line "behavioral strategy" (Shukla et al, 2017) that holds promise for weight control by way of glucose regulation. ■

THE "DISFIGURING MAYHEM" OF CANCER CACHEXIA

Withering Away: The Deadly Consequences of Fatigue and Emaciation

97—Posted August 10, 2018

"I don't have a body. I am a body," wrote Christopher Hitchens, diagnosed with esophageal cancer and having lost 14 pounds that no amount of tube feeding could correct, in his final memoir, *Mortality.* (2014) Walking to the refrigerator became "like a forced march." He continued, "You lose weight but cancer is not interested in your flab. It wants your muscle." With ruthless brutality, his cancer gave him "gut-wringing nausea on an utterly empty stomach" and the "double-cross of feeling acute hunger while fearing even the scent of food." Likewise, Dr. Paul Kalanithi, neurosurgeon and himself dying of Stage IV lung cancer, wrote, in his book *When Breath Becomes Air* (2016), how he suffered from a "profound bone-weariness" after chemotherapy and felt "withered:" "I could see my bones against my skin, a living X-ray."

These are descriptions of *cancer cachexia*--from the Greek for "bad condition." Cachexia causes a wasting of the body and stems from an imbalance of the body's anabolic and catabolic processes. It can occur in other chronic, debilitating diseases such as AIDS, chronic obstructive pulmonary disease (COPD), and cardiac disease, (Penet and Bhujwalla, *Cancer Journal*, 2015) but is associated most commonly with advanced cancer, where it affects 60 to 80 percent of patients and is "directly implicated" in 20 percent of the "spectacularly wretched deaths" (Sontag, *Illness as Metaphor*, 1989) from cancer. (Vaitkus and Celi, *Experimental Biology and Medicine*, 2017) Half of all the cancer deaths worldwide—about 8.2 million a year—are attributed to cancers (e.g. pancreatic, esophageal, pulmonary, hepatic, colorectal) that are most often associated with cachexia. (Baracos et al, *Nature Reviews*, 2018) Almost any kind of cancer, though, can be associated with cachexia. (Vaitkus and Celi, 2017)

There have been many definitions of cachexia over the years, but the one commonly, but still not universally used now, was delineated by Fearon and colleagues (*Lancet Oncology*, 2011): a multifactorial syndrome defined by ongoing loss of skeletal muscle with or without accompanying loss of fat that *cannot* be reversed by conventional nutritional support and can lead to progressive functional impairment. Essentially, cachexia creates what some researchers have called *metabolic mayhem.* (Tsoli and Robertson, *Trends in Endocrinology and Metabolism*, 2013) With the accompanying fatigue and anorexia, i.e., reduced food intake, a patient can experience a dramatic decline, both physically and psychologically, in his or her quality of life. Several factors can lead to anorexia, including nausea, changes in taste and smell, GI obstruction, malabsorption, vomiting, diarrhea, effects of chemotherapy, and even anxiety and depression. (Ryan et al, *Proceedings of the Nutrition Society*, 2016) Furthermore, anorexia can be caused by substances released by the tumors, i.e., pro-inflammatory cytokines, sometimes referred to as "tumorkines" (Tsoli and Robertson, 2013), such as interleukin 1 or 6 and tumor necrosis factor alpha, among many others. Some tumors directly affect zinc metabolism, with accompanying negative effects on food intake. (Ezeoke and Morley, *Journal of Cachexia, Sarcopenia, and Muscle*, 2015)

Game Piece with Zodiac Sign of Cancer, mid-12th century, France. The symbol of the crab has been associated with tumors: swollen veins are like the legs of a crab.

Metropolitan Museum of Art, NYC, Public Domain. Credit: Pfeiffer Fund, 2012.

Salvador Dalí, *Cancer, from Signs of the Zodiac*, 1967, lithograph from original gouaches. Dalí seems to be taking some "poetic license" here since his crab looks more like a lobster.

Collection of The Dalí Museum, St. Petersburg, Florida.. Copyright 2018, Salvador Dalí, Fundació Gala-Salvador Dalí, Artists Rights Society, NYC. Used with generous permission.

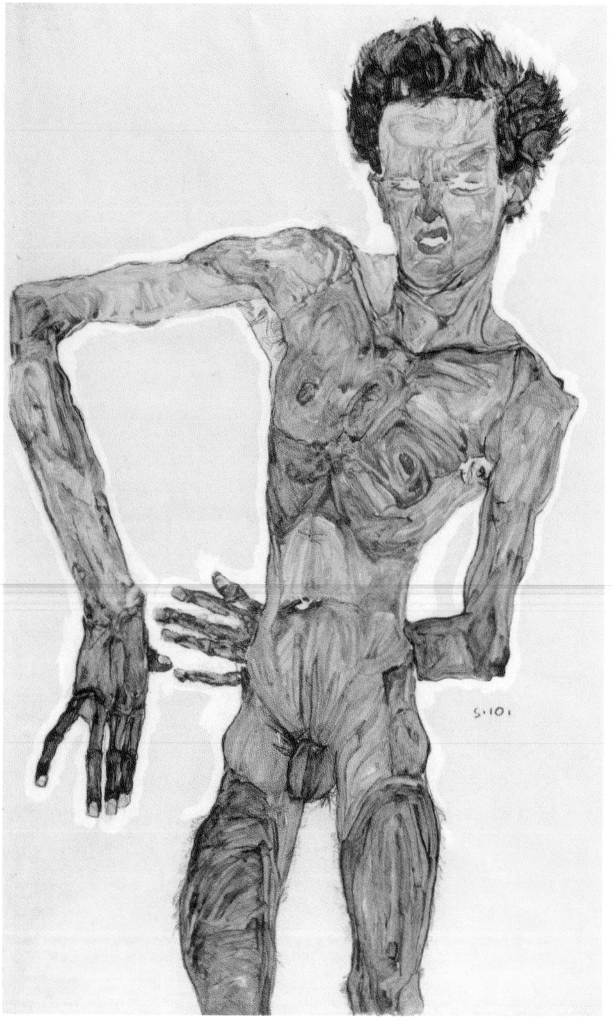

Egon Schiele, *Nude Self-portrait, Grimacing*, 1910, Albertina Museum, Vienna. Many of Schiele's paintings have that emaciated, cachectic look.

Wikipedia Commons/Public Domain.

Fearon et al (2011) noted that cachexia can occur on a continuum, and it represents "a spectrum of conditions." (Ryan et al, 2016) Sometimes, "substantial" unintentional weight loss occurs even before the diagnosis of cancer in a *pre-cachexia* syndrome, with early clinical and metabolic signs, including impaired glucose tolerance, anemia, and general inflammation. If *refractory cachexia* develops, there is often a life expectancy of less than 3 months. The mechanisms involved that lead to mortality include cardiac arrhythmia, electrolyte imbalances, respiratory difficulties due to muscle weakness, aspiration pneumonia, sepsis, etc. (Baracos et al, 2018)

The risk of progression can vary from one patient to another (and some patients are genetically more resistant to developing cachexia than others), and can depend on the type of cancer, stage, time of diagnosis, body composition, presence of systemic inflammation, reduced food, particularly protein, intake, and even the gut microbiome. (Loumaye and Thissen, *Clinical Biochemistry*, 2017) Researchers question whether all cachexia is the same because cachexia can develop from disease-specific, treatment-specific, or tissue-specific mechanisms. (Kays et al, *Journal of Cachexia, Sarcopenia, and Muscle*, 2018) Other aspects associated with the diagnosis include an *unintentional* weight loss of greater than 5 percent within six months or a body mass index (BMI) of less than 20 $kg/m.^2$ Fearon et al, 2011, though, emphasize that clinicians should

Terracotta statuette of a mime with a shield, Hellenistic, 2nd–1st century B.C.

Metropolitan Museum of Art. Credit: Gift of Gordian Weber, 2000. Public Domain.

not just focus on weight loss, but rather on *sarcopenia*, i.e., loss of muscle, with reduced muscle mass and loss of strength. Further, with the loss of muscle, patients are more susceptible to the toxic effects of treatment, with the possibility of being "overdosed."

Detecting cachexia, though, may be difficult because muscle mass can vary considerably across populations, and there is no biomarker (to date) that is highly specific for detecting skeletal muscle wasting. (Loumaye and Thissen, 2017) Furthermore, BMI is not a precise measure of body composition and does not measure muscle: people with the same BMI can have very different body compositions. (Caan et al, *Cancer Research*, 2018) CT or DXA (used to

Terracotta statuette of an emaciated woman, Late Hellenistic, 1st century B.C.

Metropolitan Museum of Art. Credit: Gift of Joseph W. Drexel, 1889. Public Domain.

Statuette of emaciated youth, Roman, 1st Century Common Era. Those with cachexia, secondary to cancer or a chronic debilitating disease, present with weakness, muscle loss, and a "withered" look.

Dumbarton Oaks, Byzantine Collection, Washington, DC. Used with permission.

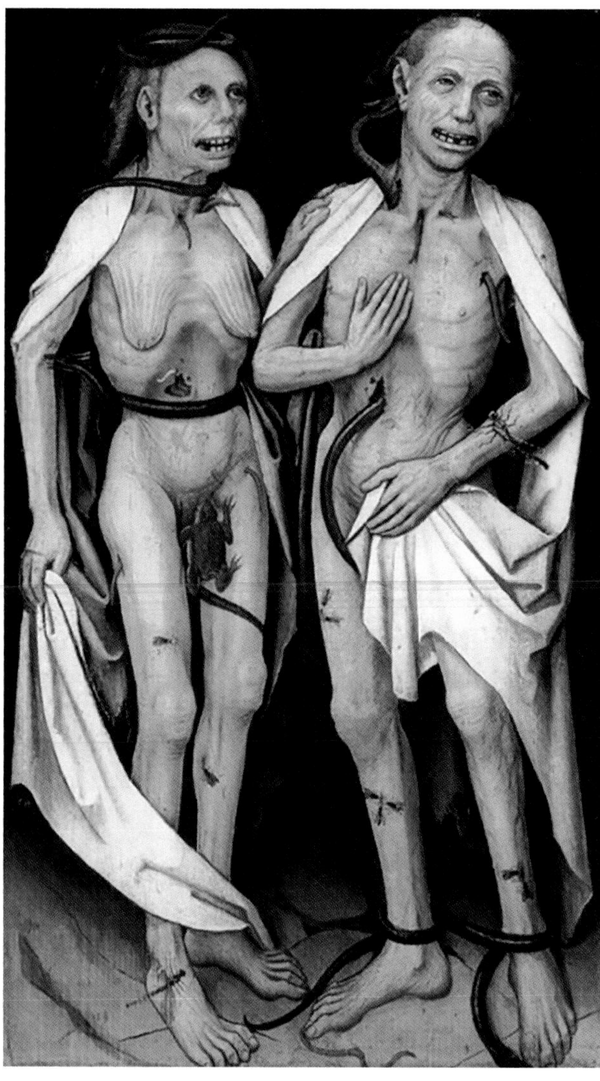

Master of the Upper Rhine, *The Deceased Lovers, Death and Lust,* 16th century.

Musée du Strasbourg, Public Domain.

evaluate bone density) scans can assess muscle, but these can subject patients to additional radiation exposure. (Loumaye and Thissen, 2017)

Ironically, what also makes the diagnosis particularly challenging in recent years is that 40 to 60 percent of patients are overweight or obese at the time they are diagnosed with cancer. (Prado et al, *Proceedings of the Nutrition Society*, 2016; Schwarz et al, *BMC Cancer*, 2017) In fact, many cancers—perhaps between 15 to 45 percent—are considered directly related to obesity. (Ryan et al, 2016) As a result, patients can have "severe depletion" of skeletal muscle that remains undetected, and with this "upward shift" in BMI, the diagnosis of cachexia becomes much less clear. (Baracos et al, 2018) Further, those with increased adipose tissue may be at risk for chemotherapy toxicity because some drugs are lipophilic, with a longer elimination half-life. (Ryan et al, 2016)

Flemish painter Pieter Bruegel the Elder, *The Fight Between Carnival and Lent,* (detail) 1559, 16th century, Kunsthistorisches Museum, Vienna.

Bridgeman Images, used with permission.

The goal of treatment is to preserve muscle and stabilize weight. Early diagnosis and careful monitoring of weight are particularly important because unlike in starvation where increased food intake reverses the condition, cachexia is not completely reversible. (Schwarz et al, 2017) Patients often require "active nutritional management," including use of tube feeding and/or intravenous feeding, either by central or peripheral vein. (Baracos et al, 2018) Unfortunately, though, clinicians often don't know enough about specific nutritional requirements for cancer patients and know even less about those requirements for those who are also obese. (Ryan et al, 2016) An exercise regimen involving aerobic and resistance exercises is also recommended but clinicians appreciate that patients are often too debilitated to comply. (Baracos et al, 2018) Some orexigenic medications/drugs have been tried but there are no conclusive data. A recent scathing review of the use of the "hunger hormone" ghrelin found the research for outcomes relating to body weight and food intake of "very low quality," due to poor methodology, inadequate reporting of data, and small sample sizes. (Khatib et al, *Cochrane Database of Systematic Reviews,* 2018) Other medications that can be appetite stimulants

American artist Benjamin Day, Illustration for *Crab Story*, drawing, 19th–20th century, Metropolitan Museum of Art, NYC. Whittelsey Fund, 1967.

Hans Memling, *Triptych, The Last Judgement,* 1467–71.
Gdańsk, Poland Muzeum Narodowe, Public Domain.

have adverse effects, as for example, corticosteroids that can increase the risk of thromobembolism or even cause muscle atrophy. (Baracos et al, 2018)

Bottom line: There are still many unanswered questions about the complex mechanisms involved in cachexia. Diagnosis should be made as early as possible but is often made more difficult because there is no universally agreed upon definition nor any biomarkers specific to cachexia, and many patients diagnosed with cancer, ironically, are overweight or obese and have "cryptic cachexia." (Tsoli and Robertson, 2013) Cancer cachexia leaves the patient physically and psychologically debilitated as it creates its metabolic mayhem. To date, there are no evidence-based treatments.

Note: My title, *Disfiguring Mayhem*, comes from a quotation by Mark Twain, 1870, writing in a Virginia newspaper (and having nothing to do with cancer or cachexia.) *Mayhem* is from the root "to maim." ■

THE BODY AS METAPHOR: SOCIAL CLASS AND OBESITY

Subjective social status and a 'mentality of scarcity'

98 — Posted September 5, 2018

"The evil of poverty is not so much that it makes a man suffer as it rots him physically and spiritually," wrote George Orwell in *Down and Out in Paris and London* (1933) considered by some to be Orwell's own account of the excruciating circumstances and utter destitution he had experienced as a *plongeur*—the lowest of menial restaurant workers. Mostly, Orwell focuses on the abject poverty that resulted in days without food: "...a man who has gone even a week on bread and margarine is not a man any longer, only a belly with a few accessory organs." His description, though, of the filth of many restaurants, with cats, rats, and cockroaches roaming the floors where raw meat lies among the garbage, sinks become clogged with grease, and the workers have no time to clean pots, plates or utensils, may make even the hardiest want to forego eating altogether.

Ironically, though, through the years, lower socioeconomic status is more often associated with obesity. One of the first to make the connection between socioeconomic status (SES) and weight was Albert Stunkard back in the 1960s. (Moore, Stunkard, and Srole, *JAMA*, 1962) Using a representative cross-section of 1,900 people from the Midtown Manhattan Study, a survey originally focused on the epidemiology of mental illness, they found a surprising relationship between social class (based on the father's occupation and education when the subject was entering adulthood) and obesity. The prevalence of obesity was found to be seven times higher among women from the lowest social class than those reared in the highest class. Later, Stunkard speculated that there were three possible relationships: obesity influences SES; SES influences obesity; or some other factor influences both. (Stunkard and Sørensen, *NEJM*, 1993) For Stunkard and his colleagues, this effect had "profound implications for theory and therapy": Essentially, it meant that obesity, despite "its genetic and biochemical determinants," is also potentially "susceptible to an extraordinary degree of control by social factors." (Moore, Stunkard and Srole, 1962) Significantly, as well, and much later in twin studies, Stunkard et al (*NEJM*, 1990) explored the genetics of obesity and were among the first to find that body mass index (BMI) has as much as a 70 percent chance of being genetically determined in some populations.

Stunkard and his colleagues continued to study the relationship between SES and obesity (Goldblatt, Moore, and Stunkard, *JAMA*, 1965). They found, among 144 published studies, that "no matter what measures of SES or obesity were used or what population group (in developed countries), the results were monotonously similar": SES was inversely (and strongly) related to obesity in women. In developing countries, they found the opposite to be true, so that possibly due to a lack of food, obesity was seen as a sign of health and wealth. Studies with men and children of either gender were more inconsistent. Their theory was that obesity is more "severely stigmatized" among women, and these attitudes develop at a very young age (Sobal and Stunkard, *Psychological Bulletin*, 1989). Obesity is a "visual defect," and unlike most other chronic diseases, represents a "greater social disability" because of its "public nature." (Stunkard and Sørensen, 1993) They also speculated that women of higher SES would have more leisure time for exercise, greater access to resources that would "facilitate" dieting (e.g. more nutritious choices and ability to afford expensive foods; interest in dieting programs); and even have greater knowledge about nutrition and diet. The work of Stunkard was groundbreaking at the time, even though Allison and his colleagues (Pavela et al, *Current Obesity Reports*, 2016) note there were "limitations in the original focus," such as reliance on a cross-sectional

Luigi Nono, *Abandoned*, 1903. Galleria Internazionale d'Arte Moderna Di Ca' Pesaro, Venice, Italy.

DeAgostini Picture Library/M. Carrieri/Bridgeman images, used with permission.

Vincent van Gogh, *A Woman with a Spade, Seen from Behind* (also called *Peasant Woman Digging*) 1885, Oil on canvas on wood panel.

Art Gallery of Ontario, Gift of Ann and Lawrence Heisey, 1997. Image copyright 2018 Art Gallery of Ontario. Used with permission.

Edward Thompson Davis, *A Beggar on the Path*, 1856. Hamburger Kunsthalle, Hamburg, Germany.

Bridgeman Images. Used with permission.

Thomas B. Kennington, *The Pinch of Poverty*, 1889, Art Gallery of South Australia.

Wikimedia Commons/Public Domain.

design, self-reports of height and weight, and little discussion of racial and ethnic differences that can have significant effects on weight and its relationship to SES.

McLaren (*Epidemiologic Reviews*, 2007) extended Stunkard et al's research to 333 published studies through to 2004. As per the hypothesis, McLaren found these new studies resembled the original group but with the a qualification: Because of large-scale societal changes and nutritional changes as a result of globalization of food markets, modernization, and economic growth, the differences between developed and developing countries were not as prominent. Furthermore, in more recent years, "virtually all social groups are increasingly affected by obesity" such that even though women of a higher SES in developed countries may still value being thin, "our obesogenic environment may make it increasingly difficult for women of any class group" to maintain that ideal. McLaren also noted that the body, incorporating appearance, type, and behavior, can be seen as a *social metaphor* for a person's status, an idea borrowed from the French sociologist Pierre Bourdieu. Social class, therefore, is not just about wealth, but "a constellation of attributes" (e.g. accent, body shape) that become highly valued.

How do we measure socioeconomic status? Most commonly, education and income are the key parameters. Galea and colleagues (Bor et al, *The Lancet*, 2017), though, explain that all measures of SES have limitations. For example, a person's earnings may be assessed cross-sectionally and may fluctuate and not necessarily reflect earnings over a lifetime. Further, educational data are "coarse," and there is not a "clear ordinal ranking" for education beyond college nor an appraisal of the quality of an education. And of course, use of self-reported data for education and income may suffer from reporting bias.

C.A. Ferrier, *A Slum* (engraving), 19th century, Private Collection.
Bridgeman images, used with permission.

In more recent years Dhurandhar (*Physiology & Behavior*, 2016) has hypothesized that obesity is more common in women because women must maintain an adequate weight for successful reproduction and nursing; those with lower SES, though, may actually have an increased desire for food and even anticipate that a food supply may be or will become inadequate, whereas those with a higher SES may be "resistant" to that perception. She further speculates that obesity is more likely present in lower-class women because increased weight may be a "strategic response" to even a *perceived* insecurity about food (Dhurandhar, 2016).

Laraia et al (*American Journal of Preventive Medicine*, 2017) note that living in poverty, with its effects on stress levels, sleep schedules (due to irregular shifts, for example), and general uncertainties regarding employment, housing, and food can contribute to a *mentality of scarcity*. These researchers, though, emphasize that it is not poverty alone that contributes to poor diets: "the majority of Americans, regardless of income, eat poorly," as judged by statistics on

Are You First Class? Color lithograph by the English School, 20th century. Postcard. This conductor has some obvious concern about this woman's subjective social status.

Bridgeman images, used with permission

healthy dietary intake (e.g., sufficient daily quantities of fruits and vegetables), "even though the reasons for eating a poor diet" may be different for one social class from another.

Further, giving lower-class people surplus money to purchase healthier food does not necessarily lead to decreased caloric intake: Instead of substituting the unhealthier food for healthier, less caloric choices, many supplement, rather than displace, their meals with additional calories (Caldwell and Sayer, *Appetite*, 2018).

Sometimes it is not just our actual social class and environment that can affect weight and even health in general, but even the *perception* of that environment (Pavela et al, 2016; Dhurandhar et al, *Obesity*, 2018)—the so-called *subjective social status*, a term described in 1950s: "a person's belief about his location in a status order ... that may or may not be congruent with his objective status." (Davis, *Sociometry*, 1956) This notion of a subjective social status echoes Ralph Waldo Emerson: "...the poor are only they who feel poor..." (*Domestic Life*, Chapter V, in *The Complete Works*, 1904).

Subjective social status is measured by a pictorial image of a 10-rung ladder, the MacArthur Scale. (Dhurandhar et al, 2018; Wijayatunga et al, *Appetite*, 2018) In recent years, Allison and his colleagues have designed studies in which they have manipulated the environments of their subjects to create the impression that they have a lowered social status. (Wijayatunga et al, 2018; Cardel et al, *Physiology & Behavior*, 2016; Kaiser et al, *Annals of the NY Academy of Sciences*, 2012) In other words, it may be the "socio" as much or more than the "economic" that may lead to increased obesity. (Kaiser et al, 2012) In one small study, rigged Monopoly games appeared to result in increased caloric consumption and decreased feelings of powerfulness and pride in those with subjectively lowered status (Cardel et al, 2016). Another, using remote food photography, found that individuals with perceived lower subjective status had increased caloric intake and a reduced ability to compensate for those additional calories. Over time, this failure to compensate may potentially lead to weight gain (Wijayatunga et al, 2018).

Pablo Picasso, *Poverty*. 1903.

Whitworth Art Gallery, The University of Manchester, UK. Source: Bridgeman images and Artists Rights Society, NYC, used with permission.

Carefree in Poverty, 1510–20, glass-stained, The Cloisters Collection, 1999, from the Metropolitan Museum of Art, NY. There is nothing particularly "carefree" about poverty.

Metropolitan Museum of Art, NYC, Public Domain.

Isidre Nonell, *A Beggar in Paris,* 1897. In his book, *Down and Out in Paris and London* (1933), George Orwell depicts the life of excruciating poverty for those who are forced to beg. Writes Orwell, (p. 174) "A beggar...has merely made the mistake of choosing a trade at which it is impossible to grow rich."

Wikimedia Commons/Public Domain.

Bottom line: Most researchers acknowledge that obesity rates have been increasing in every population worldwide, sometimes regardless of socioeconomic status. Many factors, including both genetic and environmental ones, may contribute to increased weight. Since the original work by Stunkard and his colleagues in the 1960s, researchers have appreciated that our socioeconomic status, usually assessed by income and education levels, can affect body weight. Even the *perception* of our status, such as when it is artificially manipulated and subjects, may have an effect on caloric consumption and a subsequent failure to compensate for additional caloric intake, though experimental manipulation is not the same as real-world conditions. Lower socioeconomic status has been associated with increased weight, particularly in females, but the relationship is complex, not entirely understood, and differs among subgroups of populations by sex, race, and ethnic origin.

Note: For more detailed information on actual prevalence rates of obesity in different populations, see the papers by Ogden et al, in *Morbidity and Mortality Weekly Report* (MMWR), December 22, 2017 (adults) and February 16, 2018 (youth), from the Dept. of Health and Human Services/ Centers for Disease Control and Prevention. Special thanks to Drs. Ogden and Flegal for providing me with these references for the most recent statistics (2011–2014.) ■

MATHEMATICAL MODELS: OBESITY BY THE NUMBERS

Toward greater precision in obesity research

99 — Posted September 29, 2018

How to be Thin.

EAT LESS!

How to be Thin. Eat Less! Color lithograph, English School, 20th century, Private collection.

"...when you can measure what you are speaking about, and express it in numbers, you know something about it; but when you cannot measure it...your knowledge is of a meager and unsatisfactory kind; it may be the beginning of knowledge, but you have scarcely advanced to the stage of science..." So wrote 19th century Irish mathematician and physicist Lord Kelvin, famous for his experiments on heat energy (and for whom the Kelvin scale of absolute temperature was named.) (Lecture, 1883, in *Popular Lectures and Addresses*, 1889, Volume 1.) Kelvin's words are particularly relevant to the study of obesity.

Essentially obesity is a "disorder of energy balance," based on the *First Law of Thermodynamics,* that occurs when we consume more calories than we expend. But, explain researchers Kevin Hall and Juen Guo (*Gastroenterology*, 2017), this is merely a "useful framework" and does not provide any "causal explanation" for why some people are more prone to obesity than others. They clarify, "...obesity prevention is often erroneously portrayed as a simple matter of bookkeeping," i.e., that 3500 kcal equals a pound, but this widespread view, quoted on websites and throughout the media, is "naïve and incorrect" because it does not take into consideration the complex and dynamic *interdependent* relationship between intake and expenditure. (Hall and Chow, *International Journal of Obesity,* 2013; Thomas et al, *International Journal of Obesity*, 2013)

Although this relationship has been known for years, it stubbornly persists in the scientific literature as well. For example, in response to a *Viewpoint* (Guth, *JAMA*, 1/16, 2018), Hall et al (*JAMA* 6/12, 2018) take issue with the author's simplistic understanding based on the 3500 kcal per pound rule, that small decreases in caloric intake will "progressively" lead to "substantial" weight losses over time. In other words, the 3500-kcal rule "overpromises," (despite sustained efforts by dieters), and there is not a "slow steady weight loss," specifically because of the body's physiological adaptations.

Even in a recently published professional journal, researchers had used this "invalid" rule for their projected calculations of weight change in their study of type 2 diabetes. (See the critical comments by Andrew W. Brown and colleagues in *BMJ Open Diabetes Research and Care,* 11 September 2018.)

The 3500 kcal rule had been established in the late 1950s by Max Wishnofsky (*American Journal of Clinical Nutrition*, 1958). Wishnofsky acknowledged that weight change was complex, with an irregular "series of ups

French artist Robert Delaunay, *The Runners*, c.1924 (oil on canvas), Musee des Beaux-Arts, Troyes, France. Self-reports of physical exercise are as inaccurate as those of caloric intake, height, and weight.

De Agostini Picture Library / G. Dagli Orti / Bridgeman images, used with permission.

and downs, with frequent periods of no apparent weight change," and the 3500-kcal rule did depend on several variables, including the length of time of observation, whether someone was fasting, and the body's hydration. His oversimplified conclusion, though, was based on a small sample and short-term observations, as well as a "limited understanding of fundamental metabolic processes," and its inaccuracy has led "many patients wondering why their prescribed weight loss is less than expected." (Thomas et al, *Journal of the American Academy of Nutrition and Dietetics*, 2014) Nevertheless, the 3500 kcal rule, with its "fatal flaws," became gospel, with the result that patients have been blamed and potentially stigmatized or blame themselves for a lack of willpower and motivation. (Hall and Kahan, *Medical Clinics of North America*, 2018)

It was Wilbur O. Atwater, an early 20th century researcher, who confirmed the law of conservation of energy in humans and established the energy values, still in use today, for

Spanish Energy by Penny Warden.

Source: Bridgeman Images, Private collection, used with permission.

protein (4 cal/gram); carbohydrate (4 cal/gram); fat (9 cal/gram); and alcohol (7 cal/gram). (Heymsfield et al, *European Journal of Clinical Nutrition*, 2017) But, say Allison and his colleagues, "Few dietary components are surrounded by more misinformation and myths than the calorie," in part related to our "lack of accurate and practical methods" for assessing caloric intake and for establishing our caloric requirements over time. (Heymsfield et al, *American Journal of Clinical Nutrition*, 1995)

For years, researchers have accepted the self-reports of subjects for their caloric intake, as well as for their height and weight measurements and even their physical activity. What has now been established unequivocally is that these self-reports are often grossly inaccurate and lead to "questions of data validity." Dhurandhar and colleagues (*International Journal of Obesity*, 2015), in response to the use of self-reports, have emphasized, "...something is *not* better than nothing." Reasons for this misreporting include

A Race in Ancient Greece (colour lithograph), French School, (19th century) / Bibliotheque des Arts Decoratifs, Paris, France/Archives Charmet.

Bridgeman Images, used with permission.

inaccurate food labeling; inaccurate estimates of portions; psychological denial and self-deception; faulty memory; or even wanting to present themselves more favorably. (Heymsfield et al, 1995)

As a replacement for self-reports, we now have better measurement tools, such as the doubly labeled water method, a technique developed for "free-living" humans in the early 1980s to assess our metabolic rate and total daily expenditure over a few weeks, and the dual-energy X-ray absorptiometry—the DXA scan—developed in the 1970s to measure body composition, but even these are not completely free from error. (Thomas et al, *American Journal of Clinical Nutrition*, 2014; Thomas et al, *European Journal of Clinical Nutrition*, 2018; Heymsfield et al, *Obesity Reviews*, 2018) Further, in free-living situations, using our current methods, we are limited to observation periods—"snapshots"—of about two weeks. (Hall et al, *American Journal of Clinical Nutrition*, 2012)

More recently, though, researchers are now using "validated mathematical models," and they have established, for example, that the so-called *diet plateau* that occurs after six months of dieting, in which weight loss seems to stop, is actually much more a function of patients "experiencing an exponential decay of diet adherence." (Freedhoff and Hall, *The Lancet*, 2016; Hall and Guo, 2017) For reasons that are not fully understood, diet adherence, including "sustaining dietary choices and behaviors," is "so challenging that it is poor even in short-term studies when all food is provided."

Retro Food: Counting Calories, 1957 (screen print), American School, (20th century). Researchers have debunked the simplistic rule that 3500 kcal equals one pound. Nevertheless, calories do count since weight occurs when our caloric intake is greater than our caloric expenditure.

Photo copyright GraphicaArtis/ Bridgeman Images, used with permission.

(Freedhoff and Hall, 2016) The researchers, though, do emphasize that there is considerable individual variability and while weight loss and maintenance are difficult, there are many anecdotal stories of successful dieters. Further, although there is some metabolic adaptation to weight loss, and losing weight can slow down metabolism for complex reasons, including that caloric expenditure declines with weight loss (Hall, *Obesity*, 2018), hormonal "mediators" of appetite change (Heymsfield et al, *Obesity*, 2017) and considerable physical activity is required to maintain the loss, (Hall, 2018) this does not explain the six-month plateau described by many dieters. It is the "seemingly innocuous intermittent loss" of diet adherence—i.e., the "behavioral fatigue" (Hall and Kahan, 2018)—that is mostly responsible. (Thomas et al, *American Journal of Clinical Nutrition*, 2014)

Mathematical models can also be particularly useful in simulating experimental designs that would not be ethical or practical. Another "novel application" is their use to predict, from short-term results, which patients will be more successful in their weight loss over the long-term, i.e., to determine who are the "responders" to a particular intervention and if not, to suggest a change in strategy, according to Allison and colleagues (Dawson et al, *Advances in Nutrition*, 2014.) Further, researchers have used a mathematical model to evaluate the complex interactions between genetic and non-genetic modes of transmission of obesity and to assess prevalence rates. (Ejima et al, *Obesity*, 2018) This model has predicted that obesity prevalence unfortunately will reach 41% of the U.S. population by 2030.

Wilbur O. Atwater, early 20th century researcher in metabolism established the First Law of Thermodynamics in humans as well as established the caloric values of protein, fat, carbohydrate, and alcohol that are still in use today.

Wikimedia Commons/Public Domain

Investigation of Respiration During Exercise. Subject breathes into a mask and his exhalations are measured by a gas meter and then analyzed. Researchers have come a long way from this primitive way of measuring metabolic functioning after exercise. From *The Science of Life* (London, 1929–30).

Investigation of respiration during exercise/Universal History Archive/UIG/ Bridgeman Images, used with permission.

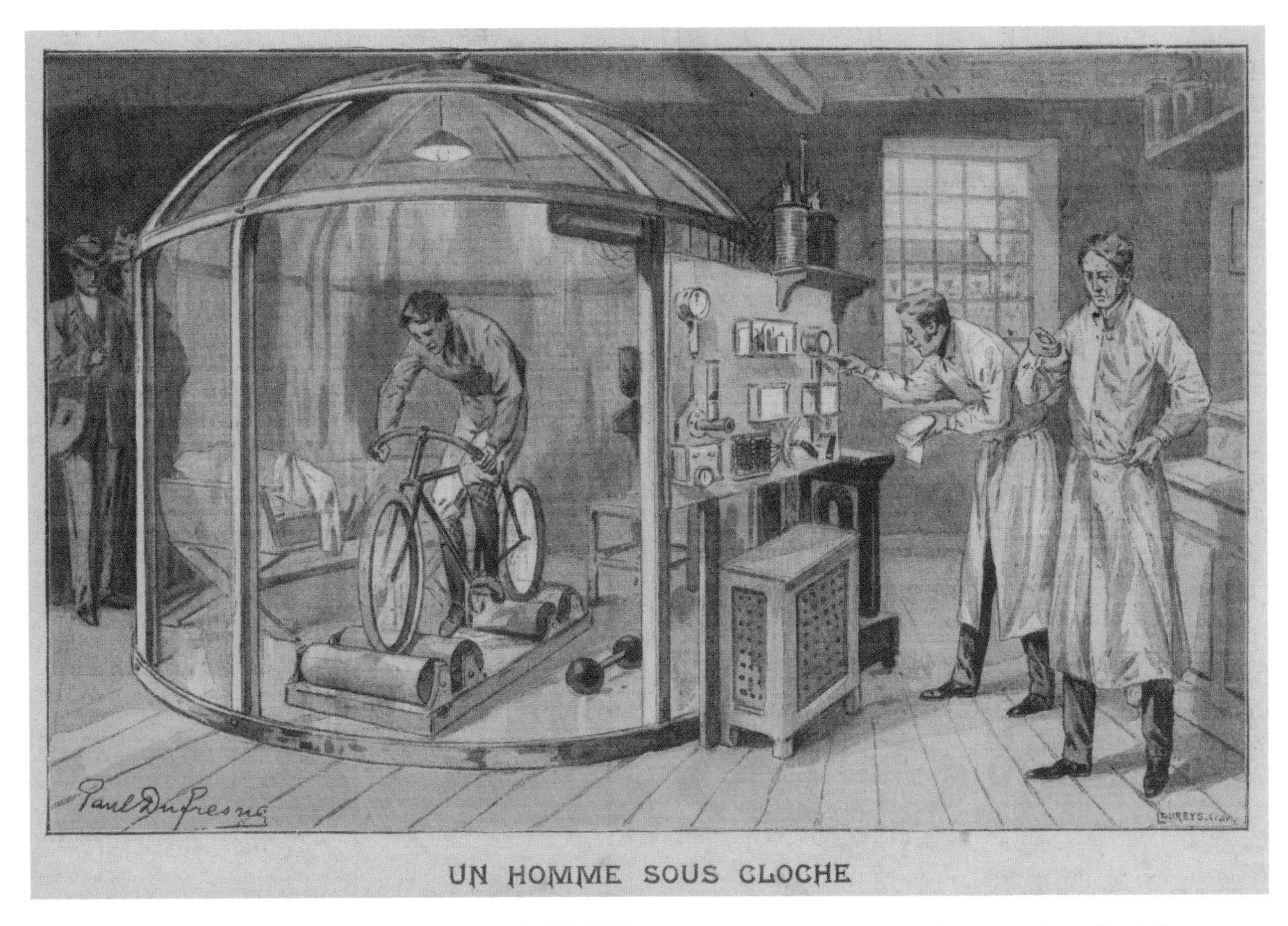

▲ Paul Dufresne, *A Man Riding a Bicycle Inside a Bell Jar in a Scientific Experiment* (color lithograph)

Bridgeman Images. Private Collection, copyright Look and Learn. Used with permission.

◀ Hubert von Herkomer, portrait of Lord Kelvin, before 1907, famous for experiments in heat energy and for emphasizing the importance of measurement in science. Glasgow Museum, Scotland.

Wikimedia Commons/Public Domain.

American artist Henry Brevoort Eddy, *New York Ledger: Bicycle Number,* 1895, print.
Metropolitan Museum of Art, NYC. Public Domain. Gift of Bessie Potter Vonnoh, 1941.

Allison and colleagues (Ivanescu et al, *International Journal of Obesity*, 2016), though, appreciate the importance of validating their mathematical models. For example, a model derived from one finite sample will not always predict as well on the overall population from which the sample was taken or on a new sample from a different population—a phenomenon they call *validity shrinkage*. To validate a model, new datasets are required. For an example of how researchers used four different classical studies, including those by Claude Bouchard and Ancel Keys, to validate their model, see Thomas et al, *American Journal of Clinical Nutrition*, 2014.

Bottom line: For years, obesity research has suffered from severe methodological deficiencies that have led some investigators to emphasize the need for better, not necessarily more, research. For example, when people think they are adhering to diets, they may, even unconsciously, not be doing so. Further, countless studies have relied, not on accurate measurement, but on notoriously

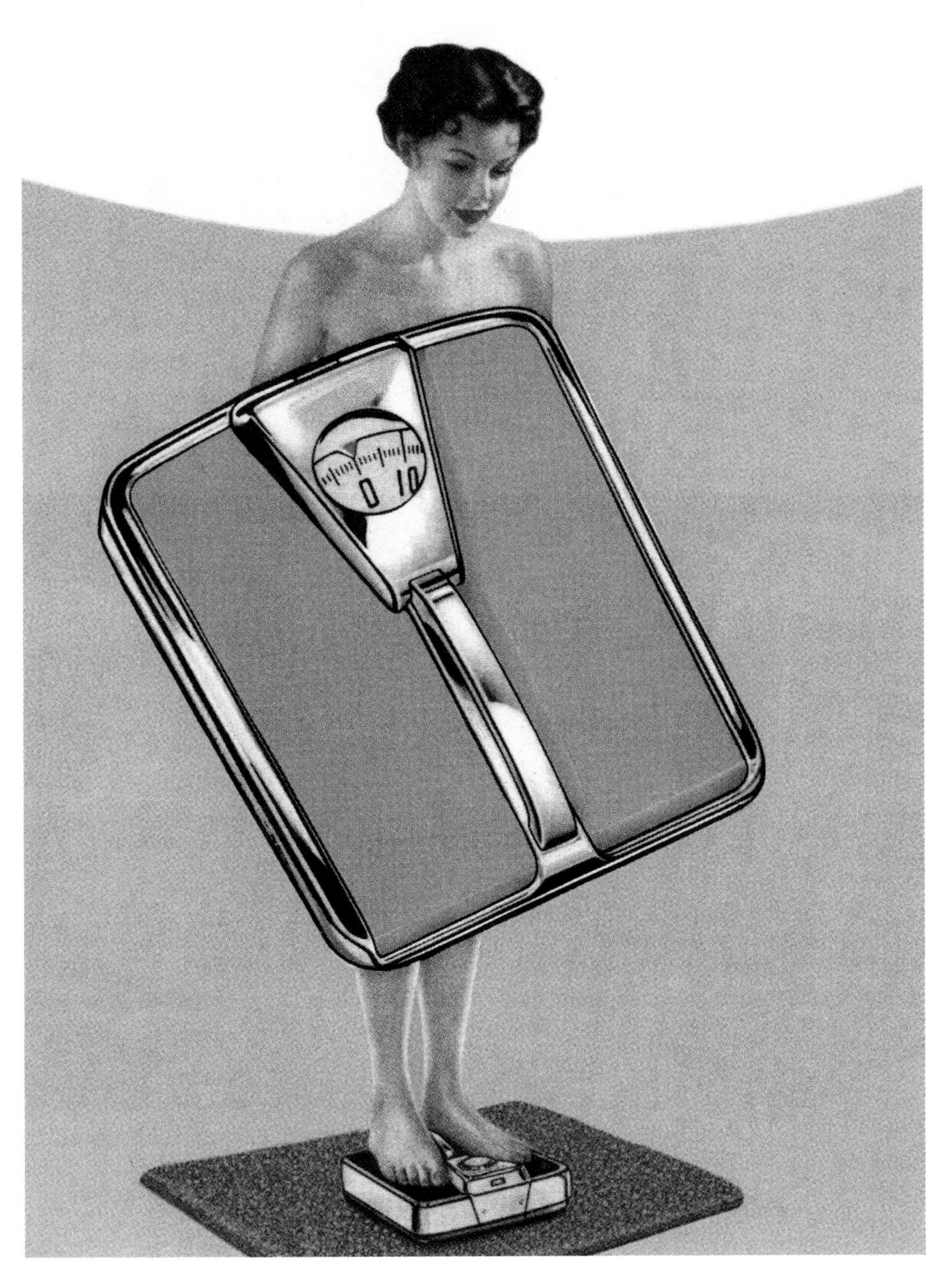

1950s Woman on Weight Scale, 1958. Screen print from photograph. People are notoriously inaccurate when asked by researchers to report their weight. Nevertheless, weighing ourselves frequently, from weekly to daily, keeps us accountable and for most, more likely to maintain a diet, though "behavioral fatigue" and an "exponential decay in dietary adherence."

Private collection, used with permission of GraphicaArtis and Bridgeman Images.

inaccurate self-reports of caloric intake, caloric expenditure, height and weight. Even the rule of thumb in which 3500 kcal equals one pound is simplistic and erroneous. To remedy these defects, researchers have begun to generate mathematical models in an attempt to create greater precision in measurement and clarify why it is so difficult for people to lose weight and particularly, to maintain that loss over time. No model is perfect: models may suffer from what investigators have called *validity shrinkage*, and even mathematical models are only as good as the data from which they are originally derived. All research is a work-in-progress. Nevertheless, these mathematical models are a vast improvement over years of inaccurate data, and obesity investigators will benefit considerably from incorporating them into their research. Lord Kelvin would approve.

Note: For one web-based tool to estimate weight loss over time, refer to the *Body Weight Planner* at http://BWplanner.niddk.nih.gov. (Hall and Kahan, 2018) ■

"ON THE MARGIN OF THE IMPOSSIBLE"

Navigating through the flood of information on obesity

100 — Posted November 5, 2018

French artist Jean Dubuffet, *Node in the Hair (The Impossible)*, oil on canvas, 1955, Private Collection.

Harry, Lord Monchensey, has just returned to his family home after an absence of eight years to celebrate his elderly mother's birthday in *The Family Reunion,* T.S. Eliot's 1939 play in verse. Alas, he is veering toward insanity due to his own guilt: on an ocean voyage a year previously, Harry's wife has been "swept off the deck in the middle of a storm" and vanishes among the waves. "You would never imagine anyone could sink so quickly," says Harry. He torments himself that he may have pushed her into the sea and been responsible for her death, "...the wish to get rid of her / Makes him believe he did..." His relatives, at a loss to help and concerned about Harry's troubled mental state, consider inviting their local physician over for a consultation. Harry's aunt Agatha, somewhat skeptical, says, "Not for the good it will do / But that nothing be left undone / On the margin of the impossible." (T.S. Eliot, *The Family Reunion: The Centenary Edition,* 1888–1988.)

It was Archibald L. Cochrane, an advocate for the use of randomized controlled trials and for whom the Cochrane Library database was named, who called attention to the Eliot quote. Writing about scientific research in general, in his now classic *Effectiveness & Efficacy: Random Reflections on Health Services,* (1971), Cochrane called for clinicians to "...abandon the pursuit of the 'margin of the impossible'..." (p. 85)

The study of obesity borders on that "margin of the impossible" because, "Despite decades of research into the causes of the obesity pandemic, we seem to be no nearer to a solution now than when the rise in body weights was first chronicled decades ago." (Hebert et al, *Mayo Clinical Proceedings,* 2013) "The circle of our understanding / Is a very restricted area," says the Chorus near the end of Eliot's play. How, though, can we make sense of where we are now?

Back in the mid-1950s, a Johns Hopkins researcher, studying the ease of abstracting information from his biological journals, wrote, "Perhaps no problem facing the individual scientist today is more defeating than the effort to cope with the flood of published scientific research, even within one's own narrow specialty." (Glass, *Science,* 1955) Tools were primitive and unsophisticated by 21st century standards: today we have capabilities of retrieval far beyond anything scientists then might have imagined, but the "effort to cope with the flood of scientific research, even within one's own narrow specialty" has grown exponentially worse. By one account, as I had written eight years ago in my first blog, over 250 different professional journals, without even including, for example, journals in the fields of economics or consumer affairs, include articles relevant to obesity. (Baier et al, *International Journal of Obesity,* 2010) We are at risk of being inundated and like Henry's wife, lost at sea, or rather, "lost in publication." (Garg et al, *Kidney International,* 2006) What are some of the general issues involved in navigating this flood of information?

For one, "Not all scientific information is created equal." (Ioannidis, *PLOS Medicine* 2018) For example, in their review of the current "medical misinformation mess,"

French artist Henri de Toulouse-Lautrec, *Acrobat on Tightrope,* pastel, 19th century, Musée Toulouse-Lautrec, La Berbie, France. An acrobat balancing on a tightrope is a metaphor for the "margin of the impossible" from which obesity researchers work.

De Agostini Picture Library/ G. Dagli Orti/Bridgeman Images, used with permission.

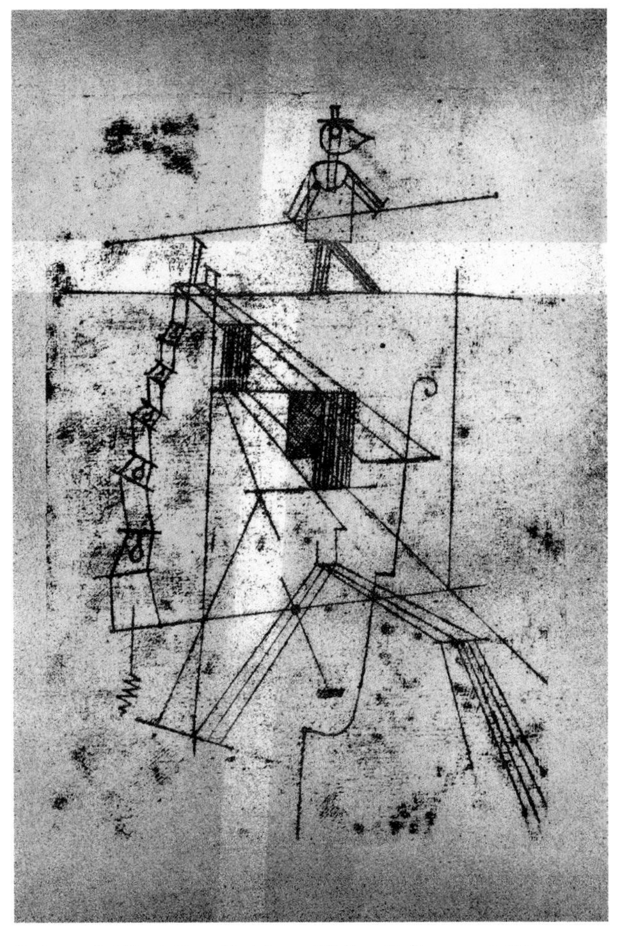

Swiss artist Paul Klee, *Tightrope Walker: Seiltanzer,* 1923, lithograph, Private Collection.

Photo copyright Christie's Images/Bridgeman Images. Copyright 2018 Artists Rights Society (ARS), New York, used with permission of both Bridgeman Images and ARS.

Ioannidis and colleagues (*European Journal of Clinical Investigation*, 2017) found that there are about 17 million articles within the search engine of PubMed that involves humans, and apparently about 1 million articles are added each year. This is not particularly good news, though, since much of the information contained in these articles is misleading, unreliable, or of "uncertain reliability." Furthermore, say Ioannidis et al (2017) most of those who read these studies are not even aware of this situation, and even if they are, most do not have sufficient proficiency necessary to evaluate the research studies they are reading.

Ioannidis (*European Journal of Epidemiology*, 2018) also called attention to the so-called *Matthew Effect*: those papers that are heavily cited continue to be cited. Merton (*Science*, 1968) had described this effect, named for the Bible's *Book of Matthew* (25.9): "For to everyone who has will more be given, and he will have abundance; but from him who has not, even what he has will be taken away. In other words, explains Merton, scientists "of considerable repute" keep getting greater recognition while those "who have not yet made their mark" have that recognition denied to them.

The media contribute to the problem, often by barraging the public with medical information, sometimes from those popular "authorities" on television who offer "evidence," much of which is "incomplete and wildly

English, *There was an Old Man of Coblenz, the Length of Whose Legs was Immense,* 1846, color lithograph, Private Collection. From *A Book of Nonsense,* published by Frederick Warne and Co., London, c. 1875; Creator, Edward Lear.

inaccurate." (Ioannidis et al, 2017) Since science is, after all, public, it must be communicated to others: "that is what we mean by a contribution to science—something given to the common fund of knowledge. In the end, then, science is a socially shared and socially validated body of knowledge." (Merton, 1968) And it is "the best method we have of coming to an impartial knowledge of the world." (Kroeger et al, *American Journal of Clinical Nutrition,* 2018) The media and even researchers themselves, though, ostensibly for well-meaning, even righteous aims, sometimes misrepresent or exaggerate, either consciously or unconsciously, scientific claims, i.e., what Cope and Allison have aptly labeled *white hat bias.* (*Acta Paediatrica,* 2010; *International Journal of Obesity,* 2010) (For more on *white hat bias,* see my blog 53) Clinicians, patients, and their families, as a result, are often left without the ability to evaluate treatment options.

One major impediment has been the persistence within the literature of thinking of obesity as a single disease with a single etiology. (Hebert et al, 2013; SR Karasu, *American Journal of Lifestyle Medicine,* 2013), though Stunkard and Wolff, as early as the 1950s, (*Psychosomatic Medicine,* 1958) noted that there was no need to presume a common etiology. Furthermore, instead of appreciating the enormous complexities of obesity, many researchers categorize obesity in the language of their own discipline. For example, physicians view obesity as a pathological state, i.e., a disease to be treated; sociologists may view it as an example of body diversity; the clergy, as an example of moral corruption and self-indulgence; anthropologists, as a disease of civilization; geneticists, as a genetic disorder; evolutionary biologists, as either appropriate or inappropriate adaptation to an obesogenic environment, with contributions from

Weighing Bars of Camphor, from *Tractatus de Herbis* by Dioscorides, Italian School, 15th century, Biblioteca Estense, Moderna, Emilia-Romagna, Italy. Sometimes, data derived from self-reports may seem as inexact and primitive as obtained from 15th century measuring devices.

bacteria, viruses, endocrine-disrupting toxins, among others; physicists, as an energy imbalance following the laws of thermodynamics; and psychiatrists and psychologists, as a disorder of self-regulation or even addiction. (SR Karasu, 2013; SR Karasu, *American Journal of Lifestyle Medicine,* 2014.) (For more on the different "languages" see my blog 26, *A Towering Babel.*)

There are also methodological difficulties, some general to science and some specific to obesity studies. Particularly prevalent in obesity studies is that non-randomized observational research far outnumbers randomized controlled studies, and there is a careless use of causal language, particularly from these observational studies. (Trepanowski and Ioannidis, *Advances in Nutrition,* 2018)

Statistical errors are unusually common among obesity studies. "If you torture your data enough, they will tell you whatever you want to hear," and "like other forms of torture, it leaves no incriminating marks when done skillfully…and may be difficult to prove even when there is incriminating evidence." (Mills, *NEJM,* 1993) Allison and his colleagues (George et al, *Obesity,* 2016) identified 10 of the most common statistical errors seen in obesity

Genesis 6: 11–24, *Noah's Ark,* from the *Nuremberg Bible,* colored woodcut, German School, 15th century, *Biblia Sacra Germanica,* Private Collection. God told Noah to gather his family and all the animals, two by two, to board the ark. It rained for 40 days and 40 nights

The Stapleton Collection/Bridgeman Images, used with permission.

research. One of the most common errors in obesity literature is assuming an intervention is effective when the study itself does not support that conclusion. (Brown et al, *Proceedings of the National Academy of Sciences*, 2018) Some other common errors include a mishandling of or even ignoring missing data or not dealing correctly with those subjects who don't complete a study, ignoring confirmation bias, and ignoring regression to the mean. *Confirmation bias* is the tendency for researchers to evaluate their results differently or even somewhat less critically when their results match their initial expectations or conform to their initial hypotheses. *Regression to the mean* is a statistical phenomenon that occurs when repeated measurements are made on the same subject, and there is no control group to compare any difference from baseline. When the measurements change on repeated exam, (and often when subjects deviate less extremely from the mean) researchers can erroneously assume the change was due to their intervention. In other words, regression to the mean can "masquerade as a treatment effect." (Kahathuduwa et al, *Diabetes, Obesity and Metabolism*, 2018)

Further, obesity research has been plagued by the complexities of inaccurate measurement, including those related to self-report of body weight, height, food intake, and exercise. "While one is either obese or not, the cutoff between the two states is arbitrary." In other words, population health "manifests itself as a continuum…(and) "We can predict the health in populations with much more certainty than we can predict health in individuals." (Galea, *The Milbank Quarterly*, 2018) "We call it health when we find no symptom/Of illness. Health is a relative term," says the physician in Eliot's play.

These measurement inaccuracies have led to what some researchers have called "pseudoscience." (Trepanowski

Noah and the Ark, 3rd–5th century A.D., Roman fresco, from Catacomb of the Giordani, Rome. Researchers have created a "Data Ark" to preserve raw data and enable greater transparency and reproducibility of studies.

De Agostini Picture Library/Bridgeman Images, used with permission.

So Near and Yet So Far, color lithograph, postcard, English School, 20th century, Private Collection. Research in obesity can seem both "so near and yet so far" in explaining the exponential rise in obesity rates over the past 40 years.

Copyright: Look and Learn/Bridgeman Images, used with permission.

and Ioannidis, 2018; Archer et al, *Current Problems in Cardiology*, 2016; Archer et al, *PLOS One*, 2013) For example, attempts at nutritional surveillance, i.e., the systematic collection of data to detect trends in consumption and assess the connection between caloric intake and obesity rates over the past 40 years, have resulted in "pseudo-quantitative" data that are "physiologically implausible." Data collected by the Centers for Disease Control from the NHANES population of civilian, non-institutionalized in the U.S. have relied on inaccurate and grossly misleading self-reports of food intake that have also excluded huge swaths of the U.S. population, including undocumented aliens, the homeless, and those institutionalized. Schoeller et al, in a letter signed by 17 leaders in the field of obesity research, documented how it has been over 20 years since Schoeller himself had found "substantial biases and inaccuracies," i.e., "fatal flaws"—particularly gross under-reporting of caloric intake in obesity research. Unbelievably, the practice of self-report remains rampant in obesity studies. (Schoeller et al, *American Journal of Clinical Nutrition*, 2013; Dhurandhar et al, *Journal of Nutrition*, 2016)

Trust in nutritional science further diminishes when one study implicates a nutrient as harmful and then another labels the same nutrient as beneficial. Ioannidis names this extreme alternation the *Proteus phenomenon*, after the Greek god who could change his shape easily. (*PLoS Medicine*, 2005) Further, obesity research presents challenges because almost all nutritional variables are correlated with one another (Ioannidis, *JAMA* 2018): not only do we eat our carbohydrates, fats, and proteins

Detail of the south spandrel showing Matthew, fresco, Italian School, 13th century, St Johann Church, Tubre, South Tyrol, Italy. The so-called *Matthew Effect*, named after a verse in the *Book of Matthew* in the Bible, describes how heavily cited articles by prominent researchers continue to be cited while those researchers "who have not yet made their mark" are less likely to be cited in the literature.

Hirmer Fotoarchiv/Bridgeman Images, used with permission.

in various combinations, but our foods expose us to thousands of chemicals, contaminants, and toxins that make it impossible to disentangle the potential influence of one component from others, as well as isolate environmental exposures and other variables such as lifestyle, education, socioeconomic status, etc. Further, adherence to a dietary protocol is often poor or the control group may adopt the experimental protocol. (Trepanowski and Ioannidis, 2018)

The public should be skeptical, writes Marion Nestle, in her book *Unsavory Truth: How Food Companies Skew the Science of What We Eat* (2018) whenever any study singles out any food, beverage, supplement, or specific ingredient that causes or reduces risk of obesity, heart disease, type 2 diabetes or cancer. (p. 228) Nestle calls the sensational findings on the benefits of single foods when they are removed from their "dietary context," *nutrifluff.* (p. 54) Since we eat all foods in combination with others, it makes no sense to accept that one food has unusual and special benefits for our health.

In a particularly innovative study, Schoenfeld and Ioannidis (*American Journal of Clinical Nutrition*, 2013) raised the provocative question whether everything we eat is related to cancer. These researchers selected 50 common ingredients from random pages in a popular cookbook

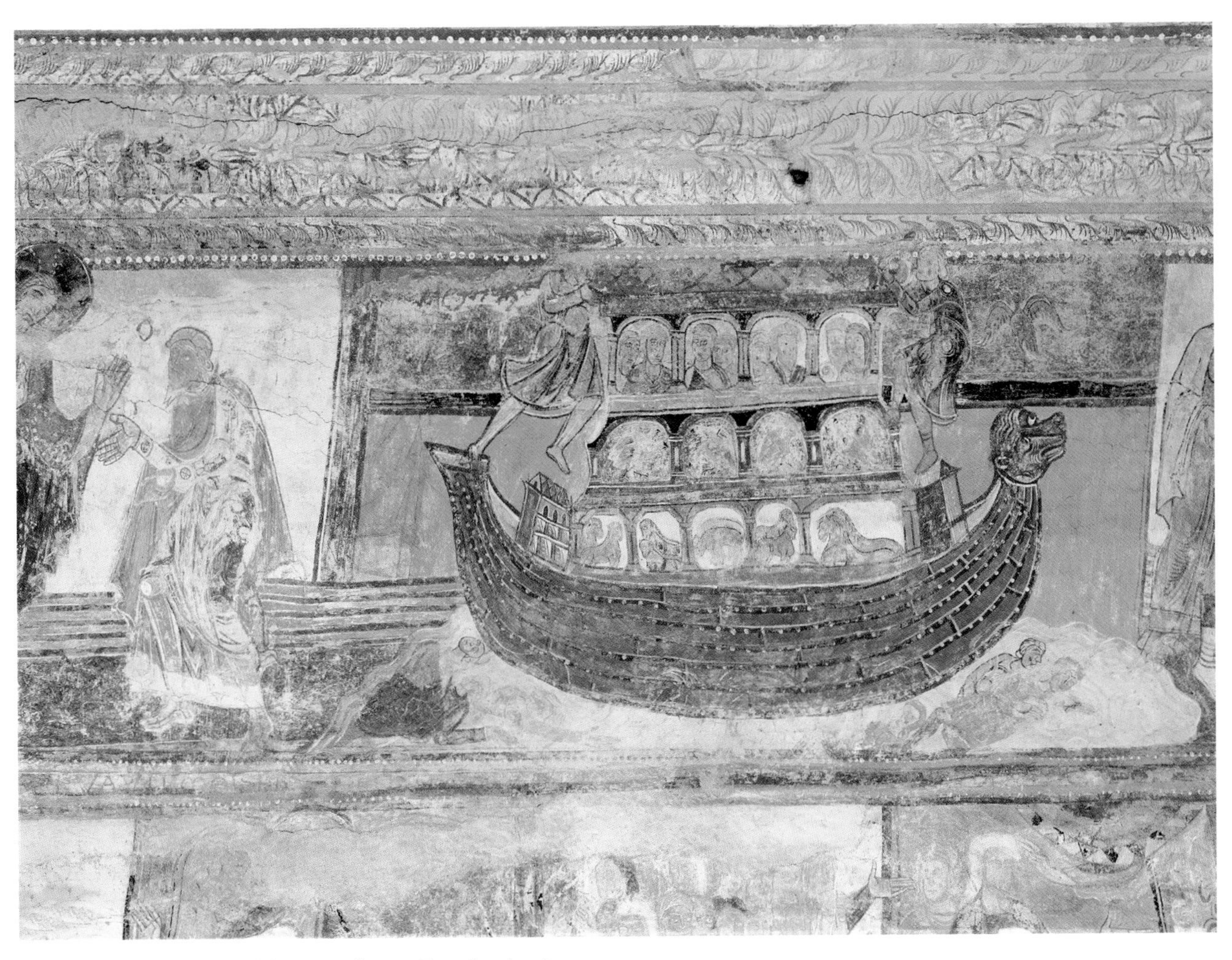

The Floating Ark, ceiling of the nave, fresco, French school, 11th century, Abbey Church of Saint-Savin-sur-Gartempe, Poitou, France.

Hirmer Fotoarchiv/Bridgeman Images, used with permission.

and found 40 of these ingredients (80%) were featured in articles that offered evidence for either an increased or decreased risk of cancer, despite weak statistical evidence. Gastrointestinal cancers, highlighted in 45% of the research, were the most commonly studied. Further, randomized controlled trials often failed repeatedly to find treatment effects for nutrients in which observational studies had previously reported strong associations, and even meta-analyses were sometimes biased and subject to misinterpretation. (Schoenfeld and Ioannidis, 2013) "If taken literally, if we increase or decrease intake of any of several nutrients by two servings a day, cancer will almost disappear worldwide." (Brown et al, *Advances in Nutrition*, 2014)

Whatever its primary focus, nutrition research has been called "among the most contentious fields of science" (Ioannidis and Trepanowski, *JAMA*, 2018) because of the potential financial conflicts of interest from industry or other sources of funding, as well as researchers' own potential biases and preferences (e.g. vegan, gluten-free, etc.) in what they eat or what causes they support. (Brown et al, 2014) Many researchers do believe it is a "puritanical and outdated view" that accepting funding from industry necessarily biases results. (Ioannidis, 2018) In fact, Allison et

The Flood, from the Atrium, detail of Noah releasing the white dove (mosaic), by Veneto-Byzantine School, 13th century, San Marco, Venice, Italy. In *Genesis* 8.6, Noah sends out a dove three times to see if the flood waters had receded after the 40 days and 40 nights of rain: the first time the dove returns for it could not find dry land; the second time, seven days later, Noah again sends forth the dove that now returns with an olive branch in its mouth to indicate the waters were beginning to recede; the third time, still another week later, the dove does not return, and Noah realizes there is dry land. What will be our sign that the flood waters of publications have receded?

Bridgeman Images, used with permission.

al found, in looking at top-tier medical journals, that randomized controlled trials were of equal quality regardless of the funding source. (Kaiser et al, *International Journal of Obesity,* 2012)

Nestle, though, offers a caveat, "Let me state for the record that financial ties with food companies are not necessarily corrupting; it is quite possible to do industry-funded research and retain independence and integrity. But food-company funding often does exert undue influence." (Nestle, 2018, p. 6) She adds, " (and)... it does suggest that the research question and interpretation require more than the usual level of scrutiny." (p. 71) For Nestle, there should be a clear distinction between marketing by food companies and science. Further, Nestle sees *financial conflicts* of interest as categorically different from *non-financial conflicts* that can depend on individual beliefs, desires, and hypotheses that vary enormously from one investigator to another.

Though clearly not specific to those conducting nutrition studies, researchers have not been required to be transparent in releasing their raw data, with the result that many studies are unable to be replicated. Nestle (2018, p. 169) remembers the joke from years ago when she was a graduate student in molecular biology at Berkeley, "Never repeat an experiment that works on the first try." In an effort to rectify the situation and preserve and make accessible these retrospective data, Hardwicke and Ioannidis (*PLOS One,* 2018) have launched an initiative—the *Data Ark*—an online repository for preserving raw data, encouraging scientific rigor, and increasing transparency among studies.

Bottom line: Just as Harry's wife in T.S. Eliot's play, is swept overboard and drowns, we are all drowning in a contaminated sea of publication. So much of the research within obesity yields pseudo-scientific data due to poor methodology, erroneous and unreliable measurement, and biases due to conflicts of interest. Ioannidis has suggested research itself needs its own study, what he and his colleagues have called *meta-research* as a way of verifying, evaluating, and rewarding research. (*PLOS Biology,* 2018) In science, there is sometimes a fine line between healthy skepticism and misrepresenting and exaggerating scientific uncertainty. (Allison et al, *American Scientist,* 2018) Though the "circle of our understanding" often seems "a very restricted area," researchers have no choice other than to navigate as best they can through the flood of information and away from that "margin of the impossible." ■

EPILOGUE

"THE MARROW OF ZEN" AND A BEGINNER'S MIND

The evolution of an obesity narrative

"In the beginner's mind, there are many possibilities; in the expert's mind, there are few," wrote Japanese master Shunryu Suzuki, who brought Zen teachings to America in the early 1960s, in his book *Zen Mind, Beginner's Mind* (1970, p. 1) "A mind should be an empty and ready mind, open to everything,"(p. 2) whereas a mind full of preconceived ideas, subjective intentions, or habits is not open to things as they are," (p. 77) he explained. For Suzuki, "It is readiness of mind that is wisdom." (p. 103) He encouraged a "smooth, free-thinking way of observation—without stagnation." (p. 105) And it is "...under a succession of agreeable or disagreeable situations, you will realize the marrow of Zen." (p. 24) While essential to any branch of science, this philosophical perspective is particularly applicable to the study of obesity.

It was during Suzuki's 12 years in San Francisco that he became a Spiritual Master to psychiatrist Albert (Mickey) Stunkard, a renowned pioneer in obesity research, during Stunkard's time on the West Coast at Stanford. Stunkard wrote of Suzuki's influence on his own thinking in his paper, aptly entitled "Beginner's Mind." (*Annals of Behavioral Medicine,* 1991) For him, the beginner's mind meant a particular magic and joy in discovery that allowed his mind to "follow whatever leads seemed most promising" whether or not he knew anything about the subject. This open accessibility led Stunkard to develop creative insights into obesity, particularly in the realm of specific eating disorders and the relationship of obesity to social class and to influences from both nature and nurture, that had not been identified previously and that are still relevant almost 60 years later.

The Medicine Buddha from the Zen Temple Ryumonji in Weiterswiller, France

Dean David B. Allison, who writes of the enormous personal impact that Stunkard's encouragement had early on in his own career development, appreciated Mickey's contagious enthusiasm, humility, "wide-eyed" inquisitiveness, and a genuine willingness to learn from anyone. (Pavela et al, *Current Obesity Reports,* 2016) In other words, Stunkard was one of those rare charismatic scientists who emphasize the importance of *problem-finding*, rather than just problem solving, and had that unique ability, not only to achieve excellence themselves, but to evoke excellence in others. (Merton, *Science*, 1968)

That excellence originates from an ability to shift perspective and develop that *free-thinking way of observation.* In a particularly original exploration of changing perspectives, Chang and Christakis (*Sociology of Health & Illness,* 2002) explored the evolving narrative of obesity through the lens of five editions, from its first publication in 1927 to 2000, of the *Cecil Textbook of Medicine*, "one of the most prominent and widely consulted" medical texts, still in circulation as *Goldman-Cecil, with its 25th Edition* most recently published in 2016.

In each edition, Chang and Christakis found that authors consistently accepted that obesity results from an imbalance of greater caloric intake than caloric expenditure. What they found, though, is that over the seven decades, the cause of this imbalance shifted "dramatically:" the obese were "initially cast as societal parasites," but "later transformed into societal victims" For example, in the 1927 edition, obesity is seen as "aberrant individual activity"—the result of specific behaviors over which the individual had control. By 1967, the focus had now shifted, and obesity had "changed from being the result of something that individuals do, to being the result of something that individuals experience" within a social context: it was society that was seen "as a source of harm," predominantly from the food industry. By 1985, the chapter's author introduces the

Portrait of the Buddhist Leader and eminent monk, Great Master Seosan. Unidentified artist, Joseon Dynasty, 1392–1910, Hanging scroll, Korean, late 17th–18th century.

Metropolitan Museum of Art, NYC, Public Domain. Seymour Fund, 1959.

Seiryū Gongen, Shinto goddess believed to be the avatar of two Esoteric Buddhist deities, from Nanbokuchō period, 1336–92, Japanese, hanging scroll, mid 14th century.

Metropolitan Museum of Art, NYC, Public Domain. Mary Griggs Burke Collection, Gift of the Mary and Jackson Burke Foundation, 2015.

Sketch by Japanese artist, Shibata Zeshin (1807–1891).

Metropolitan Museum of Art, NYC, Public Domain. Purchase, Gifts, Bequest, and Funds from various donors, by exchange, 1952.

Albert (Mickey) J. Stunkard, MD, renowned obesity researcher, who believed in the importance of a 'beginner's mind' and was mentored by Shunryu Suzuki.

Photo from the collection of the University of Pennsylvania, used with permission.

Seon or Zen Master
Cheongheo Hyujeong,
1520–1604, Korean.
Seon means *meditation*.

Pictures from History/Bridgeman Images, used with permission.

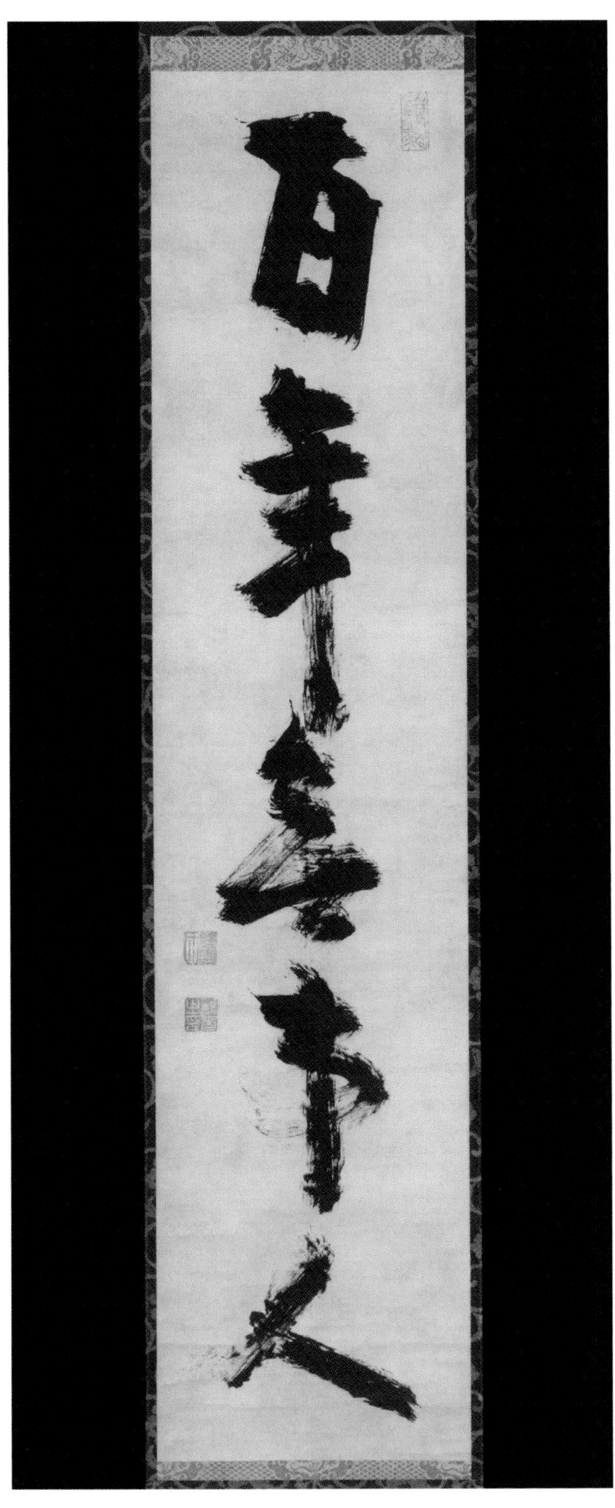

Zen calligraphy reflecting monk's priorities, "For a hundred years (I have been) a person with no attachments" by Jiun Sonja, Japanese, late 18th century.

Metropolitan Museum of Art, NYC, Public Domain. Gift of Morton Berman, in honor of Sylvan Barnet and William Burto, 2015.

disease model of obesity, albeit tentatively. Here, as well, there is an emphasis on both cultural and socioeconomic factors as contributory, while a genetic contribution "may play a role, but its mechanism remains unknown." Further, the obese now need "sympathetic attention rather than admonition" because of their "multiple failures in weight reduction."

In the last edition, published in 2000, that Chang and Christakis evaluated, obesity, with a strong emphasis on its genetic roots, is now referred to as a "complex polygenic disease," but with environmental contributions from the availability of highly palatable foods and decreases in physical activity. This edition also emphasizes the "frustrating condition for patient and physician alike" since the treatment of obesity "is fraught with difficulty and failure." Further, patients may have a psychological burden resulting from experiencing discrimination. Chang and Christakis summarize, "...we have moved from early models, which invoke the psychological causes of obesity, to contemporary models, which emphasize the psychological consequences of obesity." They add, "Over the seven decades, (the narrative) changes from one in which the individual is detrimental to society to one in which society is detrimental to the individual." Importantly, Chang and Christakis emphasize that these explanatory shifts over the years did not follow from any experimental studies.

In expanding on the work of Chang and Christakis, I read the most recent edition, the *Goldman-Cecil 25th edition* (Jensen, pp. 1458–1466, 2016) The obesity chapter emphasizes the biological: both genetic and now epigenetic contributions are noted, as are the many biological modulators involved in food intake and energy balance. There is also a reference to psychological differences among people with respect to dietary restraint and feelings of hunger. Further, there are sections on secondary causes of weight gain, such as the contribution from medications, and on the multiple potential medical complications (e.g. type 2 diabetes, sleep apnea, cancer) associated with obesity. Obesity is clearly referred to as a "chronic disease" that requires long-term treatment; there is an emphasis that "without approaches to ensure behavioral changes, body fat is invariably regained." Ironically, the author repeats the popular assumption that a deficit of 500 kcal/day will "theoretically" result in a pound of weight loss per week. (See my blog 99, *Mathematical Models: Obesity by the Numbers* for a discussion of the so-called 3500 kcal rule.)

In some ways, in this most recent edition, though, the pendulum has swung back to placing more of the onus on the patient and his or her readiness to engage with a physician: "Before a patient enters a weight management program, it is helpful to ensure that the patient is interested and

Hotei Admiring the Moon, Edo Period, 19th c., Japanese. Hotei was one of most beloved characters in Zen Buddhism.

Metropolitan Museum of Art, NYC, Public Domain. Charles Stewart Smith Collection, Gift of Mrs. Charles Stewart Smith, Charles Stewart Smith, Jr., and Howard Caswell Smith, in memory of Charles Stewart Smith, 1914.

ready to make lifestyle changes and has realistic goals and expectations. Patients who expect to lose large amounts of weight in a short time are virtually doomed to disappointment." (Jensen, 2016) If patients are not ready, the chapter's author recommends delaying treatment.

More recently, some researchers (Ralston et al, *The Lancet*, 2018) maintain it is time for a "new obesity narrative." They acknowledge that an "established narrative" relied on what they consider a "simplistic causal model" that generally blamed individuals for their obesity. Further, this model tended to disregard "all those complex factors for which an individual has no control." These authors suggest re-framing obesity to include a context for those who have "physiological limitations" from their obesity but exist within the much larger obesogenic environment. "Obesity is not simply about weight or body image. It is about human vulnerability arising from excess fat, the origins of which lie in multiple determinants ranging from molecular genetics to market forces." (Ralston et al, 2018)

The shifting obesity narratives reinforce the importance of having a willingness to continue learning from anyone, to see possibilities everywhere without preconceived notions, to be a *problem-finder.* There is a Zen story, a version of which is told by journalist and Zen scholar George Leonard, in the *Epilogue* to his book *Mastery: The Keys to Success and Long-term Fulfilment* (pp. 175–76): Jigoro Kana, the man who invented Judo and began the practice of wearing white and black belts in martial arts, was quite old and near death. Calling his students around him, he told them he wanted to be buried in a white belt, the "emblem of a beginner." Leonard's explanation is that at the moment of death, we are all beginners, even those who have reached the highest renown and accomplishment. Another interpretation, though, perhaps somewhat more philosophical, is that Kana, eager for knowledge, wanted to wear that white belt as indicative of his wish to continue learning for all of eternity. To wear a white belt humbly, to appreciate that everything, including information we learn, is transient and time-limited and ultimately subject to change, are all qualities of the beginner's mind and the "marrow of Zen." "Science," after all, "issues only interim reports." (Smith et al, *The Benefits of Psychotherapy*, 1980, p. 189) ■

ACKNOWLEDGMENTS

Both the initial and now this expanded edition of my book, *Of Epidemic Proportions: The Art and Science of Obesity,* evolved within an atmosphere of unconditional love and support I receive from my husband now of almost 43 years, T. Byram Karasu, MD, Distinguished Professor Emeritus, University Chairman Emeritus, and Dorothy and Marty Silverman Professor Emeritus in Psychiatry at Albert Einstein College of Medicine. Byram had not only encouraged me to write our textbook *The Gravity of Weight*, but also, after our book's publication, to begin writing monthly blogs for the psychologytoday.com website and continue writing them over these past many years. Byram remains my first and most important reader.

Alan Barnett, whose enormous talents as a graphic designer and whose remarkable and intuitive understanding of my own vision, has shaped, with an astonishing pace, our book's beautiful layout, including its front and back covers. Alan has created for me the book I had always imagined whenever anyone mentioned compiling my blogs into one volume and has seamlessly worked now to graft onto my first edition the additional seven blogs now included in this version. I thank Rudi Wolff, himself a graphic designer, who suggested publication with Blurb and through Blurb's website, I fortuitously found Alan.

Robert Swanson has assisted me in generating a comprehensive index that enables my readers greater access to and retrieval of all the material within this volume. I always felt that an index, perhaps the most important part of a book, was missing from my first edition.

I also want to thank John McDuffie, Publisher, of American Psychiatric Publishing, Inc, who had been our publisher on *The Gravity of Weight* and who has always welcomed my continuing to send him my blogs long after we had much hope of increasing sales for *The Gravity*.

Though always fascinated by diet and particularly by my father's lifelong struggle with his own excessive weight, I have come late to a more scholarly approach to obesity. Though there are many prominent researchers in this field, the person who has had the greatest influence on my thinking and for whom I have the greatest respect is David B. Allison, PhD, formerly Distinguished Professor and Associate Dean for Science at the University of Alabama at Birmingham and now, since August 2017, the Dean and Distinguished Professor and Provost Professor at the School of Public Health, Indiana University at Bloomington. David's intellectual integrity, discipline, and insistence on a rigorous scientific approach to the study of obesity have been exemplary and a model for me. Though we still have never met, I consider David a treasured mentor whose curiosity and enthusiasm are contagious. References to David's own writings, to date now about 600 academic papers and counting, appear either directly or indirectly on every page of my compendium. I was fortunate to begin writing my blogs shortly before David established *Obesity and Energetic Offerings,* a weekly public service free of charge to hundreds of thousands of subscribers worldwide, that provides links to publications on all aspects of obesity and general scientific research. I am always honored to have my blogs included on his prestigious list.

I thank Ms. Lybi Ma, my editor at psychologytoday.com, who has enabled me to continue writing my increasingly esoteric blogs, that have gotten longer and longer, with fewer and fewer readers, for the past eight years.

I am indebted to Drs. Barry and Bobbi Coller, whose appreciation and sensitivity to both art and medicine, inspired and nurtured my own interest in art. It was Barry's and Bobbi's course, primarily for medical students at the Icahn School of Medicine at Mount Sinai, *The Pulse of Art*, that motivated me to search for some of the extraordinary art images I have now included. Barry is the Physician-in-Chief and Vice President for Medical Affairs at Rockefeller University. Bobbi, with her PhD in art history, is an independent art curator and art educator; I am grateful to Judith A. Salerno, MD, MS, President of the New York Academy of Medicine, who shares an interest in obesity research and was involved in the several part- series HBO series, *The Weight of the Nation.* (2012)

Several museum collections, including the National Portrait Gallery, the Imperial War Museum, and the Royal Collection Trust, all in the United Kingdom, have been extraordinarily responsive. I am particularly appreciative, of the Metropolitan Museum of Art of New York City, especially because it is so severely short-staffed, for consigning, under the aegis of Julie Zeftel, Senior Manager of Rights and Permissions, so many images in its vast collection to the Public Domain.

Many individuals, as well, have been generous in allowing me to use their images: Eric Nestler, MD, PhD for use of his slide *Rodent Models of Depression?;* Erika Benincasa, representative of the Vik Muniz Studio and artist Vik Muniz; Martin Langenberg, of the Clemens Sels Museum, Neuss, Germany; Victoria and Chris Jordan, of the Chris Jordan Studios; David Leopold, representative of Matthew and Eve Levine, of the Estate of David Levine; Oda Siqveland, of Springer/*Nature* for facilitating my use of a 1991 cover of *Nature;* Ms. Carolyn Cruthirds, Coordinator of Image Licensing of the Museum of Fine Arts in Boston; Frans Lanting Studio; The Jewish Museum; Mr. William Kentridge and the Marian Goodman Gallery; Joyce Faust, of Art Resource; and Todd Leibowitz, of the Artists Rights Society, for their assistance in obtaining required permissions. I highlight, as well, Shaina Buckles Harkness, Collection Database & Image Manager and Librarian of the Dali Museum in St. Petersburg, Florida; Sue Hartke of GraphicaArtis of Washington State; Thomas Haggerty, of Bridgeman images; Joni Joseph, Collection Manager/Registrar of Dumbarton Oaks in Washington, DC; Eva Athanasiu, Copyright, Rights and Reproductions Assistant of the Art Gallery of Ontario; and Rebecca Toov, Collections Archivist of the University of Minnesota Archives, all of whom facilitated my obtaining rights and permissions for images not in the Public Domain.

Further, I acknowledge Anne Jacobson, MD, MPH, on the Editorial Board of *Hektoen International,* for her encouragement and appreciation of my writings, whether for *Hektoen* or for psychologytoday.com; and Anne Machalinski, a contributing editor for *Weill Cornell Medicine Magazine,* who followed my monthly blogs, asked to interview me, and then wrote such a glorious article detailing my interest in obesity for that publication.

I am also particularly indebted to my Weill Cornell Medicine family, including Jack D. Barchas, MD, the Barklie McKee Henry Professor and Chairman of the Department of Psychiatry and Psychiatrist-in-Chief of the NY-Presbyterian Hospital, who has had a major influence on my career at Weill Cornell; Mason L. Essif, Executive Director of Communications & Public Affairs; Anna Sokol, Marilynn Bonilla, and Krystle Lopez, all of the Office of External Affairs at Weill Cornell Medicine, who have willingly accepted my scholarly blog submissions and have featured them prominently on the Weill Cornell Medicine's Facebook page as well as re-tweeted announcements of my monthly blogs. Kevin J. Pain, Library Research Specialist at Weill Cornell Medicine's Samuel J. Wood Library, has always willingly and expeditiously found me any reference I could not retrieve myself.

Finally, I thank my brother Joseph A. Rabson, MD, whom I consulted on anything to do with plastic surgery; my sister-in-law, Mrs. Barbara Rabson, docent at the James A. Michener Art Museum in Doylestown, Pennsylvania, who alerted me to the glorious images of Chris Jordan and the whimsical images of Wayne Thiebaud, and of course, to my readers, many of whom have graciously embraced my monthly offerings and encouraged me to release this second and now expanded edition.

Sylvia R. Karasu, MD
January 2019

INDEX

ABOUT THE AUTHOR

Sylvia R. Karasu, MD is a Clinical Professor of Psychiatry at Weill Cornell Medicine, an Attending Psychiatrist at NY-Presbyterian Hospital, and a member of the Institutional Review Board of The Rockefeller University. The senior author of *The Gravity of Weight: A Clinical Guide to Weight Loss and Maintenance* (2010) and *The Art of Marriage Maintenance* (2005), Dr. Karasu is a *cum laude* graduate of the University of Pennsylvania and has her MD degree from Einstein College of Medicine. She is a Distinguished Life Fellow of the American Psychiatric Association, a graduate of the New York Psychoanalytic Institute, and an elected Fellow of the New York Academy of Medicine. Dr. Karasu is a contributor to *Hektoen International*, an online medical humanities journal, and has been writing scholarly blogs monthly on obesity for psychologytoday.com now for the past eight years. She has a private psychiatric practice in New York City.

blurb